Basic Algebra, Differential Calculus, Statistics and Probability

A Well Simplified Math Book for high Schools and Colleges

Kingsley Augustine

Printed by Amazon KDP

Table of Contents

PREFACE

Basic Algebra, Differential Calculus, Statistics and Probability is a book that contains topics under algebra, differential calculus, statistics and probability. This book serves as a useful companion for students in high schools and higher institutions of learning. It is a valuable textbook for students who want to write entrance test or examination into colleges and universities. This book consists of step-by-step explanation of topics presented in a way that is easy for students to understand. It contains very many worked examples and many self-assessment exercises to ensure that students get a mastery of each topic covered. The answers to the exercises are provided at the end of the book.

What makes this book a unique mathematical asset, is its detailed step by step approach in explaining the topics covered in these branch of mathematics. Instead of solving questions by going straight to the point, leaving you confused and frustrated, this textbook teaches you in simple English, explaining each step taken at a time. Thus, allowing anyone, regardless of their experience in algebra, differential calculus, statistics and probability, to understand each topic with ease, and hence make mathematics more interesting.

I give all thanks and Glory to God Almighty, for giving me the grace to write this book. I also wish to express my deep appreciation to my wife Mrs. Mercy Augustine for her patience, understanding and encouragement when I was writing this book. I also thank my children, Dora, Merit and Elvis for their moral support.

Kingsley Augustine.

CHAPTER 1
BASIC ARITHMETIC OPERATIONS

In order to fully understand the content of this book, it is important to be familiar with basic arithmetic operations. The following topics should remind us of these basics.

Addition and Subtraction of Integers (Directed Numbers)

a. -11 + 7 = -4
Since the two numbers have different signs, one way of solving this problem is to subtract the smaller number from the larger number. Then take the sign of the higher number in the question as the sign of the answer. In this example, 7 is the smaller number while 11 is the larger number. Hence, subtract 7 from 11 to get 4. The sign of the higher number (i.e. 11) is minus, so the answer 4 will carry a minus sign. This will finally give us -4

b. -6 + 15 = 9
Since their signs are different, subtract the smaller number from the larger number. Finally, take the sign of the higher number as the sign of the answer. In this example, 6 subtracted from 15 is 9. The sign of the higher number (i.e. 15) is plus, so the answer 9 will carry a plus sign. This will finally give us +9 or 9 as the answer as shown above.

c. 3 - 10 = -7
Their signs are different. So, subtract 3 from 10. This gives 7. The sign of the larger number (i.e. 10) is minus, so the answer 7 will carry a minus sign. This will finally give us –7.

d. -4 - 8= -12
In this case, their signs are the same. So, add the two numbers, then give the sign that they have, to the answer. In this example, 4 added to 8 gives 12. Their sign is minus, so the answer 12 will carry a minus sign. This will finally give us -12.

e. 5 - (-6) = 5 + 6 = 11
Two negative signs that are closed to each other multiply to become positive.

f. -6 - (-2) = -6 + 2 = -4. (This is similar to (a) above where the larger number has a negative sign.

g. -14 - 17 = -31
Since their signs are the same, add the two numbers to get the answer, and remember to give the sign that they have to the answer. In this example, 14 added to 17 gives 31. Their sign is minus, so the answer 31 will carry a minus sign. This will finally give us -31.

Multiplication and Division of Integers (Directed Numbers)
a. -2 x 3 = -6

Carry out multiplication and division as the usual way of multiplying and dividing numbers. However, when one of the numbers is negative, the answer will be negative. When the two numbers are negative, the answer will be positive.

b. $5 \times -4 = -20$

c. $-2 \times -6 = 12$

d. $-7 \times -3 = 21$

e. $-10 \div 2 = \dfrac{-10}{2} = -5$

f. $18 \div -6 = \dfrac{18}{-6} = -3$

g. $-48 \div -12 = \dfrac{-48}{-12} = 4$

h. $-24 \div -6 = \dfrac{-24}{-8} = 3$

Precedence (Rule of PEMDAS or BODMAS)

Precedence is the order which arithmetic operations are followed during simplifications. The order of operation is easily remembered by using the acronym below:

PEMDAS: Parenthesis, Exponent, Multiplication, Division, Addition, Subtraction.

This means that parenthesis is done first before exponents (powers and roots), then multiplication or division, and addition or subtraction. Multiplication and division are at the same level, hence deal with anyone that comes first from left to right. Similarly, addition and subtraction are at the same level, hence deal with anyone that comes first from left to right. Another acronym that can be used for precedence is:

BODMAS: Bracket, Order, Division, Multiplication, Addition, Subtraction.

This means that bracket is done first before order (powers and roots), then multiplication or division, and addition or subtraction.

Note that the 'O' in BODMAS does not mean 'Of' as popularly used in some books. However, the operation, 'Of', which can be used in some arithmetic processes, should be treated as multiplication.

Between the two acronyms i.e. PEMDAS and BODMAS, I prefer the use of BODMAS because I think it is easier to remember.

Examples

1. $8 \div 2 \times 5$

Here, the rule of BODMAS has to be applied. We have to carry out the division first before the multiplication. This is done as follows:

$8 \div 2 \times 5 = (8 \div 2) \times 5$ (The brackets show the operation that will be done first)

$$= 4 \times 5 = 20$$

2. $15 \times 2 \div 5$

Since division and multiplication are at the same level, we simplify anyone that comes first. We can work from left to right and carry out the multiplication first before division. This is done as follows:

$$15 \times 2 \div 5 = (15 \times 2) \div 5$$
$$= 30 \div 5$$
$$= 6$$

3. $6 + 24 \div 3 \times (5 - 3) \text{ of } \dfrac{3}{4}$

By applying the rule of BODMAS, the part in the bracket will be handled first. This gives:

$$6 + 24 \div 3 \times (5 - 3) \text{ of } \frac{3}{4} = 6 + 24 \div 3 \times 2 \text{ of } \frac{3}{4}$$

The next step will be division. 24 divided by 3 gives 8. This is written as follows:

$$6 + 8 \times 2 \text{ of } \frac{3}{4}$$

The 'Of' is regarded as multiplication. This now simplifies to give

$$6 + 8 \times 2 \times \frac{3}{4} = 6 + 16 \times \frac{3}{4}$$
$$= 6 + 4 \times 3 \qquad \text{(Note that 4 divides into 16 to give 4)}$$
$$= 6 + 12$$
$$= 18$$

4. $4 \times 5 - 9 \div [3 + 2^3 - (3 \times 2)]$

The inner bracket simplifies to give 6. This is given below.

$$4 \times 5 - 9 \div (3 + 2^3 - 6)$$

The exponent in the bracket gives 8 (i.e. $2^3 = 8$) as shown below.

$$4 \times 5 - 9 \div (3 + 8 - 6)$$

The addition and subtraction in the bracket simplifies to give $3 + 8 - 6 = 5$. This gives:

$$4 \times 5 - 9 \div 5$$

The multiplication and division simplify to give:

$$20 - \frac{9}{5} = \frac{100 - 9}{5}$$
$$= \frac{91}{5} = 18\frac{1}{5}$$

5. $4 + 6 - 3^2 \times (2 \times \sqrt{16})7 + 1$

The root in the bracket simplifies to give 4 as shown below:

$$4 + 6 - 3^2 \times (2 \times 4)7 + 1$$

The term in the bracket gives 8 as follows:

$$4 + 6 - 3^2 \times (8)7 + 1$$

The exponent or order or power simplifies to give 9 as shown below:

$$4 + 6 - 9 \times (8)7 + 1$$

The term outside the bracket multiplies the term in the bracket to give:

$$4 + 6 - 9 \times 56 + 1$$

We simplify the multiplication operation to give:

4 + 6 − 504 + 1

Starting from left to right we carry out the addition and subtraction as follows:

$$10 − 504 + 1 = − 494 + 1$$
$$= − 493$$

Exercise 1

1. Evaluate the following:

a. −20 + 5
b. −32 + 21
c. 2 − 7
d. −3 − 9
e. 8 − (−11)
f. −11 − (−17)
g. −56 − 26

2. Find the values of the following:
a. −5 x 2
b. 2 x −11
c. −4 x −9
d. −6 x −5
e. −24 ÷ 6
f. 42 ÷ −7
g. −72 ÷ − 9
h. −48 ÷ −12

3. Apply precedence in evaluating the following:
a. 22 ÷ 11 x 6
b. 8 x 15 ÷ 5
c. $13 + 30 ÷ 15 \times (16 − 8) \text{ of } \frac{2}{3}$
d. $12 \times \frac{5}{24} − 9 ÷ [1 + 3^4 − (2 \times 5)]$
e. $2 + 11 − 4^3 \times (3 \times \sqrt{4})5 − 3$

CHAPTER 2
SIMPLIFICATION, FACTORIZATION AND SUBSTITUTION IN ALGEBRA

The use of letters to represent numbers is known as algebra. In algebra, we imagine the letters to stand for numbers in order for operations to be easy.

Examples
1. What is y kilograms in grams

Solution
 Since 1kg = 1000g
 and 2kg = 2 x 1000g
 Then ykg = y x 1000g
    = 1000yg

Note that y x 1000 = 1000 x y = 1000y. In algebra, when a number is multiplied by a letter, the number is usually written before the letter even if the letter comes first in the multiplication. Also, when two or more letters are multiplied together, the letters are usually written in alphabetical order. For example x x d x f is written as df$x$ (in alphabetical order).

Also note that m x n is written as mn, m ÷ n is written as $\frac{m}{n}$, m + n remains m + n, while m − n remains m − n.

2. A boy was 13 years old y years ago. How old is he now? How old will he be in c years time?

Solution
It is easy to evaluate this question if the letters are imagined to be numbers. For example, if the boy was 13 years old 4 years ago, then he will be 17 years old now (i.e. 13 + 4 = 17). Therefore, the boy's present age is obtained by addition. So, his present age is (13 + y) years. Similarly, in c years time, his age will be obtained by addition. So, in c years time, the boy's age will be his present age + c. This gives:
(13 + y + c)years (Note that his present age is 13 + y)

3. A man is x years old. How old was he p years ago? How old will he be in 5 years time?

Solution
His present age is x years. His age p years ago will be obtained by subtracting p from his present age.
So, p years ago, the man's age was (x − p) years
Also, in 5 years time, his age will be:
(x + 5)years

4(a) What is $a in cents?
(b) Express x minutes in hours.

Solution

(a) Since $1 = 100 cents
and $2 = 2 x 100 cents
Then, $a = a x 100 cents
= 100a cents

(b) Since 60 minutes = 1 hour

$$1 \text{ minute} = \frac{1}{60} \times 1 = \frac{1}{60} \text{ hours}$$

$$2 \text{ minutes} = \frac{1}{60} \times 2 \text{ hours}$$

$$10 \text{ minutes} = \frac{1}{60} \times 10 \text{ hours}$$

Then, x minutes $= \frac{1}{60} \times x$ hours

$\therefore$ x minutes $= \frac{x}{60}$ hours

5. The perimeter of a rectangle is 50cm and the breadth is $y$cm. Find the area of the rectangle in terms of y.

Solution

Perimeter of a rectangle = $2(l + b)$, where l = length and b = breadth.

$50 = 2(l + y)$ (since breadth = y)

$\frac{50}{2} = \frac{2(l+y)}{2}$ (When you divide both sides by 2 in order to eliminate the 2)

$\therefore$ $25 = l + y$

$\therefore$ $25 - y = l$ (Note that when y is taken to the other side of the equation, its sign will change. Hence +y becomes –y).

$\therefore$ length, $l = 25 - y$

Area of rectangle = l x b

$\therefore$ Area = $(25 - y)$ x y

$= 25y - y^2$

Note that $y \times y = y^2$. Also, when a number or letter multiplies an expression in a bracket, the letter or number is used to multiply each term in the bracket.

6. One side of a rectangle is $(2x + 6)$cm. If the perimeter of the rectangle is $(6x + 8)$cm, find the area of the rectangle in terms of x.

Solution

Let $l = (2x + 6)$

$\therefore$ perimeter = $2(l + b)$

$6x + 8 = 2(2x + 6 + b)$

Divide both sides by 2. This gives:

$$\frac{6x + 8}{2} = \frac{2(2x + 6 + b)}{2}$$

$$\frac{6x}{2} + \frac{8}{2} = 2x + 6 + b \qquad \text{(2 has canceled out)}$$

$$3x + 4 = 2x + 6 + b$$

Collecting terms in x and constant terms on the same side of the equation gives b as follows:

$$3x + 4 - 2x - 6 = b$$

$$3x - 2x + 4 - 6 = b$$

$$x - 2 = b$$

$\therefore$ Breadth, b $= x - 2$

The area of the rectangle is given by:

Area = l x b

$$= (2x + 6) \times (x - 2)$$

$$= (2x + 6)(x - 2)$$

Area $= 2x^2 - 4x + 6x - 12$

Area $= 2x^2 + 2x - 12$

Note that in expanding the bracket above, each term in the first bracket is used to multiply each term in the second bracket. Ensure you carry the sing of each term along with the term when carrying out the expansion.

Expansion of Brackets

The expression a$(x + y)$ means a x $(x + y)$, and it is expanded as follows:

a$(x + y)$ = ax + ay

The expression (a + b)$(x + y)$ means (a + b) x $(x + y)$, and it is expanded as follows:

(a + b)$(x + y)$ = ax + ay + bx + by. Each term in the first bracket is used to multiply each term in the second bracket.

Examples

1. Expand 2$(x - 3y)$

<u>Solution</u>

2$(x - 3y)$ = 2(x) +2(-3y)

$\qquad$ = 2x - 6y

2. Expand and simplify the expression, 2x - [3 - (a + b)]

<u>Solution</u>

2x - [3 - (a + b)] = 2x - (3 - a - b)

$\qquad$ = 2x - 3 + a + b

Note that the inner bracket was expanded first before the outer bracket. Also, a negative sign outside a bracket is taken as -1 outside the bracket. For example

-(a+b) = -1(a + b) = (-1 x a) + (-1 x b) = -a - b as expressed above.

3. Remove bracket and simplify a + 2(a − b) − (a + 3b)

<u>Solution</u>
a + 2(a − b) − (a + 3b) = a + 2a − 2b − a − 3b
Collecting like terms together gives:
$\qquad$ a + 2a − a − 2b − 3b
$\qquad$ = 2a − 5b
Note that terms that have the same letters are like terms. Numbers without letters are also like terms.

4. Simplify $2x(x − 3) − 2(x^2 + 2x)$

<u>Solution</u>
$\quad 2x(x − 3) − 2(x^2 + 2x) = 2x(x) + 2x(-3) − 2(x^2) − 2(+2x)$
$\qquad\qquad\qquad\qquad = 2x^2 − 6x − 2x^2 − 4x$
Collecting like terms together gives:
$\qquad 2x^2 − 2x^2 − 6x − 4x$
$\quad = −10x \qquad$ (since $2x^2 − 2x^2 = 0$)

5. Expand (a − b)(a − c)

<u>Solution</u>
$\quad$ (a − b)(a − c) = a x a + (a x −c) + (−b x a) + (−b x −c)
$\qquad\qquad\qquad = a^2 − ac − ab + bc \qquad$ (This cannot simplify further)
Take note of how the negative sign was used in the expansion.

6. Expand and simplify $(2x + 5y)(3x − 2y)$

<u>Solution</u>
$(2x + 5y)(3x − 2y) = (2x$ x $3x) + (2x$ x $−2y) + (5y$ x $3x) + (5y$ x $−2y)$
$\qquad\qquad\qquad = 6x^2 − 4xy + 15xy − 10y^2$
$\qquad\qquad\qquad = 6x^2 + 11xy − 10y^2$
-Note that −4xy and +15xy are like terms which simplify to +11xy. Note that x x x = x^2. Similarly, x^2 x x = x^3. This is done by adding their powers. Note that x has a power of 1 which is not usually written.

7. Expand and simplify $(3a − 3)^2$

<u>Solution</u>
$(3a − 3)^2 = (3a − 3)(3a − 3)$
$\qquad\quad = (3a$ x $3a) + (3a$ x $−3) + (−3$ x $3a) + (−3$ x $−3)$
$\qquad\quad = 9a^2 − 9a − 9a + 9$
$\qquad\quad = 9a^2 − 18a + 9$

Note : that $-x- = +$ (i.e. minus multiplied by minus gives plus), $-x+ = -$, $+x- = -$. For example $-2a \times -a = +2a^2$, and $2 \times -3 = -6$

8. Find the coefficient of ab in the expansion and simplification of: $(2a-3b)^2 - (a-b)^2$

Solution
$(2a-3b)^2 - (a-b)^2 = (2a-3b)(2a-3b) - [(a-b)(a-b)]$
$$= 4a^2 - 6ab - 6ab + 9b^2 - (a^2 - ab - ab + b^2)$$
$$= 4a^2 - 6ab - 6ab + 9b^2 - a^2 + ab + ab - b^2$$
Collecting like terms together gives
$$4a^2 - a^2 - 6ab - 6ab + ab + ab + 9b^2 - b^2$$
$$= 3a^2 - 10ab + 8b^2$$
$\therefore$ The coefficient of ab = -10
Note that coefficient is a number multiplying a particular variable or letter/letters

9. Simplify and find the coefficient of d^2 in $3d(2d+3) - (3d+1)(2d+1)$

Solution
$3d(2d+3) - (3d+1)(2d+1) = 3d(2d+3) - [(3d+1)(2d+1)]$
Note that the outer bracket is introduced due to the negative sign multiplying the bracket
$$= 6d^2 + 9d - (6d^2 + 3d + 2d + 1)$$
$$= 6d^2 + 9d - 6d^2 - 3d - 2d - 1$$
$$= 6d^2 - 6d^2 + 9d - 3d - 2d - 1$$
$$= 0d^2 + 4d - 1$$
$$= 4d - 1$$
$\therefore$ The coefficient of $d^2 = 0$

Lowest Common Multiple (L.C.M) of Algebraic Terms

The LCM of a given set of numbers is the first multiple of the largest number which can divide all the other numbers without a remainder .
For letters, the LCM is obtained by multiplying all the letter together. However, when a particular letter is repeated in the various terms, the LCM is the letter that has the highest power.

Examples
Find the LCM of the following:
1. x, y, z
2. $2ab, 3ax, ac$
3. $5ax^2, 3a^2y, 3xy$
4. $6a$ and $3b$
5. $(x-y)$ and $(2x+y)$
6. $(a-2b)$ and $(4a-8b)$

<u>Solutions</u>

1. The LCM of x, y, $z = xyz$ (Their multiplication)

2. 2ab, 3ax, ac. Here, first find the LCM of 2 and 3. The larger number is 3. The multiples of 3 are 3, 6, 9, 12, … e.t.c. Out of these multiples the first one that can divide 2 and 3 without a reminder is 6. So the LCM of 2 and 3 is 6.
For the letters, the highest power of a is 1, so the LCM of a in all the terms is a. Similarly, the LCM of b, c, and x are b, c, and x, since they all have power of 1. We now multiply all these LCMs to obtain the overall LCM. Therefore the LCM of 2ab, 3ax and ac is 6abc$x$.

3. 5ax^2, 3a^{2}y, 3xy. For the numbers, the largest is 5. The multiples of 5 are 5, 10, 15, 20 … e.t.c. The first multiple that can divide 3, 3 and 5 is 15. Hence, the LCM of 5, 3, 3 is 15.
For the letters, a in its highest power is a^2, for x is x^2 and for y is y. So the LCM of 5ax^2, 3a^{2}y and 3xy is 15a$^2x^2$y.

4. 6a and 3b. For the numbers, the larger value is 6. The multiples of 6 are 6, 12, 18, … e.t.c. The first multiple that can divide 6 and 3 is 6. The letters are single in the two terms. So the LCM of 6a and 3b is 6ab.

5. $(x - y)$ and $(2x + y)$. Here, the terms in the bracket are not the same and they cannot be factorized to look alike, so the LCM is obtained by multiplying them together.
$\therefore$ The LCM of $(x - y)$ and $(2x + y)$ is $(x - y)(2x + y)$

6. $(a - 2b)$ and $(4a - 8b)$. If we look at the second expression carefully, we will observe that it can be factorized to have an expression like the first one. If we factorize the second expression it will give $4(a - 2b)$. Hence, we can say that we are looking for the LCM of $(a - 2b)$ and $4(a - 2b)$. Since $(a - 2b)$ is common, we simply pick one as an LCM and then we take 4 as an LCM since it is the only number present. When we multiply 4 and $(a - 2b)$ (i.e. one of the common expression) it gives $4(a - 2b)$.
$\therefore$ The LCM of $(a - 2b)$ and $(4a - 8b)$ is $4(a - 2b)$.

Addition and Subtraction of Algebraic Fractions

Examples

1. Simplify the following:

(a) $\dfrac{3x}{8} + \dfrac{x}{8}$

(b) $\dfrac{1}{5x} + \dfrac{3}{4x}$

(c) $\dfrac{2a}{3b} - \dfrac{3}{4x}$

(d) $\dfrac{3x+2}{5a} + \dfrac{2b-1}{10b}$

(e) $\dfrac{a+3b}{a} - \dfrac{2a-b}{4b} - 3$

Solutions

(a) $\dfrac{3x}{8} + \dfrac{x}{8}$

The LCM of 8 and 8 is 8. So, divide 8 (i.e. LCM) by each of the denominators and multiply the answer by the corresponding numerator. This gives:

$$\dfrac{3x}{8} + \dfrac{x}{8} = \dfrac{(1 \times 3x) + (1 \times x)}{8} = \dfrac{3x+x)}{8} = \dfrac{4x}{8}$$

$$= \dfrac{x}{2} \quad \text{(in its lowest term)}$$

(b). $\dfrac{1}{5x} + \dfrac{3}{4x}$

The LCM of $5x$ and $4x$ is $20x$. Divide $20x$ by each denominator and multiply the value obtained by the corresponding numerator. This gives:

$$\dfrac{1}{5x} + \dfrac{3}{4x} = \dfrac{(4 \times 1) + (5 \times 3)}{20x} = \dfrac{4+15}{20x} = \dfrac{19}{20x}$$

Note that $\dfrac{20x}{5x} = 4$, while $\dfrac{20x}{4x} = 5$. These values are used to multiply their corresponding numerators as shown in the solution above.

(c) $\dfrac{2a}{3b} - \dfrac{3}{4x} = \dfrac{(4x \times 2a) - (3b \times 3)}{12bx} = \dfrac{8ax - 9b}{12bx}$

Note that the answer above cannot be simplified further. Also, $\dfrac{12bx}{3b} = 4x$, while $\dfrac{12bx}{4x} = 3b$, as b and x cancels out respectively.

(d) $\dfrac{3x+2}{5a} + \dfrac{2b-1}{10b} = \dfrac{2b(3x+2) + a(2b-1)}{10ab} = \dfrac{6bx + 4b + 2ab - a}{10ab}$ (This cannot be simplified further)

(e) $\dfrac{a+3b}{a} - \dfrac{2a-b}{4b} - \dfrac{3}{1}$ (The 3 is expressed as $\dfrac{3}{1}$)

This gives: $\dfrac{a+3b}{a} - \dfrac{2a-b}{4b} - \dfrac{3}{1} = \dfrac{4b(a+3b) - a(2a-b) - 4ab(3)}{4ab}$

$$= \dfrac{4ab + 12b^2 - 2a^2 + ab - 12ab}{4ab}$$

$$= \dfrac{4ab + ab - 12ab + 12b^2 - 2a^2}{4ab}$$

$$= \dfrac{-7ab + 12b^2 - 2a^2}{4ab}$$

19

$$= \frac{12b^2 - 7ab - 2a^2}{4ab}$$

2. Express each of the following as a single fraction in its simplest form.

(a) $\dfrac{3x + 2}{3} - \dfrac{x - 1}{4} - \dfrac{5}{12}$

(b) $\dfrac{5}{a + 4} - \dfrac{2}{a - 2}$

(c) $\dfrac{a - 3}{a + 2} - \dfrac{a - 2}{a - 5}$

Solutions

(a) $\dfrac{3x + 2}{3} - \dfrac{x - 1}{4} - \dfrac{5}{12} = \dfrac{4(3x+2) - 3(x - 1) - (1 \times 5)}{12}$

$$= \frac{12x + 8 - 3x + 3 - 5}{12}$$

$$= \frac{12x - 3x + 8 + 3 - 5}{12}$$

$$= \frac{9x + 6}{12}$$

$$= \frac{3(3x + 2)}{12} \qquad \text{(After factorizing the numerator)}$$

$$= \frac{3x + 2}{4} \qquad \text{(After equal division by 3)}$$

(b) $\dfrac{5}{a + 4} - \dfrac{2}{a - 2}$

The LCM of a + 4 and a − 2 is (a + 4)(a − 2)

Also, if we use this LCM to divide the first denominator, it gives: $\dfrac{(a + 4)(a - 2)}{a + 4} = a - 2$ (Since a + 4 will cancel out).

Similarly, the LCM divided by the second denominator gives: $\dfrac{(a + 4)(a - 2)}{a - 2} = a + 4$ (Since a − 2 will cancel out)

$$\therefore \frac{5}{a + 4} - \frac{2}{a - 2} = \frac{(a-2)5 - (a+4)2}{(a + 4)(a - 2)} = \frac{5a - 10 - (2a + 8)}{(a + 4)(a - 2)}$$

$$= \frac{5a - 10 - 2a - 8}{(a + 4)(a - 2)}$$

$$= \frac{3a - 18}{(a + 4)(a - 2)}$$

$$= \frac{3(a-6)}{(a+4)(a-2)}$$

(c) $\dfrac{a-3}{a+2} - \dfrac{a-2}{a-5}$

The LCM of a + 2 and a − 5 is (a + 2)(a − 5)

$$\therefore \quad \frac{a-3}{a+2} - \frac{a-2}{a-5} = \frac{(a-3)(a-5) - (a-2)(a+2)}{(a+2)(a-5)}$$

$$= \frac{a^2 - 5a - 3a + 15 - (a^2 + 2a - 2a - 4)}{(a+2)(a-5)}$$

$$= \frac{a^2 - 8a + 15 - (a^2 - 4)}{(a+2)(a-5)} \qquad \text{(Note that +2a − 2a = 0)}$$

$$= \frac{a^2 - 8a + 15 - a^2 + 4)}{(a+2)(a-5)}$$

$$= \frac{a^2 - a^2 - 8a + 15 + 4}{(a+2)(a-5)}$$

$$= \frac{-8a + 19}{(a+2)(a-5)}$$

Highest Common Factor (H.C.F) of Algebraic Terms

The HCF of a given set of numbers is the number which is just lower than or equal to the smallest number in the given set, and which all the numbers in the given set can divide without a remainder.

For letters, the HCF of a particular letter is obtained by taking the letter with the lowest power in the given set.

Examples

Find the HCF of the following:

1. 4, 6, 8
2. 3, 9, 15
3. $8a^2b$, $20ab^2$
4. a^2b^4c, $a^3b^2c^3$
5. $36xy^3z^5$, $54xy^5z^3$, $72xy^4z^5$

Solutions

1. The H.C.F of 4, 6, 8 is 2. The smallest number in the set is 4. So, a number that is just lower than or equal to 4 which 4, 6, and 8 can divide without a remainder is 2.

So, their H.C.F is 2.

2. The H.C.F of 3, 9, and 15 is 3. This is because the lowest number is 3, and all the numbers can divide 3 without remainder.

3. For the terms, $8a^2b$ and $20ab^2$, the H.C.F of 8 and 20 is 4. The lower number is 8, and a number which is just lower than 8 or equal to 8, and which 8 and 20 can divide without remainder is 4. Concerning the letters, between a^2 and a, we take "a", since it has the lower power. Similarly, between b and b^2 we take b since it has the lower power. Therefore, combining all these answers together gives the H.C.F of $8a^2b$ and $20ab^2$ as 4ab.

4. a^2b^4c, $a^3b^2c^3$
For the letter a, we take a^2 since it has the smaller power. For the letter b, we take b^2, while for the letter c, we take c since it has the lower power.
Therefore, the H.C.F of a^2b^4c and $a^3b^2c^3$ is a^2b^2c.

5. $36xy^3z$, $54xy^5z^3$, $72xy^4z^5$
For the numbers, 36, 54 and 72, the smallest is 36. All the numbers can divide 2, 3, 6, 9 and 18. Note that these factors must not be greater than 36 which is the smallest in the set. So, among 2, 3, 6, 9 and 18, the highest is 18. So, 18 is the number which is just lower than or equal to 36, and which all the numbers (i.e. 36, 54, 72) can divide without remainder.
For the letters, the lowest power of x is 1 (i.e. x), the lowest power of y is 3 (i.e. y^3), while the lowest power of z is 3 (i.e. z^3). So, the H.C.F of the letters is xy^3z^3
∴ The HCF of $36xy^3z^5$, $54xy^5z^3$, and $72xy^4z^5$ = $18xy^3z^3$.

Factorization of Algebraic Expression

An algebraic expression can be expressed as a product of its factors. When factorizing expressions we simply take the HCF of the terms in the expression and put it outside a bracket. Then we divide each individual term in the expression by the HCF in order to obtain the new terms that will be inside the bracket.

Examples
1. Factories 3a – 12b

Solution
The HCF of 3 and 12 is 3. For the letters, since the letters are not repeated, then their HCF which is 1 will not be used here. When there is no common letter or number to a set of terms, then their HCF is taken as 1.
So, factorize the expression by taking the HCF and then introducing a bracket. Then, divide each term of the expression by the HCF. This is done as follows:
$3a - 12b = 3(\frac{3a}{3} - \frac{12b}{3})$ (The HCF of 3a and 12b is 3, which is used outside the bracket, and also used to divide each term in the expression)

= 3(a – 4b)   (After simplifying the division above)

∴ 3a – 12b = 3(a – 4b)

(Note that the expansion of the factorized expression will give the original expression)

2. Factorize 15y + 10.

<u>Solution</u>

15y + 10

The HCF of 15y and 10 is 5. The letter is not considered since it appears in only one of the terms

∴ $15y + 10 = 5(\dfrac{15y}{5} + \dfrac{10}{5})$

$= 5(3y + 2)$ (This is obtained after the division of 15 by 5, and 10 by 5)

∴ 15y + 10 = 5(3y + 2)

3. Factorize: –6m + 15mn

<u>Solution</u>

–6m + 15mn

The HCF of –6m and 15m is –3m. Always take the sign of the first term.

Note that n is not present in both terms, so it is not considered in the HCF.

∴ $-6m + 15mn = -3m(\dfrac{-6m}{-3m} + \dfrac{15mn}{-3m})$

$= -3m(2 – 5n)$

(Note that –6m divided by –3m is 2, while 15mn divided by –3m is –5n)

4.   Factorize $-4a^5 + 2a^3b - 10a^2b^2$

<u>Solution</u>

$-4a^5 + 2a^3b - 10a^2b$

The HCF of $-4a^5$, $2a^3b$ and $10a^2b$ is $-2a^2$. Note that b is not present in all the three terms, so it is not considered in the HCF. Since $-2a^2$ is the HCF of the terms, it also means that $-2a^2$ is what is common to the three terms in the expression.

∴ $-4a^5 + 2a^3b - 10a^2b^2 = -2a^2(\dfrac{-4a^5}{-2a^2} + \dfrac{2a^3b}{-2a^2} - \dfrac{10a^2b^2}{-2a^2})$

$= -2a^2(2a^3 – ab + 5b^2)$

Note that $\dfrac{a^5}{a^2} = a^{5-2} = a^3$ (simply subtract the powers). Also note that a negative sign outside a bracket changes the sign of every term in the bracket.

5. Factorize $3m^3 – 2m^2 + m$

<u>Solution</u>

$3m^3 – 2m^2 + m$

The terms m^3, m^2 and m have m in common. This means that m is their HCF. 3 and 2 have nothing but 1 in common. So, their HCF is not considered here.

$$\therefore \quad 3m^3 - 2m^2 + m = m(\frac{3m^3}{m} - \frac{2m^2}{m} + \frac{m}{m})$$
$$= m(3m^2 - 2m + 1)$$

Factorization by Grouping

In factorization by grouping four terms are usually given in the expression. The first two of the terms will be factorized before the other two terms are factorized. After this, the common term obtained is used to carry out the final factorization of the terms.

Examples

1. Factorize $ax + ay + 3bx + 3by$

Solution

The terms ax and ay have 'a' in common. This also means that their HCF is 'a'. The terms $3bx$ and $3by$ have 3b in common. This also means that their HCF is 3b. We now factorize each pair as follows:

$ax + ay + 3bx + 3by = a(x + y) + 3b(x + y)$

The final step is to take the factors outside each of the bracket (i.e. a and 3b) and enclose them in a bracket. Then, one of the two equal bracket terms is also taken along. This gives:

$a(x + y) + 3b(x + y) = (a + 3b)(x + y)$

This also means that $(x + y)$ is common to the terms above. So, it can be factorized as follows:

$$a(x + y) + 3b(x + y) = (x + y)[\frac{a(x + y)}{(x + y)} + \frac{3b(x + y)}{(x + y)}]$$

$(x + y)$ cancels out to give: $(x + y)(a + 3b)$ as obtained above.

Hence, $ax + ay + 3bx + 3by = (x + y)(a + 3b)$

2. Factorize $2ab - 5a + 2b - 5$

Solution

'a' is common to the first two terms. The next two terms have nothing in common, so we can take 1 to be the common term between them.

$\therefore \quad 2ab - 5a + 2b - 5 = a(2b - 5) + 1(2b - 5)$

Taking one of the bracket (since they are equal) along with the factors outside each of the bracket gives:

$a(2b - 5) + 1(2b - 5) = (2b - 5)(a + 1)$

3. Factorize $2ax - 2ay - 3bx + 3by$

Solution

2a is common to the first two terms, while −3b is common to the last two terms. Note that the sign of the first of the two terms must be taken. For example, $-ab + 2ax$ has −a as the common term. Also, when a negative term is used as a common term, the inner sign of the bracket

24

formed will change. For example, $-2a - 3ax$ will give $-a(2 + 3x)$. Notice the change of sign when the bracket is formed, especially as $-3ax$ becomes $+3x$ in the bracket.

$$\therefore \quad 2ax - 2ay - 3bx + 3by = 2a(x - y) - 3b(x - y)$$
$$= (x - y)(2a - 3b)$$

Notice the change in sign of +3by to −y after factorizing.

4. Factorize $15 - xy + 5y - 3x$

Solution

Since the first two terms have nothing in common, and the last two terms have nothing in common, then we regroup the expression. To regroup the expression means to rearrange the terms such that the first two terms will have a common factor and the last two terms will have a common factor. This gives:

$15 - xy + 5y - 3x = 15 + 5y - xy - 3x$

∴ The expression to be factorized is:

$15 + 5y - xy - 3x$

The first two terms have 5 as a factor while the last two terms have $-x$ as a factor. The expression is now factorized as follows:

$$15 + 5y - xy - 3x = 5(3 + y) - x(y + 3)$$
$$= (3 + y)(5 - x)$$

Note that $(3 + y)$ is also equal to $(y + 3)$.

Difference of Two Squares

$$(a + b)(a - b) = a^2 - ab + ab - b^2$$
$$= a^2 - b^2 \quad \text{(since } -ab + ab = 0\text{)}$$
$$\therefore \quad a^2 - b^2 = (a + b)(a - b)$$

This can be used to factorize the difference of the squares of two quantities.

Examples

1. Factorize $81 - w^2$

Solution

$81 - w^2$

We can express 81 as a square in order to make the expression a difference of two squares. This gives $81 - w^2 = 9^2 - w^2$

$$= (9 + w)(9 - w) \quad \text{[This is similar to } a^2 - b^2 = (a + b)(a - b) \text{ as explained above]}$$

2. Factorize $16y^2 - 4z^2$

Solution

$$16y^2 - 4z^2 = 4^2y^2 - 2^2z^2 \quad \text{(Since } 16 = 4^2 \text{ and } 4 = 2^2\text{)}$$
$$= (4y)^2 - (2z)^2 \quad \text{(This is a difference of two squares)}$$

$$= (4y + 2z)(4y - 2z)$$

3. Factorize $5 - 5m^2$

Solution
The common factor is 5. This is used to factorize the expression as follows:
$$5 - 5m^2 = 5(1 - m^2)$$
But, $1 - m^2$ is a difference of two squares because it can be expressed as
$$1^2 - m^2 \quad \text{(Since } 1^2 = 1\text{)}$$
$$\therefore \ 5(1 - m^2) = 5(1^2 - m^2)$$
We now factorize the terms in the bracket (i.e. difference of two squares) as follows:
$$5(1^2 - m^2) = 5(1 + m)(1 - m)$$
$$\therefore \ 5 - 5m^2 = 5(1 + m)(1 - m)$$

4. Factorize $25x^2 - 9$

Solution
$$\begin{aligned}
25x^2 - 9 &= 5^2 x^2 - 3^2 \\
&= (5x)^2 - 3^2 \\
&= (5x + 3)(5x - 3)
\end{aligned}$$

5. Find the value of $62^2 - 38^2$ without using calculator.

Solution
$$\begin{aligned}
62^2 - 38^2 &= (62 + 38)(62 - 38) \\
&= 100 \times 24 = 2400
\end{aligned}$$
Notice that what was obtained from one bracket was used to multiply what was obtained from the other bracket.

6. Find the value of $84^2 - 16$ without using calculator.

Solution
$84^2 - 16 = 84^2 - 4^2$ (Both terms must be expressed in squares).
$$\begin{aligned}
\therefore \quad 84^2 - 4^2 &= (84 + 4)(84 - 4) \\
&= 88 \times 80 \\
&= 7040
\end{aligned}$$

Factorization of Simple Quadratic Expression

A quadratic expression is an expression in which 2 is the highest power of the unknown. For example, $2x^2 - 5x + 8$ is a quadratic expression.

Examples

1. Factorize $x^2 + 8x + 12$

Solution

Since the co–efficient of x^2 is 1 (i.e. $1x^2$ also means x^2), then we can apply a short method as follows:

$$x^2 + 8x + 12 = (x \quad)(x \quad)$$

Find two numbers such that their product is +12 and their sum is +8. These two numbers must be factors of 12 since their product has to give 12. An easy way to find the factors of a number is represented below:

$$
\begin{array}{c|c}
2 & 12 \\
2 & 6 \\
3 & 3 \\
  & 1 \\
\end{array}
$$

∴ The factors are 12 and 1, 6 and 2, 3 and 4. These are the numbers (or multiplication of numbers) on the right side and left side of the line above, along with a number that must be used to multiply them in order to obtain 12.

∴ Among the three pairs of factors of 12 above, the factors whose product is +12, and sum is +8 are +6 and +2. Note that the correct sign must be taken along with the factors. These two numbers are now used to complete the two brackets above. This gives:

$$x^2 + 8x + 12 = (x + 6)(x + 2)$$

2. Factorize $c^2 - 8c - 20$

Solution

The co–efficient of c^2 is 1. So let's apply the direct method.

$$c^2 - 8c - 20 = (c \quad)(c \quad)$$

Find two numbers such that their product is −20 (i.e. the last term) and their sum is −8 (i.e. the middle term). These two numbers must be factors of 20. We can obtain the factors as follows:

$$
\begin{array}{c|c}
2 & 20 \\
2 & 10 \\
5 & 5 \\
  & 1 \\
\end{array}
$$

∴ The factors are 20 and 1 (since 20 x 1 = 20). The second factors are 10 and 2 (since 10 x 2 = 20). The third factors are 5 and 4 (since 5 x 4 = 20). Note that 4 is obtained from 2 x 2. Among these factors, it is −10 and +2 that will give a product of −20 and a sum of −8. Note that you have to try different signs with the factors in order to get the right sign. The two numbers are now used to factorize the expression by putting them in the bracket above. This gives:

$$c^2 - 8c - 20 = (c - 10)(c + 2)$$

3. Factorize $m^2 - 11m + 24$

<u>Solution</u>

$$m^2 - 11m + 24 = (m \quad)(m \quad)$$

Two numbers whose product is +24 and sum is −11 are −8 and −3. This is because −8 x (−3) = +24,

and −8 + (−3) = −11. These two numbers are now used to complete the brackets above.

∴ $m^2 - 11m + 24 = (m - 8)(m - 3)$

Note that when these brackets are expanded, the original quadratic expression will be obtained.

4. Factorize $2a^2 - 10a + 12$

<u>Solution</u>

The co–efficient of a^2 is 2. It is not 1, so the direct method used above cannot be applied here. Hence the first step in this case is to multiply the first and last terms (i.e. $2a^2$ and +12). This gives: $2a^2$ x 12 = $24a^2$

We now find two numbers as coefficients of 'a' such that their product is $24a^2$ and their sum is −10a (i.e. the middle term in the original expression). These two numbers must be factors of $24a^2$. They are −6a and −4a. We now substitute these two terms for −10a in the original expression. This gives:

$$2a^2 - 10a + 12 = 2a^2 - 6a - 4a + 12$$

We now factorize by grouping. Between the first two terms 2a is common while between the last two terms −4a is common. This now factorizes as follows:

$$2a^2 - 10a + 12 = 2a^2 - 6a - 4a + 12$$
$$= 2a(a - 3) - 4(a - 3)$$
$$= (a - 3)(2a - 4)$$

Substitution

Examples

1. Evaluate the following if $x = 2$, and y = 3
(a) $2x - 3y$
(b) $x + 2(3x - y)$

<u>Solutions</u>
(a) $2x - 3y$

Substitute 2 for x and 3 for y. Note that $2x$ means 2 x x. Similarly, 3y means 3 x y

∴ $2x - 3y$

   = (2 x 2) − (3 x 3)

$= 4 - 9$

$= -5$

(b) $x + 2(3x - y)$ (Substitute 2 for x and 3 for y)

   $= 2 + 2[(3 \times 2) - 3]$

   $= 2 + 2(6 - 3)$

   $= 2 + 2(3)$

   $= 2 + 6 = 8$

Note that the term in the bracket is first evaluated. Also, a term outside a bracket is multiplying the contents of the bracket.

2. Evaluate the following given that a = –1, b = 3 and c = –5

(a) $2(a - 2c) - 4b$

(b) $(b - a)(c + 4a)$

(c) $\dfrac{a + bc}{b - ac}$

Solutions

(a) $2(a - 2c) - 4b$

   $= 2[-1 - (2(-5)] - (4 \times 3)$ (Substitute the values of the letters as given in the question)

   $= 2[-1 - (-10)] - 12$

   $= 2(-1 + 10) - 12$

   $= 2(9) - 12$

   $= 18 - 12$

   $= 6$

b. $(b - a)(c + 4a)$

   $= [3 - (-1)][-5 + (4(-1)]$

   $= (3 + 1)[-5 + (-4)]$

   $= 4(-5 - 4)$

   $= 4 \times (-9)$

   $= -36$

Note that it is advisable to enclose each multiplication in a bracket. For example 4b is expressed as (4 x 3), and 4a is expressed as [4 x (–1)]. Always evaluate the contents of inner brackets before the outer brackets and then outside the brackets.

(c) $\dfrac{a + bc}{b - ac}$

   $= \dfrac{-1 + [3 \times (-5)]}{3 - [(-1) \times (-5)]}$

   $= \dfrac{-1 + (-15)}{3 - (5)}$

$$= \frac{-1 - 15}{3 - 5}$$

$$= \frac{-16}{-2}$$

= 8   (The negative sign cancels out)

Note that the numerator and denominator have to be reduced to a single value before division is possible.

3. Given that p = 3, and q = –1, find the value of x when $x = p^2q - q^2p$.

Solution
$x = p^2q - q^2p$
$x = [3^2 \times (-1)] - [(-1)^2 \times 3]$ (Simply substitute the given values of p and q)
$= [9 \times (-1)] - (1 \times 3)$
$= -9 - 3 = -12$

Note that the terms having powers must be evaluated first before further evaluations.

4. Evaluate: $ab\sqrt{c^2 + b^2}$, given that a = 2, b = –3, c = 4

Solution
$ab\sqrt{c^2 + b^2}$
$= [2 \times (-3)] \times \sqrt{4^2 + (-3)^2}$
$= (-6) \times \sqrt{16 + 9}$
$= -6 \times \sqrt{25}$
$= -6 \times 5 = -30$

5. If $x = -7$ and y = 3, calculate the values of:

(a) $(\frac{x + y}{x - y})^2$

(b) $2x^2y + y^2x$

Solutions

(a) $(\frac{x + y}{x - y})^2$

$= (\frac{-7 + 3}{-7 - 3})^2$

$= (\frac{-4}{-10})^2$

$= (\frac{2}{5})^2$ (After simplifying the term in the bracket)

$$= \frac{2^2}{5^2}$$

$$= \frac{4}{25}$$

Note that the contents of the bracket must be evaluated to a single digit before the use of the power. The power is used for both the numerator and denominator.

(b) $2x^2y + y^2x$

$= [2(-7)^2(3)] + [(3^2)(-7)]$

$= (2 \times 49 \times 3) + [9 \times (-7)]$

$= 294 + (-63)$

$= 294 - 63 = 231$

Exercise 2

1. What is y kilometers in meters?
2. A man was 38 years old m years ago. How old is he now? How old will he be in p years time?
3. A man is b years old. How old was he c years ago? How old will he be in 12 years time?
4.(a) What is $y in cents?
(b) Express n seconds in minutes. What is n seconds in hours?
5. The perimeter of a rectangle is 44cm and the breadth is $x$cm. Find the area of the rectangle in terms of x.
6. One side of a rectangle is $(x + 10)$cm. If the perimeter of the rectangle is $(4x + 18)$cm, find the area of the rectangle in terms of x.
7. Expand $5(2x - y)$
8. Expand and simplify the expression, $11x - [5 - 3(x + y)]$
9. Remove bracket and simplify $3a + (a - 2b) - (2a + b)$
10. Simplify $3x(7x + 2) - (10x^2 - 9x)$
11. Expand $(2a + 3b)(a - 5b)$
12. Expand and simplify $(3x - 2y)(5x - y)$
13. Expand and simplify $(4a - 5)^2$
14. Find the coefficient of xy in the expansion and simplification of: $(3x + 2y)^2 - (2x - 3y)^2$
15. Simplify and find the coefficient of y^2 in $5y(y - 2) - (2y - 3)(y - 4)$
16. Find the LCM of the following:
a. l, m, n
b. 6cd, 2ed, ec
c. $2a^3x$, $5ay^2$, $6x^2y$
d. 20m and 10n
e. $(x + y)$ and $(3x - y)$
17. Simplify the following:
a. $\frac{2x}{5} + \frac{x}{5}$

31

b. $\dfrac{4}{9x} + \dfrac{1}{3x^2}$

c. $\dfrac{a}{10b} - \dfrac{7}{15x}$

d. $\dfrac{2x+5}{8a} + \dfrac{b-3}{5b}$

e. $\dfrac{5a-b}{3} - \dfrac{a-5b}{5a^2} - 1$

18. Express each of the following as a single fraction in its simplest form.

a. $\dfrac{x-4}{10} - \dfrac{3x-2}{5} + \dfrac{4}{15}$

b. $\dfrac{2}{3a+7} - \dfrac{1}{6a+14}$

c. $\dfrac{2a-1}{3a-1} \cdot \dfrac{3a+4}{a-2}$

19. Find the HCF of the following:
a. 5, 20, 30
b. 6, 14, 24
c. $12a^3b^2$, $30a^2b^3$
d. ab^2c^3, ab^5c^2
e. $24x^2yz^3$, $72x^2y^2z^4$, $36x^3yz^2$
20. Factorize $4m - 8n$
21. Factorize $12y - 8$.
22. Factorize: $-20y - 50yz$
23. Factorize $5m^4 - 10m^2 - m^3$
24. Factorize $2x + 6y + 5mx + 15my$
25. Factorize $3ab + 7a - 3b - 7$
26. Factorize $5bx - 5by + nx - ny$
27. Factorize $15 - xy + 3y - 5x$
28. Factorize $100 - m^2$
29. Factorize $25x^2 - 9y^2$
30. Factorize $7 - 7b^2$
31. Factorize $36p^2 - 4$
32. Find the value of $63^2 - 37^2$ without using calculator.
33. Find the value of $92^2 - 64$ without using calculator.
34. Factorize $x^2 + 6x + 9$
35. Factorize $b^2 - b - 20$
36. Factorize $n^2 - 14n + 48$
37. Factorize $5a^2 - 11a - 12$

38. Evaluate the following if $x = -3$, and y = -2

a. $x^2y - y^2x$

b. $(y^2 - x^2)(y - x)$

39. Evaluate the following given that a = -1, b = 3 and c = -5

a. $2(4a - c) + 7b$

b. $(3a - b)(b - 2c)$

c. $\dfrac{ab - bc}{a - bc}$

40. Given that p = 1, and q = -3, find the value of x when $x = pq - q^2q$.

41. If x = -4 and y = 2, calculate the values of:

a. $(\dfrac{x + 2y}{x - 6y})^2$

b. $2y^2 - x^2y$

CHAPTER 3
LAWS OF INDICES

The following are the laws of indices. They are true for all values of a, b and $x \neq 0$

Law 1: $x^a \times x^b = x^{a+b}$

Law 2: $x^a \div x^b = x^{a-b}$

Law 3: $x^0 = 1$

Law 4: $x^{-a} = \dfrac{1}{x^a}$ or $bx^{-a} = \dfrac{b}{x^a}$ or $\left(\dfrac{b}{x}\right)^{-a} = \left(\dfrac{x}{b}\right)^{a}$

Examples
Simplify the following:

1. $10^5 \times 10^4$

2. $m^8 \div m^5$

3. $\dfrac{a^{-8}}{a^3}$

4. $5x^2 \times 4x^0 \times 2x^{-6}$

5. $y^{-5} \div b^0$

<u>Solution</u>

1. $10^5 \times 10^4 = 10^{5+4} = 10^9$

2. $m^8 \div m^5 = m^{8-5} = m^3$

3. $\dfrac{a^{-8}}{a^3} = a^{-8-3} = a^{-11} = \dfrac{1}{a^{11}}$

4. $5x^2 \times 4x^0 \times 2x^{-6} = (5 \times 4 \times 2)x^{2+0+(-6)} = 40x^{2-6} = 40x^{-4} = \dfrac{40}{x^4}$

5. $y^{-5} \div b^0 = y^{-5} \div 1 = y^{-5} = \dfrac{1}{y^5}$

Product of Indices
In applying product of indices, the following are true:

$(x^a)^b = x^{ab}$

Similarly, $(x^a y^b)^c = x^{ac} y^{bc}$ and $\left(\dfrac{x}{y}\right)^a = \dfrac{x^a}{y^a}$

Examples
Simplify the following:
1. $(h^4)^{-5}$
2. $(2^{-3})^2$
3. $(-c^3)^2$
4. $(-4u^2v)^3$

Solution

1. $(h^4)^{-5} = h^{4 \times (-5)} = h^{-20} = \dfrac{1}{h^{20}}$

2. $(2^{-3})^2 = 2^{-3 \times 2} = 2^{-6} = \dfrac{1}{2^6} = \dfrac{1}{64}$

3. $(-c^3)^2 = -c^{3 \times 2} = -c^6 = c^6$ (A negative number that is raised to an even number power will give a positive value).

4. $(-4u^2v)^3 = -4^{1 \times 3}\, u^{2 \times 3}\, v^{1 \times 3} = -4^3 u^6 v^3 = -64u^6 v^3$

Fractional Indices

In applying fractional indices, the following are true:

$$x^{1/a} = \sqrt[a]{x} \quad \text{and} \quad x^{a/b} = \sqrt[b]{x^a} \quad \text{or} \quad x^{a/b} = \left(\sqrt[b]{x}\right)^a$$

In all cases, $x \ne 0$

Examples
Simplify the following:
1. $27^{\frac{1}{3}}$
2. $9^{-\frac{1}{2}}$
3. $(25a^2)^{\frac{1}{2}}$
4. $\sqrt{1\dfrac{9}{16}}$
5. $\left(\dfrac{16}{54}\right)^{-\frac{2}{3}}$

Solutions

1. $27^{\frac{1}{3}} = \sqrt[3]{27} = 3$

2. $9^{-\frac{1}{2}} = \dfrac{1}{9^{\frac{1}{2}}} = \dfrac{1}{\sqrt{9}} = \dfrac{1}{3}$ (Note that $\sqrt[2]{9}$ is $\sqrt{9}$ since 2 is not usually written with the square root sign).

3. $(25a^2)^{\frac{1}{2}} = 25^{\frac{1}{2}}a^{(2 \times \frac{1}{2})} = 25^{\frac{1}{2}}a^1 = (\sqrt{25}) \times a = 5a$

4. $\sqrt{1\frac{9}{16}} = \sqrt{\frac{25}{16}} = \frac{5}{4}$

5. $\left(\frac{16}{54}\right)^{-\frac{2}{3}} = \left(\frac{8}{27}\right)^{-\frac{2}{3}}$ (When the fraction is expressed in its lowest term)

$\left(\frac{8}{27}\right)^{-\frac{2}{3}} = \left(\frac{27}{8}\right)^{\frac{2}{3}} = \frac{27^{2/3}}{8^{2/3}} = \frac{(\sqrt[3]{27})^2}{(\sqrt[3]{8})^2} = \frac{3^2}{2^2} = \frac{9}{4}$

Note that by taking the inverse of the term in the bracket, the negative power becomes positive.

Equations in Indices

Examples

Solve the following equations:

1. $4^{x-1} = 64$
2. $n^{-\frac{2}{3}} = 9$
3. $2a^{-3} = -16$
4. $9^x = 27$
5. $5x = 40x^{-\frac{1}{2}}$

<u>Solutions</u>

1. $4^{x-1} = 64$

This is solved by expressing both sides of the equation in the same base and then equating the powers. This gives:

$\quad 4^{x-1} = 4^3$ (Note that $64 = 4^3$)

Equating the powers gives:

$\quad x - 1 = 3$

$\quad x = 3 + 1$

$\quad x = 4$

2. $n^{-\frac{2}{3}} = 9$

In this case, the unknown letter is the base. To solve this, make the power of n to 1 by multiplying this power by its inverse and using the same sign of the power. The other side of the equation should also be raised to the same power. This gives:

$\quad (n^{-\frac{2}{3}})^{-3/2} = 9^{-3/2}$

$\quad n^{(-2/3 \times -3/2)} = \frac{1}{9^{3/2}}$ (The inverse of $\frac{2}{3}$ is $\frac{3}{2}$. The sign of the original power is also used with the inverse)

$\quad n^1 = \frac{1}{(\sqrt{9})^3} = \frac{1}{3^3} = \frac{1}{27}$ (Note that $-\frac{2}{3} \times -\frac{3}{2} = 1$. This gives n^1)

$\quad n = \frac{1}{27}$

3. $2a^{-3} = -16$

Divide both sides by 2.

$$a^{-3} = \frac{-16}{2}$$
$$a^{-3} = -8$$

Now make the power of 'a' to be 1 by multiplying this power by its inverse. This gives:

$$(a^{-3})^{-\frac{1}{3}} = (-8)^{-\frac{1}{3}}$$
$$a^1 = \frac{1}{(-8)^{\frac{1}{3}}} = \frac{1}{\sqrt[3]{-8}} = \frac{1}{-2}$$
$$a = -\frac{1}{2}$$

4. $9^x = 27$

Expressing both sides of the equation in the same base gives:

$$(3^2)^x = 3^3$$
$$3^{2x} = 3^3$$

Equating the powers gives

$$2x = 3$$
$$x = \frac{3}{2}$$

5. $5x = 40x^{-\frac{1}{2}}$

Divide both sides by 5.

$$x = \frac{40x^{-\frac{1}{2}}}{5}$$
$$x = 8x^{-\frac{1}{2}} \quad \text{(Since } 40 \div 5 \text{ gives 8)}$$

Divide both sides by $x^{-\frac{1}{2}}$

$$\frac{x}{x^{-\frac{1}{2}}} = \frac{8x^{-\frac{1}{2}}}{x^{-\frac{1}{2}}}$$

Cancelling out $x^{-\frac{1}{2}}$ on the right hand side gives,

$x^{1-(-\frac{1}{2})} = 8$ (Note that x can be expressed as x^1. Also, from the law of indices, $x \div x^{-\frac{1}{2}} = x^{1-(-\frac{1}{2})}$)

$$\therefore x^{1+\frac{1}{2}} = 8$$
$$x^{3/2} = 8$$

Make the power of x to be 1 by multiplying it by its inverse. Also raise the power of 8 to the same inverse. This gives:

$$(x^{3/2})^{\frac{2}{3}} = 8^{\frac{2}{3}}$$
$$x^1 = 8^{\frac{2}{3}}$$
$$x = (\sqrt[3]{8})^2 = (2)^2$$
$$x = 4$$

37

Exercise 3

1. Simplify the following:
   a. $-3(te^3)^4$
   b. $(4ab^3)^3$
   c. $\dfrac{(-a)^2 \times a^7}{(-a)^5}$
   d. $(-g^4)^5$
   e. $\dfrac{(m^2)^3}{m^4 \times (-m)}$

2. Simplify the following:
   a. $(3a)^{-1}$
   b. $(a^2)^{-\frac{1}{2}}$
   c. $(49x^3)^{\frac{1}{2}}$
   d. $(27x^{3/2})^{\frac{2}{3}}$

3. Solve the following equations:
   a. $x^{-\frac{1}{2}} = 5$
   b. $a^{-2} = 9$
   c. $9^{x-2} = 27$
   d. $\dfrac{4^{2x-1}}{16^2} = 64$

CHAPTER 4
LINEAR EQUATIONS AND CHANGE OF SUBJECT OF FORMULAE

Linear Equations

Linear equations can sometimes be expressed with brackets, with fractions or both. When solving linear equations with fractions, clear the fractions by multiplying each term of the equation by the L.C.M of the denominators of the equation.

Examples

1. Solve the following equations:

a. $2x + 4(3 - x) = 11$

b. $6(a - 3) - 2(5a - 8) = -4$

c. $a - 5(2 + a) - (3a - 4) = 2(2a - 1) - 7$

<u>Solutions</u>

a. $2x + 4(3 - x) = 11$

Expanding the bracket gives:

$2x + 12 - 4x = 11$

Collect like terms. This means to collect terms in x on one side of the equation and constant terms on the other side of the equation.

$12 - 11 = 4x - 2x$ (When a number crosses the equality sign, its sign will change)

$1 = 2x$

Divide both sides by 2

$\therefore \quad x = \dfrac{1}{2}$

b. $6(a - 3) - 2(5a - 8) = -4$

Expanding the brackets gives:

$6a - 18 - 10a + 16 = -4$ (Note that $-2 \times -8 = +16$)

Collect like terms in 'a' on one side of the equation.

$6a - 10a = -4 + 18 - 16$

$-4a = -2$

Divide both sides by -4

$\dfrac{-4a}{-4} = \dfrac{-2}{-4}$

$a = \dfrac{-2}{-4}$

$a = \dfrac{1}{2}$

c. $a - 5(2 + a) - (3a - 4) = 2(2a - 1) - 7$

Expanding the brackets gives:

$a - 10 - 5a - 3a + 4 = 4a - 2 - 7$ (Note that $-5 \times (+a) = -5a$, and $- \times -4 = -1 \times -4 = +4$)

39

Collect like terms on one side of the equation.

$7 + 2 + 4 - 10 = 5a - a + 3a + 4a$

$3 = 11a$

Divide both sides by 11

$$\frac{3}{11} = \frac{11a}{11}$$

$$a = \frac{3}{11}$$

2. Solve the following equations:

a. $\frac{3}{4}x - \frac{1}{3}(x - 2) = \frac{5}{6} - (2x - 1)$

b. $\frac{1}{6}(5x - 2) - \frac{2}{3}(4 - x) = 1$

c. $\frac{5}{2m - 3} - \frac{3}{4} = \frac{1}{6} + 7$

Solutions

a. $\frac{3}{4}x - \frac{1}{3}(x - 2) = \frac{5}{6} - (2x - 1)$

In order to clear the fractions, multiply each term in the equation by 12 which is the L.C.M of the denominators, i.e. 4, 3 and 6. This gives:

$12(\frac{3x}{4}) - 12 \times \frac{1}{3}(x - 2) = 12(\frac{5}{6}) - 12(2x - 1)$

Cancelling out by using the 12 to divide the various denominators gives:

$3(3x) - 4(x - 2) = 2(5) - 12(2x - 1)$

$9x - 4x + 8 = 10 - 24x + 12$

Collect like terms:

$9x - 4x + 24x = 10 + 12 - 8$

$29x = 14$

$$x = \frac{14}{29}$$

b. $\frac{1}{6}(5x - 2) - \frac{2}{3}(4 - x) = 1$

In order to clear the fractions, multiply each term in the equation by 6 which is the L.C.M of the denominators, i.e. 6 and 3. This gives:

$6 \times \frac{1}{6}(5x - 2) - 6 \times \frac{2}{3}(4 - x) = 6 \times 1$

Cancelling out by using the 6 to divide the various denominators gives:

$(5x - 2) - 2 \times 2(4 - x) = 6$

$5x - 2 - 4(4 - x) = 6$

$5x - 2 - 14 + 4x = 6$

$5x + 4x = 6 + 2 + 14$

$9x = 22$

$\therefore \quad x = \dfrac{22}{9}$

c. $\dfrac{5}{2m-3} - \dfrac{3}{4} = \dfrac{1}{6} + 7$

Multiply each term in the equation by 12(2m − 3) which is the L.C.M of the denominators, i.e. (2m − 3), 4 and 6. This gives:

$12(2m - 3)\dfrac{5}{2m-3} - 12(2m - 3)\dfrac{3}{4} = 12(2m - 3)\dfrac{1}{6} + 12(2m - 3) \times 7$

Cancelling out gives:

$12(5) - 3(2m - 3)3 = 2(2m - 3) + 12(2m - 3)7$

$60 - 9(2m - 3) = 4m - 6 + 84(2m - 3)$

$60 - 18m + 27 = 4m - 6 + 168m - 252$

$60 + 27 + 252 + 6 = 4m + 168m + 18m$

$345 = 190m$

$m = \dfrac{345}{190}$

$\therefore \quad m = \dfrac{69}{38}$ (After equal division by 5)

Change of Subject of Formulae

If m = b + c, then m is the subject of the formula. If it is rearranged to give b = m − c, then b is now the new subject of the formula.

In changing the subject of a formula, simply solve the equation for the letter which is to become the new subject.

Examples

1. Make h the subject of the formula: $s = \dfrac{wd}{h}\left(h - \dfrac{d}{2}\right)$

<u>Solution</u>

$s = \dfrac{wd}{h}\left(h - \dfrac{d}{2}\right)$

Expanding the bracket gives:

$s = \dfrac{wd}{h}(h) - \dfrac{wd}{h}\left(\dfrac{d}{2}\right)$

Canceling out the h gives:

$s = wd - \dfrac{wd^2}{2h}$

To clear fractions, multiply throughout by 2h (LCM)

$$2h(s) = 2h(wd) - 2h(\frac{wd^2}{2h})$$

$2hs = 2hwd - wd^2$ (Note that the 2h at the end on the right side has cancelled out).

Collect terms in h on one side.

$wd^2 = 2hwd - 2hs$

Factorizing the right hand side gives:

$wd^2 = h(2wd - 2s)$

Divide both sides by $(2wd - 2s)$

$$\frac{wd^2}{2wd - 2s} = \frac{h(2wd - 2s)}{2wd - 2s}$$

Cancelling out the $2wd - 2s$ on the right hand side gives:

$$h = \frac{wd^2}{2wd - 2s}$$

2. Given that $I = \dfrac{E}{\sqrt{R^2 + W^2L^2}}$, express R in terms of I, E, W and L.

Solution

$$I = \frac{E}{\sqrt{R^2 + W^2L^2}}$$

Cross multiply to obtain:

$I\sqrt{R^2 + W^2L^2} = E$

Square both sides to remove the square root sign.

$(I\sqrt{R^2 + W^2L^2})^2 = E^2$

$I^2(R^2 + W^2L^2) = E^2$ (Note that the square also affect I in the bracket)

$I^2R^2 + I^2W^2L^2 = E^2$

$I^2R^2 = E^2 - I^2W^2L^2$

Divide both sides by I^2

$$R^2 = \frac{E^2 - W^2I^2L^2}{I^2}$$

This can also be simplified as follows:

$$R^2 = \frac{E^2}{I^2} - \frac{I^2W^2L^2}{I^2}$$

Canceling out I^2 on the right side gives:

$$R^2 = \frac{E^2}{I^2} - W^2L^2$$

Take the square root of both sides in order to remove the square on R^2.

$$\therefore \quad R = \sqrt{\frac{E^2}{I^2} - W^2 L^2}$$

Note that when a term/terms in a square root sign are squared, it gives only the terms in the root sign. For example, $(\sqrt{m^2 b^2})^2 = m^2 b^2$. Take note of the absence of the root sign

3. Make x the subject of the formula $R = \sqrt{\dfrac{ax - P}{Q + bx}}$

Solution

$$R = \sqrt{\frac{ax - P}{Q + bx}}$$

Square both sides to remove the square root sign
$$R^2 = \frac{ax - P}{Q + bx}$$
By cross multiplication it gives:
$R^2(Q + bx) = ax - P$
$R^2 Q + R^2 bx = ax - P$
Collecting terms in x gives
$ax - R^2 bx = R^2 Q + P$
Factorizing the left hand side gives:
$x(a - R^2 b) = R^2 Q + P$
Divide both sides by $(a - R^2 b)$
$$\frac{x(a - R^2 b)}{(a - R^2 b)} = \frac{R^2 Q + P}{a - R^2 b}$$

$$x = \frac{R^2 Q + P}{a - R^2 b}$$

Exercise 4

1. Solve the following equations:
a. $5x + 2(3 - x) = 10$
b. $2(2a - 3) - 5(4a - 1) = -6$
c. $b - 4(1 + b) - (5b - 1) = -(b - 3) - 2$

2. Solve the following equations:
a. $\dfrac{1}{4}x - \dfrac{2}{3}(x - 1) = \dfrac{3}{4} - (5x - 2)$

b. $\frac{1}{6}(5x-2)-\frac{5}{12}(3-2x)=1$

c. $\frac{2}{2n-3}-\frac{3}{5}=2\frac{1}{2}$

3. Make p the subject of the formula: $tp = md(p-\frac{d}{3})$

4. Given that $V = \frac{P}{\sqrt{E^2+I^2C^2}}$, express C in terms of I, V, E and P.

5. Make x the subject of the formula $R = \sqrt{\frac{bx-S}{T+ax}}$

6. Make m the subject of the formula: $Amq = md(\frac{q}{5}-\frac{dA}{5})$

7. Given that $L = \frac{TC^2}{Y^3-B}$, express Y in terms of L, T, B and C.

8. Make x the subject of the formula $\sqrt{\frac{Q^2}{x}} = \sqrt{\frac{M}{2x^3}}$

CHAPTER 5
LINEAR EQUATIONS FROM WORD PROBLEMS

Examples

1. The sum of 6 and one-third of a certain number is one more than twice the number. Find the number.

<u>Solution</u>
Let the number be x

One-third of the number $= \frac{1}{3}x$

The sum of 6 and one-third of the number $= 6 + \frac{1}{3}x$.

Twice the number $= 2x$

$\quad 6 + \frac{1}{3}x$ is one more than $2x$. This means that:

$\quad 6 + \frac{x}{3} - 2x = 1$

Multiply each term by 3 which is the LCM of all the denominators. Note that all the whole numbers have 1 as their denominator. This multiplication by the LCM of their denominators is done in order to clear the fraction. Each term is now multiplied by 3 as follows:

$\quad (6 \times 3) + 3(\frac{x}{3}) - (2x \times 3) = 1 \times 3$

$\quad 18 + x - 6x = 3$ (Note that in $3(\frac{x}{3})$, the 3 cancels out to give x)

Collect terms in x on one side and the other terms on the other side of the equation. This gives:

$\quad 18 - 3 = 6x - x$ (Note the change in the signs of the terms that cross the equality sign).

$\quad 15 = 5x$

Divide both sides of the equation by 5 (i.e. the co-efficient of x which is the unknown).

$\quad \frac{15}{5} = \frac{5x}{5}$

$\therefore \quad 3 = x$ (The 5 cancels out)

$\therefore \quad x = 3$

2. When 10 is subtracted from the product of 6 and a certain number, and the result is divided by 4, the answer is equal to the number. What is the number?

<u>Solution</u>
Let the number be a

The product of 6 and the number $= 6a$

When 10 is subtracted from this product it gives: $6a - 10$

This result divided by 4 is $\frac{6a-10}{4}$

The answer is equal to the number means that:

$\quad \frac{6a-10}{4} = a$

Cross multiply to solve this equation. This gives:

$\quad 6a - 10 = 4a$

collect like terms in a on the left hand side

$$6a - 4a = 10$$
$$2a = 10$$

Divide both sides by 2

$$\frac{2a}{2} = \frac{10}{2}$$
$$\therefore \quad a = 5$$

3. A man is three times as old as his son. Five years ago, the sum of their ages was 38. Find the age of the man and his son.

Solution

Let the son's age be x

$\therefore$ The man's age = $3x$ (i.e. three times his son's age)

Five years ago, the son's age was: $x - 5$

Similarly, five years ago the man's age was: $3x - 5$

The sum of their age five years ago, was:

$$x - 5 + 3x - 5$$

Since this sum is 38, we express it as follows:

$$x - 5 + 3x - 5 = 38$$
$$x + 3x = 38 + 5 + 5$$
$$4x = 48$$
$$\therefore \qquad x = \frac{48}{4}$$
$$x = 12$$

$\therefore$ The son's age = 12 years

The man's age = $3x$ = 3 x 12 = 36years

4. A girl is 5 years old and her mother is 32 years old. In how many years time will the mother be twice as old as the girl?

Solution

Let the number of years time be y.

At that time, the girl's age will be = 5 + y

At that time the mother's age will be = 32 + y

Since the mother's age will be twice the girl's age, it means that:

$$2(5 + y) = 32 + y \qquad \text{(Note that 2(5 + y) is twice the girl's age)}$$
$$\therefore \qquad 10 + 2y = 32 + y$$
$$2y - y = 32 - 10 \qquad \text{(After collecting like terms)}$$
$$y = 22$$

$\therefore$ In 22 years time, the mother will be twice as old as the girl.

5. A cyclist takes 5hours to travel from a town A to a town B. When coming back, the journey took him 1hour less because he increased his speed by 15km/h. What was his speed when he

was going from A to B.

Solution
Let his speed from A to B be x

$$\text{Speed} = \frac{\text{Distance}}{\text{Time}}$$

∴   Distance = speed x time

$$= x \times 5 = 5x \quad \text{(When going from A to B)}$$

When coming from B to A (i.e. the return journey), his speed was $x + 15$ (since he increased His speed by 15km/h), while his time was: $5 - 1 = 4$ (Since he took 1 hour less)

∴   Distance = speed x time

$$= (x + 15) \times 4$$
$$= 4(x + 15)$$

Since the distance from A to B is equal to the distance from B to A, then it follows that:

$$5x = 4(x + 15) \quad \text{(Distance A to B is } 5x \text{, while distance B to A is } 4(x + 15)\text{. Both are equal)}$$
$$5x = 4x + 60$$
$$5x - 4x = 60$$
$$x = 60$$

His speed when going from A to B was 60km/h

6. A man walked for 3hours at 8km/h. He then cycled at 12km/h for a certain period of time. If his total distance was 54km, for how many hours did he cycle?

Solution

$$\text{Speed} = \frac{\text{Distance}}{\text{Time}}$$

When he walked, his total distance was:

Distance = speed x time

$$= 8 \times 3$$
$$= 24\text{km}$$

When he cycled, his total distance was:

Distance = speed x time

Let the time taken to cycle be x

∴   Distance cycled = 12 x x

$$= 12x$$

His total distance is given by:

$$24 + 12x$$

Since this total distance was 54km, then:

$$24 + 12x = 54$$
$$12x = 54 - 24$$
$$12x = 30$$
∴ $$x = \frac{30}{12}$$
$$x = \frac{5}{2} \quad \text{(After equal division by 6)}$$

$$x = 2\frac{1}{2}$$

∴   The man cycled for $2\frac{1}{2}$ hours

7. A sum of $324 is made up of $10 notes and $1 notes. If there are eight times as many $1 notes as there are $10 notes, find the number of each note.

<u>Solution</u>
Let there be x $1 notes.
∴   The number of $10 notes = $8x$ (i.e. eight times as many $1 notes)
∴   Total amount of $1 notes = x x 1 = x (This means $$x$)
Similarly, total amount of $10 notes = $8x$ x 10 = $80x$ (This means $$80x$)
Since this total gives the sum of $324, it means that:

$x + 80x = 324$
$81x = 324$ (Note that x means $1x$)
∴ $x = \dfrac{324}{81}$
$x = 4$

∴   Number of $1 = 4
Number of $10 notes = $8x$ = 8 x 4 = 32

8. Chloe has $30 and Peter has $186. If Chloe saves $5 a day and Peter spends $7 a day, after how many days will they have equal amounts?

<u>Solution</u>
Let the number of days for both of them to have equal amount of money be y.
If Chloe saves $5 a day, then after y days, she must have saved $5y (i.e. 5 x y)
If Peter spends $7 a day, then after y days he must have spent $7y (i.e. 7 x y)
So, total amount that Chloe will have = 30 + 5y   (i.e. her initial money plus her savings)
Total amount that Peter will have = 186 – 7y   (i.e. his initial money minus his spending)
Since they will both have equal amount of money, then:

30 + 5y = 186 – 7y
∴   5y + 7y = 186 – 30
12y = 156
∴ $y = \dfrac{156}{12}$
y = 13

∴   They will have equal amounts of money after 13 days.

9. A total of x litres of water is needed to fill 30 tanks with the same amount of water in each tank. If the size of each tank is 3 litres less, there will be enough water to fill 32 tanks. What is the value of x.

Solution

Amount of water in each of the 30 tank is: $\dfrac{x}{30}$

When the size of each of these 30 tanks is 3 litres less than its original size, then each tank will contain: $\dfrac{x}{30} - 3$

Since this amount of water can fill each of the 32 tanks, then the total amount of water in the 32 tanks is:

$32(\dfrac{x}{30} - 3)$

This total amount of water has not changed. It is also equal to x litres (i.e. the original quantity of water). This means that:

$32(\dfrac{x}{30} - 3) = x$

$\therefore \quad \dfrac{32x}{30} - 96 = x$

Multiply each term by 30 in order to clear fraction. This gives:

$30(\dfrac{32x}{30}) - 30(96) = 30(x)$

$32x - 2880 = 30x$

$32x - 30x = 2880$

$2x = 2880$

$\therefore \qquad x = \dfrac{2880}{2}$

$x = 1440$

$x = 1440$ litres

10. When a plane travels a certain journey at an average speed of 200km/h, it arrives an airport at 5am. When it travels at an average speed of 240km/h, it arrives at the airport at 3am. What is the length of the journey?

Solution

Let the duration of the journey when it travels at 200km/h be m hours.

$\therefore$ The duration when it travels at 240km/h will be:

$\qquad m - 2$ (i.e. it arrives 2 hours earlier, from 5am to 3am. This means it takes lesser time)

But, average speed $= \dfrac{\text{Total distance}}{\text{Total time}}$

$\therefore$ At 200km/h: $200 = \dfrac{\text{Total distance}}{m}$

$\therefore$ Total distance $= 200 \times m$

$= 200m$

Similarly, at 240km/h: Total distance $= 240(m - 2)$

Since the distance is the same, then:

$200m = 240(m - 2)$

$200m = 240m - 480$

$480 = 240m - 200m$

$480 = 40m$

$$\therefore \quad m = \frac{480}{40}$$

$$m = 12$$

Recall that: Total distance = 200m or 240(m − 2)

∴   Total distance = 200 x 12 = 2400

∴   The length of the journey is 2400km

11. A man walks to a village at 18km/h and returns at 12km/h. If the whole journey takes $6\frac{1}{4}$ hours, what is the total distance walked?

Solution

Let x be the time taken when he walked at 18km/h

∴   The time taken for the return journey when he walked at 12km/h will be:

$6\frac{1}{4} - x$ (since total time for the two journeys is $6\frac{1}{4}$)

Also, speed = $\dfrac{\text{Distance}}{\text{Time}}$

∴   At 18km/h: $18 = \dfrac{\text{Distance}}{x}$

∴   Distance = $18x$

Similarly,

At 12km/h: $12 = \dfrac{\text{Distance}}{6\frac{1}{4} - x}$

∴   Distance = $12(6\frac{1}{4} - x)$

Since the to and fro distances are the same, then:

$$18x = 12(6\frac{1}{4} - x)$$

$$18x = 12(\frac{25}{4} - x)$$

$$18x = 12(\frac{25}{4}) - 12(x)$$

$$18x = 3(25) - 12x$$

$$18x + 12x = 75$$

$$30x = 75$$

$$\therefore \quad x = \frac{75}{30}$$

$$x = 2\frac{1}{2} \text{hours}$$

30

∴   Distance = $18x$ or $12(6¼ - x)$

∴   Distance = $18 \times 2½$

$$= 18 \times \frac{5}{2} = 45\text{km}$$

This distance is for one part of the journey

∴   Total distance for the to and fro journey is:

$$45 \times 2 = 90$$

∴   The man walked a total distance of 90km.

12. The result of taking 3 from a number and multiplying the answer by 4 is the same as taking 3 from five times the number. Find the number.

Solution
Let the number be c
3 taken from c means: c – 3
Multiplying this by 4 gives: 4(c – 3)   (This is the first part of the statement)
Five times the number is 5c
3 taken from 5c is 5c – 3   (This is the second part of the statement)
Since the first part = the second part, then:

$$4(c – 3= 5c – 3$$
$$4c – 12 = 5c – 3$$
$$4c – 5c = 12 – 3$$
$$–c = 9$$
$$\therefore \quad –9 = c$$
$$\therefore \quad c = –9$$

Hence, the number is –9.

13. The sum of three consecutive numbers is 63. Find the numbers.

Solution
Let the first number be x.
∴   The second number will be $x + 1$, while the third number will be $x + 2$. For example, if the first number is 10, then the second number will be 11 which is (10 + 1), while the third number will be 12 which is (10 + 2).
So, the numbers are $x, x + 1$ and $x + 2$.
Since their sum is 63, then it follows that:

$$x + (x + 1) + (x + 2) = 63$$
$$3x + 3 = 63$$
$$3x = 63 – 3$$
$$3x = 60$$
$$\therefore \quad x = \frac{60}{3}$$
$$x = 20$$

∴ The numbers are 20, 21 and 22 from $x, x + 1$ and $x + 2$.

14. The sum of two numbers is 21. Five times the first number added to two times the second number is 66. Find the two numbers.

Solution
Let the first number be x.
∴ The second number is $21 – x$. Note that the value of the sum of two numbers minus the first number gives the second number. For example, if 8 + 6 = 14, then 14 – 8 = 6 or 14 – 6 = 8. This shows that

Sum – one of the numbers = the other number

∴ five times the first number is : 5 x x = $5x$

2 times the second number is: 2 x $(21 - x)$ = $2(21 - x)$

When these two expressions are added together, it gives 66 as follows:

$5x + 2(21 - x) = 66$

$5x + 42 - 2x = 66$

$5x - 2x = 66 - 42$

$3x = 24$

∴ $x = \dfrac{24}{3}$

$x = 8$

∴ The first number is 8 and the second number is 21 – 8 = 13, i.e. 8 and 13.

15. 2 is added to twice a certain number and the sum is doubled. The result is 10 less than 5 times the original number. Find the number.

Solution

Let the number be y.

Twice the number is 2 x y = 2y

2 added to twice the number = 2 + 2y

When this sum is doubled, it gives: 2 x (2 + 2y) = 2(2 + 2y)

From second sentence, 5 times the number = 5 x y = 5y

10 less than 5 times the number is: 5y – 10

Since the result from first sentence = the result from second sentence, then it follows that:

$2(2 + 2y) = 5y - 10$

$4 + 4y = 5y - 10$

$4 + 10 = 5y - 4y$

$14 = y$

∴ $y = 14$

The number is 14.

16. The sum of two numbers is 38. When 8 is added to twice one of the numbers, the result is 5 times the other number. Find the two numbers.

Solution

Let one of the numbers be n.

∴ The other number = 38 – n (Since their sum is 38)

Twice one of the numbers is: 2 x n = 2n

8 added to twice one of the numbers is: 8 + 2n

5 times the other number is: 5 x (38 – n) = 5(38 – n)

Since these two results are equal according to the question, then we equate them as follows:

$8 + 2n = 5(38 - n)$

$8 + 2n = 190 - 5n$

$2n + 5n = 190 - 8$

$$7n = 182$$
$$n = \frac{182}{7}$$
$$n = 26$$

∴   The two numbers are n = 26 and 38 – n = 38 – 26 = 12.
They are 26 and 12

17. A woman's age and her son's age add up to 45 years. Five years ago, the woman was 6 times as old as her son. How old was the woman when the son was born?

Solution
Let the woman's age be x
∴ The son's age is = 45 – x (Since their sum is 45)
Five years ago, the woman's age = $x - 5$
Five years ago, the son's age was: 45 – x – 5 = 40 – x
Since five years ago, the woman's age was 6 times that of her son, then this gives:
$$x - 5 = 6(40 - x)$$
$$x - 5 = 240 - 6x$$
$$x + 6x = 240 + 5$$
$$7x = 245$$
$$x = \frac{245}{7}$$
$$x = 35$$

∴ The woman's age is x = 35years.
The son's age is 45 – x = 45 – 35 = 10years.
∴ The woman's age when the son was born is given by the difference between their ages (i.e. how many years older is the woman than her son). This gives
$$35 - 10 = 25$$
∴   The woman was 25 years old when the son was born.

18. The numerator of a fraction is 3. If 9 is added to the numerator and 4 is added to the denominator, the fraction is doubled. What is the fraction?

Solution
Let the denominator be y
∴ The fraction is $\dfrac{3}{y}$
6 added to the numerator = 3 + 6 = 9
4 added to the denominator = y + 4
∴ The new fraction is: $\dfrac{9}{y + 4}$
This new fraction is double of the initial fraction. This gives
$$\frac{9}{y + 4} = 2\left(\frac{3}{y}\right)$$
$$\frac{9}{y + 4} = \frac{6}{y}$$

Cross multiply to give:

$$9y = 6(y + 4)$$
$$9y = 6y + 24$$
$$9y - 6y = 24$$
$$3y = 24$$
$$\therefore \quad y = \frac{24}{3}$$
$$y = 8$$

$\therefore$ The fraction is: $\frac{3}{y}$, which is $\frac{3}{8}$

Exercise 5

1. The sum of 9 and one-fourth of a certain number is five less than twice the number. Find the number.

2. When 2 is subtracted from the product of 5 and a certain number, and the result is divided by 3, the answer is equal to one and a half times the number. What is the number?

3. A man is six times as old as his daughter. Four years ago, the man was four times as old as his daughter. Find the age of the man and his daughter.

4. A boy is 16 years old and his mother is 42 years old. In how many years time will the mother be twice as old as the boy?

5. A cyclist takes 10hours to travel from a town A to a town B. When coming back, the journey took him 2hours more because he reduced his speed by 30km/h. What was his speed when he was coming from B to A.

6. A man runs for 2hours at 6km/h. He then cycled at 16km/h for a certain period of time. If his total distance was 72km, for how many hours did he cycle?

7. A sum of $186 is made up of $10 notes and $1 notes. If there are three times as many $10 notes as there are $1 notes, find the number of each note.

8. John has $20 and Rose has $200. If John saves $3 a day and Rose spends $9 a day, after how many days will they have equal amount of money?

9. A total of x litres of wine is needed to fill 21 tanks with the same amount of wine in each tank. If the size of each tank is 55 litres more, the wine will fill only 10 tanks. What is the value of x.

10. When a plane travels a certain journey at an average speed of 180km/h, it arrives an airport at 2pm. When it travels at an average speed of 150km/h, it arrives at the airport at 4.30pm. What is the length of the journey?

11. A man walks to a village at 12km/h and returns at 15km/h. If the whole journey takes 4hours, what is the total distance walked?

12. The result of taking 5 from a number and multiplying the answer by 8 is the same as adding 8 to four times the number. Find the number.

13. The sum of three consecutive odd numbers is 45. Find the numbers.

14. The sum of two numbers is 29. Two times the first number added to six times the second number is 78. Find the two numbers.

15. 6 is added to thrice a certain number and the sum is halved. The result is 3 less than 2 times the original number. Find the number.

16. The sum of two numbers is 40. When 10 is added to twice one of the numbers, the result is 4 times the other number. Find the two numbers.

17. A woman's age and her son's age add up to 52 years. 8 years ago, the woman was 17 times as old as her son. How old was the woman when the son was born?

18. The numerator of a fraction is 5. If 2.5 is added to the numerator and 8 is subtracted from the denominator, the fraction is halved. What is the fraction?

CHAPTER 6
SIMULTANEOUS LINEAR EQUATIONS

In simultaneous equations, we solve a pair of equations and determine the values of two variables. Apart from graphical method, the two methods of solving simultaneous equations are:
1. Substitution Method
2. Elimination Method

Substitution Method

In this method, we make one of the variables a subject of formula from one of the equations, and substitute it into the other equation. It is easier to use this method when one of the variables has a coefficient of 1.

Examples

1. Solve simultaneously the equations: $x + y = 4$ and $2x - y = 5$

Solution
Let us rewrite this equation and label them equation 1 and 2 for easy identification. This gives:

$$x + y = 4 \quad ....\text{Equation (1)}$$
$$2x - y = 5 \quad ....\text{Equation (2)}$$

A careful look at the two equations shows that x and y in equation 1 have coefficients of 1, while y in equation 2, has a coefficient of −1. Any of these three variables can easily be made the subject of formula in their respective equation. So, let us make x in equation 1 the subject of formula of the equation. This is done as follows:

$$x + y = 4 \quad ....\text{Equation (1)}$$
$$x = 4 - y$$

Let us represent this equation as equation 3 as follows:

$$x = 4 - y \quad ....\text{Equation (3)}$$

Since $x = 4 - y$, let us substitute $4 - y$ for x in equation 2. Note that we made x the subject of formula from equation 1, so the substitution of the expression for x should be done in the other equation (i.e. equation 2), and not in the same equation from which x was made the subject of the formula. Hence we substitute $4 - y$ for x in equation 2 as follows:

$$2x - y = 5 \quad ....\text{Equation (2)}$$
$$2(4 - y) - y = 5 \qquad \text{(Note that } x \text{ has been replaced with } 4 - y \text{ as stated above)}$$
$$8 - 2y - y = 5$$
$$-3y = 5 - 8$$
$$-3y = -3$$
$$y = \frac{-3}{-3} \qquad \text{(When we divide both sides by −3)}$$
$$y = 1$$

Now, substitute 1 for y in equation 3 in order to obtain x. Note that we can use any of the three equations, but equation 3 is the best to use since it directly gives us the value of x. This gives:

$x = 4 - y$ ....Equation (3)

$x = 4 - 1$

$x = 3$

Therefore, $x = 3$ and y = 1

2. Solve the simultaneous equations: $3x + 2y = 10$ and $4x - y = 6$

<u>Solution</u>

Let us rewrite this equation and label them equation 1 and 2 for easy identification. This gives:

$3x + 2y = 10$ ....Equation (1)

$4x - y = 6$ ....Equation (2)

A careful look at the two equations shows that only y in equation 2 has a coefficients of –1. So, let us make y in equation 2 the subject of formula of the equation. This is done as follows:

$4x - y = 6$ ....Equation (2)

Hence, $4x - 6 = y$

Or, $y = 4x - 6$ (Note that p = q also means that q = p)

Let us label this equation as equation 3 as shown below.

$y = 4x - 6$ ....Equation (3)

Since $y = 4x - 6$, let us substitute $4x - 6$ for y in equation 1. Note that we made y the subject of formula from equation 2, so the substitution of the expression for y should be done in the other equation (i.e. equation 1), and not in the same equation from which y was made the subject of the formula. Hence we substitute $4x - 6$ for y in equation 1 as follows:

$3x + 2y = 10$ ....Equation (1)

$3x + 2(4x - 6) = 10$ (Note that y has been replaced (substituted) with $4x - 6$)

$3x + 8x - 12 = 10$

$11x = 10 + 12$

$11x = 22$

$x = \dfrac{22}{11}$ (When we divide both sides by 11)

$x = 2$

Now, substitute 2 for x in equation 3 in order to obtain y. Note that we can use any of the three equations, but equation 3 is the best to use since it directly gives us the value of y. This gives:

$y = 4x - 6$ ....Equation (3)

$y = 4(2) - 6$

$y = 8 - 6$

$y = 2$

Therefore, $x = 2$ and y = 2

3. Solve simultaneously, the equations m − 6n = 1 and 9m + n = −46

<u>Solution</u>

$$m - 6n = 1 \quad ....Equation\ (1)$$
$$9m + n = -46 \quad ....Equation\ (2)$$

From equation 2:

$$n = -46 - 9m \quad ....Equation\ (3)$$

Note that we can also easily make m the subject of the formula from equation 1. Since, n = –46 – 9m, let us substitute –46 – 9m for n in equation 1. This gives:

$$m - 6n = 1 \quad ....Equation\ (1)$$
$$m - 6(-46 - 9m) = 1$$
$$m + 276 + 54m = 1$$
$$55m = 1 - 276$$
$$55m = -275$$
$$m = \frac{-275}{155}$$
$$m = -5$$

Now, substitute –5 for m in equation 3 in order to find n. This gives:

$$n = -46 - 9m \quad ....Equation\ (3)$$
$$n = -46 - 9(-5)$$
$$n = -46 + 45$$
$$n = -1$$

Therefore, n = –1 and m = –5

4. Solve simultaneously, the equations 3a – 5b = 25 and 7a + 2b = 31

<u>Solution</u>

$$3a - 5b = 25 \quad ....Equation\ (1)$$
$$7a + 2b = 31 \quad ....Equation\ (2)$$

A careful look at the two equations shows that there is no variable with a coefficient of 1. However, we can still make any of the variable a subject of its equation, and substitute it into the other equation. Hence, from equation 1, let as make b the subject of the formula as follows:

$$3a - 5b = 25 \quad ....Equation\ (1)$$
$$3a - 25 = 5b$$

Or, $5b = 3a - 25$ (Note that m = n also means that n = m)

Dividing both sides by 5 gives:

$$b = \frac{3a - 25}{5} \quad ....Equation\ (3)$$

Since $b = \frac{3a - 25}{5}$, let us substitute $\frac{3a - 25}{5}$ for b in equation 2. This gives:

$$7a + 2b = 31 \quad ....Equation\ (2)$$

$$7a + 2\left(\frac{3a - 25}{5}\right) = 31$$

$$7a + \frac{6a - 50}{5} = 31$$

Multiply each term by 5 in order to clear out the fractions. This gives:

$$5(7a) + 5\left(\frac{6a - 50}{5}\right) = 5(31)$$

$$35a + 6a - 50 = 155$$
$$41a = 155 + 50$$
$$41a = 205$$
$$a = \frac{205}{41}$$
$$a = 5$$

Now, substitute 5 for a in equation 3. This gives:

$$b = \frac{3a - 25}{5} \quad ....\text{Equation (3)}$$

$$= \frac{3(5) - 25}{5}$$

$$= \frac{15 - 25}{5}$$

$$= \frac{-10}{5}$$

$$b = -2$$

Therefore, a = 5 and b = −2

5. Solve the equations 8p − 3q = −5 and 2p − 11q = −32

Solution

$$8p - 3q = -5 \quad ....\text{Equation (1)}$$
$$2p - 11q = -32 \quad ....\text{Equation (2)}$$

A careful look at the two equations shows that there is no variable with a coefficient of 1.
However, let us make p the subject of formula in equation 2. This is done as follows:

$$2p - 11q = -32 \quad ....\text{Equation (2)}$$
$$2p = -32 + 11q$$

Dividing both sides by 2 gives:

$$p = \frac{-32 + 11q}{2} \quad ....\text{Equation (3)}$$

Substitute $\frac{-32 + 11q}{2}$ for p in equation 1. This gives:

$$8p - 3q = -5 \quad ....\text{Equation (1)}$$

$$8\left(\frac{-32 + 11q}{2}\right) - 3q = -5$$

$$4(-32 + 11q) - 3q = -5$$
$$-128 + 44q - 3q = -5$$
$$41q = -5 + 128$$
$$41q = 123$$
$$q = \frac{123}{41}$$
$$q = 3$$

Substitute 3 for q in equation 3 in order to find p. This gives:

$$p = \frac{-32 + 11q}{2} \quad \text{.....................Equation (3)}$$

$$= \frac{-32 + 11(3)}{2}$$

$$= \frac{-32 + 33}{2}$$

$$p = \frac{1}{2}$$

Hence, $p = \frac{1}{2}$ and $q = 3$

Elimination Method

Elimination method is used in solving simultaneous equations, usually when there is no variable with a coefficient of 1. In elimination method, we get rid of (eliminate) one of the variables by making its coefficient the same in both equations.

Examples

1. Solve the equation: $4x - 5y = 5$ and $2x - 3y = 2$.

Solution

$$4x - 5y = 5 \text{ ....Equation (1)}$$
$$2x - 3y = 2 \text{ ....Equation (2)}$$

When using elimination method, any operation of multiplication and division can be used to manipulate any of the equations in order to make a particular variable to have equal coefficient.

A careful look at the two equations shows that we can make the coefficient of x in equation (2) to become 4 so that the coefficient of x in the two equations will be 4. In order to make the coefficient of x to be 4 in equation (2), we multiply equation (2) by 2. This gives:

$$2(2x) - 2(3y) = 2(2) \quad \text{(Note that each term in equation (2) should be multiplied by 2)}$$
$$4x - 6y = 4 \text{ ....Equation (3)} \quad \text{(Note that our result is labelled equation 3)}$$

Now, equation 1 and 3 have the same coefficient of x. This means that x can now be eliminated by subtracting one equation from the other. Note that the coefficient of x which is 4 in equation 1 and 3 is a positive value. Hence one equation will be subtracted from the other in order to eliminate x. However, when the coefficients of a variable have opposite signs (i.e.

positive and negative), we will have to add the two equations in order to eliminate that variable.

In this case, for us to eliminate x, let us subtract equation 3 from equation 1 (we can also subtract equation 1 from equation 3) as follows:

$4x - 5y - (4x - 6y) = 5 - 4$ (Take note of the use of bracket in the subtraction)

Note that when subtracting, you subtract the left hand side of both equations separately, and subtract the right hand side of both equations separately as shown above. This simplifies to give:

$4x - 5y - 4x + 6y = 1$ (Note the changes in sign of the terms in the bracket)

$4x - 4x - 5y + 6y = 1$

$y = 1$ (Note that $4x - 4x = 0$, and $-5y + 6y = y$)

It is important to note that the terms in equation 1 come first on both sides of the equation during subtraction since we are carrying out: Equation 1 minus Equation 3.

Now that we have obtained y = 1, let us substitute 1 for y in equation 2 in order to find x. Note that the substitution can be done in any of the three equations.

$2x - 3y = 2$ ....Equation (2)

$2x - 3(1) = 2$ (Note that 1 has been substituted for y)

$2x - 3 = 2$

$2x = 2 + 3$

$2x = 5$

$x = \dfrac{5}{2}$

Hence, $x = \dfrac{5}{2}$ and y = 1

2.   Solve simultaneously, the equations $5x - 4y = -2$ and $4x + 5y = 23$.

<u>Solution</u>

$5x - 4y = -2$ ....Equation (1)

$4x + 5y = 23$ ....Equation (2)

An easy way of making the coefficient of x to be the same in both equations is to multiply one equation by the coefficient of x in the other equation. This means that we multiply equation (1) by 4 (i.e. the coefficient of x in equation 2), and multiply equation (2) by 5 (i.e. the coefficient of x in equation 1). This will make the coefficient of x to be 20 in the two equations. Hence, let us multiply equation (1) by 4 and equation (2) by 5 in order to obtain equation (3) and equation (4) respectively as shown below:

$4(5x) - 4(4y) = 4(-2)$ ....Equation (3)

$5(4x) + 5(5y) = 5(23)$ ....Equation (4)

These two equations simplifies to give:

$20x - 16y = -8$ ....Equation (3)

$20x + 25y = 115$ ....Equation (4)

Now, equation 3 and 4 have the same coefficient of x. This means that x can now be eliminated by subtracting one equation from the other. Hence, let us subtract equation 3 from equation 4.

Note that any order can be followed, i.e. we can subtract equation 4 from equation 3. Therefore, equation 4 minus equation 3 gives:

$20x + 25y - (20x - 16y) = 115 - (-8)$ (Note the subtraction of left and right side separately)

$20x + 25y - 20x + 16y = 115 + 8$ (Note the changes in sign of the terms in the bracket)

$20x - 20x + 25y + 16y = 123$

$41y = 123$ (Note that $20x - 20x = 0$, and $25y + 16y = 41y$)

$$y = \frac{123}{41}$$

$y = 3$

Now that we have obtained y = 3, let us substitute 3 for y in equation 1 in order to find x. Note that the substitution can be done in any of the four equations.

$5x - 4y = -2$ ....Equation (1)

$5x - 4(3) = -2$ (Note that 3 has been substituted for y)

$5x - 12 = -2$

$5x = 12 - 2$

$5x = 10$

$$x = \frac{10}{5}$$

$x = 2$

Hence, $x = 2$ and y = 3

3.   Solve simultaneously, the equations $11x + 3y = 4$ and $7x - 2y = -17$.

<u>Solution</u>

$11x + 3y = 4$ ....Equation (1)

$7x - 2y = -17$ ....Equation (2)

In our last two examples we made the coefficient of x to be equal. In this example let us make the coefficient of y to be equal. An easy way of making the coefficient of y to be the same in both equations is to multiply equation (1) by 2 (i.e. the coefficient of y in equation 2), and multiply equation (2) by 3 (i.e. the coefficient of y in equation 1). This will make the coefficient of y to be 6 (we are not taking the signs into consideration) in the two equations. Hence, let us multiply equation (1) by 2 and equation (2) by 3 in order to obtain equation (3) and equation (4) respectively as shown below:

$2(11x) + 2(3y) = 2(4)$ ....Equation (3)

$3(7x) - 3(2y) = 3(-17)$ ....Equation (4)

These two equations simplifies to give:

$22x + 6y = 8$ ....Equation (3)

$21x - 6y = -51$ ....Equation (4)

Now, equations 3 and 4 have the same coefficient of y. This means that x can now be eliminated by adding the two equations together. Hence, let us add equation 3 with equation 4. Note that in example 1 and 2 above, we subtracted one equation from the other. This was because the signs of the coefficients of x were the same. However, in this example (example 3), the signs of the coefficients of y (i.e. the variable to be eliminated) are not the same. Hence, we have to add them together in order to eliminate y. This summarizes the rule that when the signs of the coefficient of the variable to be eliminated are the same, we subtract one equation from the other in order to eliminate that variable. However, when the coefficients of a variable to be eliminated have opposite signs (i.e. positive and negative), we will have to add the two equations in order to eliminate that variable.

Therefore, equation 4 plus equation 3 gives:

$22x + 6y + (21x - 6y) = 8 + (-51)$ (Take note of the addition of left and right side separately)

$22x + 6y + 21x - 6y = 8 - 51$

$22x + 21x + 6y - 6y = -43$

$43x = -43$

$x = \dfrac{-43}{43}$

$x = -1$

Now that we have obtained $x = -1$, let us substitute −1 for x in equation 1 in order to find y.

$11x + 3y = 4$ ....Equation (1)

$11(-1) + 3y = 4$

$-11 + 3y = 4$

$3y = 4 + 11$

$3y = 15$

$y = \dfrac{15}{3}$

$y = 3$

Hence, $x = -1$ and y = 3

4. Solve simultaneously, the equations: 6m − 5n − 3 = 0 and 4m + 3n + 17 = 0

Solution

Each of the equations above can be rearranged by taking the constant term to the right hand side of the equations. This gives:

$6m - 5n = 3$ ....Equation (1)

$4m + 3n = -17$ ....Equation (2)

Let us make the coefficient of n to be equal in the two equations. An easy way of doing this is to multiply equation (1) by 3 (i.e. the coefficient of n in equation 2), and multiply equation (2) by 5 (i.e. the coefficient of n in equation (1). This will make the coefficient of n to be 15 in the two equations. Hence, equation (1) multiplied by 3 and equation (2) multiplied by 5 will give equation (3) and equation (4) respectively as shown below:

$3(6m) - 3(5n) = 3(3)$ ....Equation (3)

$5(4m) + 5(3n) = 5(-17)$ ....Equation (4)

These two equations simplify to give:

$18m - 15n = 9$ ....Equation (3)

$20m + 15n = -85$ ....Equation (4)

Now, equations 3 and 4 have the same coefficient of n. This means that n can now be eliminated by adding the two equations together. We have to add the two equations since the signs of the coefficients of n are different.

Another method of carrying out our operation (either addition or subtraction) is to arrange the two equations to be one above the other and then add or subtract the corresponding like terms. Hence, equation 3 plus equation 4 can be used to describe this method as shown below.

$$18m - 15n = 9 \dots\dots\dots\dots\text{Equation (3)}$$
$$20m + 15n = -85 \dots\dots\dots\text{Equation (4)}$$

Equation (3) + equation (4): $38m \quad\quad = -76$

In the addition above, the followings were carried out in each column:

$18m + 20m = 38m$

$-15n + 15n = 0,$

and $9 + (-85) = 9 - 85 = -76.$

Also, zero n was not written since n has been eliminated.

We now continue from above as follows:

$38m = -76$

$m = \dfrac{-76}{38}$

$m = -2$

Substitute –2 for m in equation (1) in order to obtain n. This gives:

$6m - 5n = 3 \dots\dots\dots\dots\text{Equation (1)}$

$6(-2) - 5n = 3$

$-12 - 5n = 3$

$-5n = 3 + 12$

$-5n = 15$

$n = \dfrac{15}{-5}$

$n = -3$

Therefore, m = –2 and n = –3

5.   Solve the simultaneous equations $5p - 8q = 16$ and $12q - p = 24$

Solution

The second equation above has been arranged differently from the first one. In the first equation the term in p is written first, but in the second equation the term in q has been written first. Therefore in order to avoid making mistake, it is advisable to have the same pattern of arrangement in the two equations.

Hence, let us now label the two equations and rearrange the second equation as shown below.

$5p - 8q = 16 \dots\dots\dots\dots\text{Equation (1)}$

$-p + 12q = 24 \dots\dots\dots\dots\text{Equation (2)}$

Let us make the coefficient of p to be equal in the two equations. In order to do this we simply multiply equation (2) by 5 so that the coefficient of p in the two equations will be 5 (we have ignored the signs). Hence, if we multiply equation (2) by 5, it will give us equation (3) as follows:

$5(-p) + 5(12q) = 5(24) \dots\dots\dots\text{Equation (3)}$

This simplifies to give:

$-5p + 60q = 120 \dots\dots\dots\text{Equation (3)}$

Now, equations 1 and 3 have the same coefficient of p. This means that p can now be eliminated by adding the two equations together. We have to add the two equations since the signs of the coefficients of p are different.

Hence, equation (1) and (3) can be brought together and added to eliminate p as shown below.

$$5p - 8q = 16 \text{ ....Equation (1)}$$

$$-5p + 60q = 120 \text{ ....Equation (3)}$$

Equation (1) + equation (3): $52q = 136$

Note that $5p + (-5p) = 0$, $-8q + 60q = 52q$, and $16 + 120 = 136$, as all obtained above when the corresponding like terms are added. Also, zero p was not written since it has been eliminated. We now continue from above as follows:

$$52q = 136$$
$$q = \frac{136}{52}$$
$$q = \frac{34}{13} \qquad \text{(After equal division by 4)}$$

Substitute $\frac{34}{13}$ for q in equation (2) in order to obtain p. This gives:

$$-p + 12q = 24 \text{ ....Equation (2)}$$
$$-p + 12(\frac{34}{13}) = 24$$
$$-p + \frac{408}{13} = 24$$
$$-p = 24 - \frac{408}{13}$$
$$-p = \frac{312 - 408}{13}$$
$$-p = \frac{-96}{13}$$
$$p = \frac{96}{13} \qquad \text{(After dividing both sides by } -1 \text{ in order to make p positive)}$$

Therefore, $p = \frac{96}{13}$, and $q = \frac{34}{13}$

6. Solve simultaneously, the following equations:
    $$2c + 5d = 0$$
    $$3c - 2d = 19$$

<u>Solution</u>
$$2c + 5d = 0 \text{ ....Equation (1)}$$
$$3c - 2d = 19 \text{ ....Equation (2)}$$

Let us make the coefficient of c to be equal in the two equations. In order to do this we multiply equation (1) by 3 (i.e. the coefficient of c in equation 2), and multiply equation (2) by 2 (i.e. the coefficient of c in equation 1). This will make the coefficient of c to be 6 in the two equations. Hence, equation (1) multiplied by 3 and equation (2) multiplied by 2 will give equation (3) and equation (4) respectively as shown below:

$$3(2c) + 3(5d) = 3(0) \text{ ....Equation (3)}$$
$$2(3c) - 2(2d) = 2(19) \text{ ....Equation (4)}$$

These two equations simplify to give:

$$6c + 15d = 0 \text{ ....Equation (3)}$$
$$6c - 4d = 38 \text{ ....Equation (4)}$$

We now eliminate c by subtracting equation (4) from equation (3). We have to subtract the two equations since the signs of the coefficients of c are the same.

Hence, equation 3 minus equation 4 is evaluated as shown below.

$$6c + 15d = 0 \text{ ....Equation (3)}$$
$$6c - 4d = 38 \text{ ....Equation (4)}$$

Equation (3) – equation (4): $19d = -38$

Note that $6c - 6c = 0$, $15d - (-4d) = 19d$, and $0 - 38 = -38$, as all obtained above.

We now continue from above as follows:

$$19d = -38$$
$$d = \frac{-38}{19}$$
$$d = -2$$

Substitute –2 for d in equation (1) in order to obtain c. This gives:

$$2c + 5d = 0 \text{ ....Equation (1)}$$
$$2c + 5(-2) = 0$$
$$2c - 10 = 0$$
$$2c = 10$$
$$c = \frac{10}{2}$$
$$c = 5$$

Therefore, c = 5 and d = –2

7. Solve simultaneously, the following equations:

$$11x - 7y = -11$$
$$5x + 3y = 29$$

<u>Solution</u>

$$11x - 7y = -11 \text{ ....Equation (1)}$$
$$5x + 3y = 29 \text{ ....Equation (2)}$$

Let us make the coefficient of x to be equal in the two equations. In order to do this, multiply equation (1) by 5 (i.e. the coefficient of x in equation 2), and multiply equation (2) by 11 (i.e. the coefficient of x in equation 1). This will give equation (3) and equation (4) respectively as shown below:

$$5(11x) - 5(7y) = 5(-11) \text{ ....Equation (3)}$$
$$11(5x) + 11(3y) = 11(29) \text{ ....Equation (4)}$$

These two equations simplifies to give:

$55x - 35y = -55$ ....Equation (3)

$55x + 33y = 319$ ....Equation (4)

We now eliminate x by subtracting equation (4) from equation (3) (note that we can also subtract equation 3 from equation 4). We have to subtract the two equations since the signs of the coefficients of x are the same.

Hence, equation 3 minus equation 4 is evaluated as shown below.

$$55x - 35y = -55 \quad ....\text{Equation (3)}$$

$$55x + 33y = 319 \quad ....\text{Equation (4)}$$

Equation (3) – equation (4): $\qquad -68y = -374$

Note that $55x - 55x = 0$, $-35y - (+33y) = -68y$, and $-55 - 319 = -374$, as all obtained above.

Hence: $-68y = -374$ (From above)

$$y = \frac{-374}{-68}$$

$$y = \frac{11}{2} \qquad \text{(When expressed in its lowest term)}$$

$$y = 5\frac{1}{2}$$

Substitute $\dfrac{11}{2}$ for y in equation (2) (we can use any equation) in order to obtain x. This gives:

$5x + 3y = 29$ ....Equation (2)

$$5x + 3\left(\frac{11}{2}\right) = 29$$

$$5x + \frac{33}{2} = 29$$

$$5x = 29 - \frac{33}{2}$$

$$5x = \frac{58 - 33}{2}$$

$$5x = \frac{25}{2}$$

$$x = \frac{\frac{25}{2}}{5}$$

$$x = \frac{25}{2} \times \frac{1}{5}$$

$$= \frac{5}{2}$$

$$x = 2\frac{1}{2}$$

Therefore, $x = 2\frac{1}{2}$ and $y = 5\frac{1}{2}$

Exercise 6

1. Solve simultaneously the equations: $2x - y = -3$ and $5x + y = -4$
2. Solve the simultaneous equations: $x + y = 5$ and $2x - y = 13$
3. Solve simultaneously, the equations $5m - n = 8$ and $3m - n = 4$
4. Solve simultaneously, the equations: $a - 3b = -5$ and $4a - 3b = 25$
5. Solve the equations: $p - q = 7$ and $p - 3q = 13$
6. Solve the equation: $10x - 5y = -30$ and $20x + 3y = -34$
7. Solve simultaneously, the equations $4x - y = 6$ and $3x + y = 15$.
8. Solve simultaneously, the equations $x + 5y = 2$ and $9x - y = 12$.
9. Solve simultaneously, the equations: $11m - 3n - 11 = 3$ and $8m + 5n - 3 = 0$
10. Solve the simultaneous equations $2p - 15q = 1$ and $2q - 11p = 5$
11. Solve simultaneously, the following equations:

    $5c + 5d = 10$

    $c - d = 5$
12. Solve simultaneously, the following equations:

    $x - 4y = -20$

    $12x + 5y = 10$

CHAPTER 7
WORD PROBLEMS LEADING TO SIMULTAENOUS LINEAR EQUATIONS

Examples

1. Three spoons and five knives cost $46. Five spoons and ten knives cost $90. Find the cost of a spoon and a knife.

Solution

Let the cost of a spoon be x, and the cost of a knife be y.

∴ From the first sentence, we have:

$3x + 5y = 46$ ....Equation (1)

Similarly, from the second sentence, we have:

$5x + 10y = 90$ ....Equation (2)

Bringing equation (1) and (2) together gives:

$3x + 5y = 46$ ....Equation (1)

$5x + 10y = 90$ ....Equation (2)

In order to make the coefficient of y to be the same in equation (1) and (2) (so as to solve by elimination method), lets simply divide equation (2) by 2. This gives:

$$\frac{5x + 10y = 90}{2}$$

This is simplified by dividing each term by 2 to give: $2.5x + 5y = 45$ .... equation (3)

Note that the coefficient of y is made to be the same in order to eliminate y (elimination method).

Therefore bringing equations (1) and (3) together gives:

$3x + 5y = 46$ ....Equation (1)

$\underline{2.5x + 5y = 45}$ ....Equation (3)

Equation (1) – equation (3): $0.5x = 1$

$$x = \frac{1}{0.5}$$

$$x = 2$$

Substitute 2 for x in any of the equations. Let us use equation (1)

∴ $3x + 5y = 46$ ....Equation (1)

$3(2) + 5y = 46$

$6 + 5y = 46$

∴ $5y = 46 - 6$

$5y = 40$

∴ $y = \frac{40}{5}$

$y = 8$

∴ A spoon cost $2, while a knife cost $8.

2. One-quarter of a boy's score in English plus one-fifth of his score in Mathematics makes 31. Half of his score in mathematics added to two-third of his score in English gives 80. What was his score in each subject?

Solution

Let his score in English be 'a' and his score in Mathematics be 'b'.

∴ From the first sentence, we have:

$$\frac{1}{4}a + \frac{1}{5}b = 31$$

Multiply each term by 20 (L.C.M of 4 and 5) in order to remove the fractions. This gives:

$$20\left(\frac{a}{4}\right) + 20\left(\frac{b}{5}\right) = 20 \times 31$$

∴ $5a + 4b = 620$ ....Equation (1)

From the second sentence of the question, we have

$$\frac{2}{3}a + \frac{1}{2}b = 80$$

Multiply each term by 6 (i.e. L.C.M of 3 and 2). This gives:

$$6\left(\frac{2a}{3}\right) + 6\left(\frac{b}{2}\right) = 6 \times 80$$

$2(2a) + 3(b) = 480$ (After 6 has been used to divide each of the denominator)

∴ $4a + 3b = 480$ ....Equation (2)

Bringing equations (1) and (2) together in order to solve them simultaneously gives:

$5a + 4b = 620$ ....Equation (1)

$4a + 3b = 480$ ....Equation (2)

Multiply equation (1) by 3 and equation (2) by 4 in order to make the coefficients of b to be the same in the two equations. This gives:

$3(5a + 4b = 620)$

$4(4a + 3b = 480)$

These simplify to:

$15a + 12b = 1860$ .... Equation (3)

$16a + 12b = 1920$ .... Equation (4)

Equation (4) – Equation (3): $a \quad = 60$

∴ $a = 60$

Substitute 60 for a in equation (2). This gives:

$4a + 3b = 480$ ....Equation (2)

$4(60) + 3b = 480$

$240 + 3b = 480$

$3b = 480 - 240$

∴ $3b = 240$

$$b = \frac{240}{3}$$

∴ $b = 80$

∴ He scored 60 in English and 80 in Mathematics

3. Divide 63 into two parts such that one part is two-fifth of the other.

Let the two parts be x and y.

∴ $x + y = 63$ ....Equation (1)   (Note that the two parts sum up to 63)

$\frac{2}{5}x = y$ ....Equation (2)   (Note that y has been made $\frac{2}{5}$ of x as stated above)

70

Substitute $\dfrac{2x}{5}$ for y in equation (1)

$\qquad x + y = 63$ ....Equation (1)

$\qquad x + \dfrac{2x}{5} = 63$

Multiply each term by 5 in order to remove the fraction. This gives:

$\qquad 5x + 5\dfrac{2x}{5} = 5 \times 63$

$\qquad 5x + 2x = 315$

$\qquad\qquad 7x = 315$

$\therefore \qquad\qquad x = \dfrac{315}{7} \qquad\qquad 7$

$\qquad\qquad x = 45$

From equation (1), $y = 63 - x$

$\qquad\qquad y = 63 - 45$

$\qquad\qquad y = 18$

$\therefore$ The two parts are 45 and 18.

4. A bicycle moves for p hours at 10km/h and q hours at 15km/h. If his total journey is 80km in 7 hours, find the values of p and q.

Solution

Recall that: Speed $= \dfrac{\text{Distance}}{\text{Time}}$

$\therefore$ At 10km/h, we have: $10 = \dfrac{\text{Distance}}{p}$ (Since time = p)

$\therefore$ Distance = 10p   (After cross multiplication)

Similarly, at 15km/hr, we have:

$\qquad$ Distance = 15q   (Since time is q in this case)

$\therefore$ Total distance is given by:

$\qquad 10p + 15q = 80$ ....Equation (1)

The total time is: $p + q = 7$ ....Equation (2)   (Note that the journey took 7 hours)

Multiply equation (2) by 10 in order to make the coefficient of p to be 10 in the two equations. This gives:

$\qquad 10(p + q = 7)$

$\qquad 10p + 10q = 70$ ....Equation (3)

Bringing equations (1) and (3) together in order to solve them simultaneously gives:

$\qquad\qquad 10p + 15q = 80$ ....Equation (1)

$\qquad\qquad \underline{10p + 10q = 70}$ ....Equation (3)

Equation (1) – equation (3): $\qquad 5q = 10$

$\qquad\qquad\qquad q = \dfrac{10}{5}$

$\qquad\qquad\qquad q = 2$

From equation (2), p can be expressed as:

$\qquad p = 7 - q$

$\qquad p = 7 - 2$ (Since q = 2)

p = 5

∴　p = 5 hours and q = 2 hours.

5.　The perimeter of an isosceles triangle is 56cm. If the two equal sides are (8y)cm and (12y − 4x + 2)cm, while the third side is (2x + 4y)cm, find the values of x and y.

<u>Solution</u>

Equating the two equal sides since they are equal gives:

　　8y = 12y − 4x + 2

∴　4x + 8y − 12y = 2　　(By collecting the unknown terms on the left hand side)

　　4x − 4y = 2

Dividing each term by 2 in order to simplify the equation gives:

　　2x − 2y = 1 ....Equation (1)

Adding the three sides of the triangle gives the perimeter as follows:

　　8y + 12y − 4x + 2 + 2x + 4y = 56

　　24y − 2x = 56 − 2

　　−2x + 24y = 54 ....Equation (2)

Bringing equations (1) and (2) together gives:

$$2x - 2y = 1 \text{ ....Equation (1)}$$
$$-2x + 24y = 54 \text{ ....Equation (2)}$$

In order to eliminate x, we add equation (1) and (2):　　22y = 55

$$y = \frac{55}{22}$$

$$y = \frac{5}{2} \qquad \text{(After equal division by}$$

11)

$$y = 2\frac{1}{2}$$

Substitute $\frac{5}{2}$ for y in equation (1)

　　2x − 2y = 1 ....(Equation (1)

$$2x - 2(\frac{5}{2}) = 1$$

　　2x − 5 = 1

　　2x = 1 + 5

　　2x = 6

∴　　$x = \frac{6}{2}$

　　$x = 3$

∴　$x = 3$ and $y = 2\frac{1}{2}$

6.　The three sides of an equilateral triangle are (3x)cm, (4y)cm and (x + y + 3)cm. Find:

a. The values of x and y.

b. The sides of the triangle.

72

Solutions

a. Since the three sides of an equilateral triangle are equal, then it follows that: $3x = 4y = x + y + 3$

Equating any two of the terms above can be used to obtain two equations as follows:

$3x = 4y$

$\therefore \quad 3x - 4y = 0$ ....Equation (1)

and $3x = x + y + 3$

$\therefore \quad 3x - x - y = 3$

$2x - y = 3$ ....Equation (2)

From equation (2), $y = 2x - 3$ ....Equation (3)

Substitute $2x - 3$ for y in equation (1)

$\therefore \quad 3x - 4y = 0$ ....Equation (1)

$3x - 4(2x - 3) = 0$

$3x - 8x + 12 = 0$ \qquad (Note that $-4 \times -3 = +12$)

$\therefore \quad -5x = -12$

$x = \dfrac{-12}{-5}$

$x = \dfrac{12}{5}$ \qquad (The negative sign has cancelled out)

$\therefore \quad x = 2\dfrac{2}{5}$

From equation (3) substitute: $x = \dfrac{12}{5}$

$\therefore \quad y = 2x - 3$ ....Equation (3)

$y = 2(\dfrac{12}{5}) - 3$

$= \dfrac{24}{5} - 3$

$= \dfrac{24 - 15}{5}$

$y = \dfrac{9}{5}$

$y = 1\dfrac{4}{5}$

Therefore, $x = 2\dfrac{2}{5}$ and $y = 1\dfrac{4}{5}$

b. Each side of the triangle is given by:

$3x$ or $4y$ or $x + y + 3$

$\therefore \quad 3x$ will give:

$3 \times \dfrac{12}{5} = \dfrac{36}{5}$

$= 7\dfrac{1}{5}$cm

$\therefore \quad$ Each side of the triangle is $7\dfrac{1}{5}$cm

7. The sum of the digits of a two digits number is 13. If the digits are interchanged, the number is decreased by 45. Find the number.

<u>Solution</u>
Let the two digits number be xy. Therefore, the number itself is obtained as follows:

$10x + y$ (For example 24 = (2 x 10) + 4 = 20 + 4 = 24)

Note that x is the tens digit while y is the unit digit. When the digits are interchanged to obtain yx, then the new number is: $10y + x$

Since the original number is decreased by 45 after the digits are interchanged, it means that the first number is greater than the second number by 45.

$\therefore$ $(10x + y) - (10y + x) = 45$

$10x + y - 10y - x = 45$

$9x - 9y = 45$

$\therefore$ $x - y = 5$ (After dividing each term by 9 in order to simplify the equation)

$\therefore$ $x - y = 5$ ....Equation (1)

Recall from the first sentence that:

$x + y = 13$ ....Equation (2)

$\underline{x - y = 5}$ ....Equation (1)   (This is simply equation 1 brought down)

Equation (2) – (1): $2y = 8$ (x has been eliminated by the subtraction. Also +y – (–y) = 2y)

$\therefore$ $y = \dfrac{8}{2}$

$y = 4$

From equation (1), substitute 4 for y. This gives:

$x - y = 5$ ....Equation (1)

$x - 4 = 5$

$x = 5 + 4$

$x = 9$

$\therefore$ The number is 94 (i.e. xy)

Check: The number is 94. When the digits are interchanged it becomes 49. Their difference is 94 – 49 = 45. Hence it is correct as given in the question.

8.   In a two digits number, the difference between the digits is 1. The number is 1 more than 5 times the sum of the digits. Find the number if the tens digit is less than the unit digit.

<u>Solution</u>
Let the number be xy (x is the tens digit)

$y - x = 1$ ....Equation (1)   (Note that the unit digit is greater than the tens digit). This equation can also be expressed as:

$-x + y = 1$

The sum of the digits is $x + y$. And 5 times the sum of the digits is $5(x + y)$

The number itself is:

$10x + y$

Since the number is 1 more than 5 times the sum of the digits, then it follows that:

$10x + y - 5(x + y) = 1$

$10x + y - 5x - 5y = 1$

$5x - 4y = 1$ ....Equation (2)

Bringing equation (1) and (2) together in order to solve them simultaneously gives:

74

$-x + y = 1$ ....Equation (1)

$5x - 4y = 1$ .... equation (2)

Multiply equation (1) by 5. This is done in order to make the coefficient of x in the two equations to be 5. This gives:

$-5x + 5y = 5$ ....Equation (3)

Bringing equation (2) and (3) together gives:

$5x - 4y = 1$ ....Equation (2)

$-5x + 5y = 5$ ....Equation (3)

Equation (2) + equation (3): $y = 6$

Substitute 6 for y in equation (1)

∴ $-x + y = 1$ ....Equation (1)

$-x + 6 = 1$

∴ $x = 6 - 1$

$x = 5$

∴   The number is 56 (i.e. xy)

9.   The sum of a man's age and his son's age is 45 years. Five years ago, the man was 6 times as old as his son. How old was the man when his son was born?

Solution

Let the man's present age be x, and his son's present age be y.

∴ $x + y = 45$ ....Equation (1)

Five years ago, the man's age was $x - 5$, while his son's age was y – 5. At that time, his age was 6 times that of his son. This means that:

$x - 5 = 6(y - 5)$

$x - 5 = 6y - 30$

$x - 6y = -30 + 5$

$x - 6y = -25$ ....Equation (2)

Bringing equations (1) and (2) together gives:

$x + y = 45$ ....Equation (1)

$x - 6y = -25$ ....Equation (2)

Equation (1) – (2) $7y = 70$ (Note that +y –(–6y) = 7y, and 45 –(–25) = 70)

$$y = \frac{70}{7}$$

∴ $y = 10$

Substitute 10 for y in equation (1). This gives:

$x + y = 45$ ....Equation (1)

$x + 10 = 45$

$x = 45 - 10$

$x = 35$

∴   The man's present age is 35 years while his son's present age is 10 years. This means that the son was born 10 years ago. Therefore, 10 years ago, the man's age was 35 – 10 = 25 years.

∴   The man was 25 years when his son was born.

10. If 1 is added to the numerator of a fraction, and 2 is added to the denominator, the fraction becomes $\frac{1}{3}$. If 3 is added to both the numerator and denominator of the fraction, the fraction becomes $\frac{1}{2}$. Find the fraction.

<u>Solution</u>

Let the numerator be x and the denominator be y

∴ The fraction is $\frac{x}{y}$

From the first sentence:

$$\frac{x+1}{y+2} = \frac{1}{3}$$

∴ $3(x + 1) = 1(y + 2)$ (When we cross multiply)

$3x + 3 = y + 2$

$3x - y = 2 - 3$

∴ $3x - y = -1$ ....Equation (1)

From the second sentence:

$$\frac{x + 3}{y + 3} = \frac{1}{2}$$

∴ $2(x + 3) = 1(y + 3)$

$2x + 6 = y + 3$

$2x - y = 3 - 6$

∴ $2x - y = -3$ ....Equation (2)

Bringing equations (1) and (2) together gives:

$3x - y = -1$ ....Equation (1)

$\underline{2x - y = -3}$ ....Equation (2)

Equation (1) − (2) x $= 2$

Hence $x = 2$

Substitute 2 for x in equation (1)

∴ $3x - y = -1$ ....Equation (1)

$3(2) - y = -1$

$6 - y = -1$

$6 + 1 = y$

∴ $y = 7$

∴ The fraction is $\frac{2}{7}$ (i.e. $\frac{x}{y}$)

Exercise 7

1. Five books and eight pens cost $90. Six books and four pens cost $80. Find the cost of a book and a pen.

2. One-third of a girl's score in Mathematics plus two-seventh of her score in Biology makes 30. One-fifth of her score in Biology added to half of her score in Mathematics gives 29. What was her score in each subject?

3. Divide 75 into two parts such that one part is one-third of the other.

4. A man walks for m hours at 12km/h and n hours at 18km/h. If his total journey is 114km in 8 hours, find the values of m and n.

5. The perimeter of an isosceles triangle is 50cm. If the two equal sides are (6y)cm and $(2y - 4x + 4)$cm, while the third side is $(2x + 6y)$cm, find the values of x and y. Hence, determine the length of each side of the triangle.

6. The three sides of an equilateral triangle are $(5x)$cm, (4y)cm and $(x + 2y + 9)$cm. Find:
a. The values of x and y.
b. Each side of the triangle.

7. The sum of the digits of a two digits number is 11. If the digits are interchanged, the number is decreased by 27. Find the number.

8. In a two digits number, the difference between the digits is 7. The number is 4 more than 8 times the sum of the digits. Find the number if the tens digit is greater than the unit digit.

9. The sum of a woman's age and her daughter's age is 65 years. Ten years ago, the woman was 8 times as old as her daughter. How old was the woman when her daughter was born?

10. If 5 is added to the numerator of a fraction, and 2 is added to the denominator, the fraction becomes $\frac{4}{5}$. If 2 is added to both the numerator and denominator of the fraction, the fraction becomes $\frac{1}{2}$. Find the fraction.

CHAPTER 8
LOGICAL REASONING

Statements

In logical reasoning, a statement is a spoken or written sentence that gives information about something. The statement may be true or false.

Some statement may be either true or false. They are called closed statements. Statements that we do not know whether they are true or false, are called open statements.

Examples

Let us classify the following as closed or open statements or neither, and also say if they are true or false.

1. X: England is a country. (Closed, and true)
2. Y: w is a prime number. (Open. Impossible to say)
3. Z: How old are you? (Neither. Not applicable)

Simple and Compound Statements

In logical reasoning a simple statement is a sentence which has only one main verb.

A compound statement is a sentence that contains two or more simple statements.

A statement can either be true or false. In logical context, questions, commands, sayings and exclamations are not considered as statement.

John is a clever and brave boy, is a compound statement. Jane is a geologist or a teacher is also a compound statement. When 'and' is used in a compound statement we call it a conjunction. When 'or' is used in a compound statement we call it a disjunction.

Symbols for Conjunction and Disjunction

Let us examine two simple statements as given below:

   Los Angeles is in USA

   USA is in North America

Let the first statement be represented by p and the second statement be represented by q. A compound statement can be obtained from the two simple statement as follows:

a. Los Angeles is in USA or USA is in North America. This can be expressed in symbol by: p or q which also means p v q, where the connective v means 'or'

b. Los Angeles is in USA and USA is in North America. This can be expressed in symbol by: p and q which also means p ∧ q, where the connective ∧ means 'and'.

Truth Table

The truth table can be used to determine the truth value of a compound statement. A compound statement p v q can be read as; either p or q or both. The truth table for p v q is illustrated below.

p	q	p v q
T	T	T
T	F	T
F	T	T
F	F	F

This shows that the statement p v q is false only when p and q are false.
The truth table for p ∧ q is illustrated below.

p	q	p ∧ q
T	T	T
T	F	F
F	T	F
F	F	F

This shows that the statement p ∧ q is true only when both p and q are true.

Negation

Negation simply means 'not'. For a statement, s, the negation of s can be written as ~s.
If s is true, then ~s is false, and if s is false, then, ~s is true.

Examples
The negation of the following statements are given in brackets below.
1. Peter is brilliant (Peter is not brilliant)
2. Jane is not a sailor (Jane is a sailor)
3. $x = 12$ ($x \neq 12$)

Implication (Conditional Statements)

Let us take a statement, s, as stated below:
    s: If a student plays football then the student takes part in sports.
Statement, s, is actually a combination of the following statements.
    p: A student in a school plays football
    q: A student in a school takes part in sports
s, states that if p is true, then q will also be true. We call a compound statement like this an implication.
Symbolically, we write the implication s as p ⇒ q, which reads as 'p implies q'. It can also be interpreted as any of the following:
1. If p then q,
2. p only if q
3. q follows from p
4. q if p

In the implication s: p ⇒ q, the substatement p is called the antecedent, while q is called the consequent. The statement p ⇒ q is sometimes called a conditional statement.
The truth table for p ⇒ q is illustrated below.

p	q	p ⇒ q
T	T	T
T	F	F
F	T	T
F	F	T

This shows that the statement p ⇒ q is always true except when the first statement is true and the second statement is false.F

Examples
1. The statements, A, B, C are given below.

    A: Lawyers are liars
    B: Lawyers are brilliant
    C: Lawyers are neat

Write the following statement as implications
a. Brilliant lawyers are liars
b. No neat lawyers are brilliant

Solution
a. B ⇒ A   (This means B implies A)
b. ~C ⇒ B   (This means ~C implies B)

2. Write the following implications using ordinary language.
a. m is negative ⇒ m² is positive
b. a² = 36 ⇒ a = +6 or –6

Solution
a. If m is negative then m² is positive
b. If a² is equal to 36 then a is equal to +6 or –6

Equivalent Statements
Two compound statements are said to be logically equivalent if they have the same truth value.
If in an implication, p ⇒ q and q ⇒ p, i.e. p implies q and q implies p, then we write p ⇔ q and say that p and q form equivalent statements.
From an implication p ⇒ q, we can deduce the equivalent implication as, ~p ⇒ ~q. The converse of p ⇒ q is q ⇒ p.

Examples
1. Write down the converse of the implication:

    A: John is smart ⇒ John plays football

Solution
The converse of A is B where B is given by:

    B: John plays football $\Rightarrow$ John is smart

2. Write statements which are equivalent to the following:
a. If you play a game the computer will be hot
b. No student who offer physics is in the technical class

Solution
The equivalent statements are:
a. If you do not play a game the computer will not be hot
b. Those students who offer physics are not in the technical class

3. Use the truth table to prove that: $\sim (p \vee q) = \sim p \wedge \sim q$

Solution
The tables are as shown below.

p	Q	p $\vee$ q	$\sim (p \vee q)$
T	T	T	F
T	F	T	F
F	T	T	F
F	F	F	T

p	q	$\sim p$	$\sim q$	$\sim p \wedge \sim q$
T	T	F	F	F
T	F	F	T	F
F	T	T	F	F
F	F	T	T	T

The last columns in each of the two tables above have the same truth values. Hence, $\sim (p \vee q) =$ $\sim p \wedge \sim q$

Valid Argument
A conclusion made from a set of statements (call the premises) can either be valid or invalid. The argument (the premises and conclusion) is valid if the fact that the premises are true, forces the conclusion to be true.

81

Use of Venn Diagrams in Logical Operations

Examples

1. Determine if the following argument is valid.

   All Asian people are kind

   Chan is an Asian

   Chan is kind

Solution

First draw the Venn diagram from the first premise. Kind people should be the overall region. Since all Asian people are kind, the region for all Asian people should go into the region for kind people. The second premise, Chan is an Asian, (represented by C) will go inside the region representing all Asian. Hence, the two premises are true since they fall in the same circle. Therefore the conclusion is true, i.e. the argument is valid. Note that the last statement is the conclusion.

The Venn diagram is as shown below.

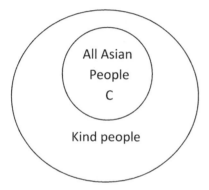

2. Is the following argument valid?

   Blue buildings are beautiful buildings

   This building is Blue

   This building is beautiful

Solution

Let B represent 'This building'. The Venn diagram below shows that B is in blue buildings and blue buildings is in beautiful buildings. Note that the overall circle represents beautiful buildings, which is the last part of the first statement.

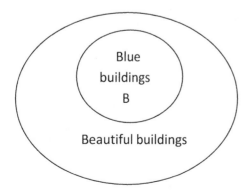

The Venn diagram shows that the two premises are true since they fall in the same circle. Therefore the argument is valid.

3. Determine if the argument below is valid.
    All workers in the factory eat bread
    Daniel eats bread
    Daniel is a worker

Solution

Let D represents Daniel. The Venn diagram below shows that D can be outside 'All workers in the factory' since we do not know if Daniel is a worker in the factory. Hence, D is in 'People who eat bread', but it can be outside 'All workers in the factory'. Therefore the conclusion that Daniel is a worker in the factory is not true. This shows that the argument is not valid.
Note that the overall region of the circle represents 'People who eat Bread' which is the last part of the first statement.

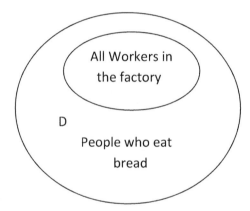

The Venn diagram shows that the two premises are not true since they do not fall in the same circle. Therefore the argument is invalid.

4. Determine if the following argument is valid.

    Smith is a lawyer
    Smith is a liar
    All lawyers are liars

Solution

This is a situation of two groups of people (i.e. lawyers and liars. Smith (represented by S) fall into the two groups. However, the two groups cannot be put together since the statements have not indicated that they can be together. The Venn diagram is as shown below.

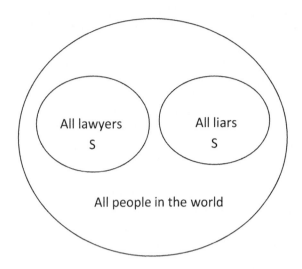

The Venn diagram above shows that these two groups of people in the world cannot be completely put together. Hence the conclusion drawn from them is not true. Therefore the argument is not valid.

Exercise 8

1. Classify the following into closed or open statements or neither, and also say if they are true or false.
a. P: New York is a state.
b. Q: x is an odd number.
c. R: What is your nationality?

2. State the negation of the following statements:
a. Stanley is neat
b. Jane is not a doctor
c. m = -5

3. The statements, X, Y, Z are given below.

       X: Politicians are liars
       Y: Politicians are smart
       Z: Politicians are rich

Write the following statement as implications
a. Smart politicians are liars
b. No rich politician is smart

4. Write the following implications using ordinary language.
a. r is negative $\Rightarrow r^3$ is negative
b. $y^3 = -64 \Rightarrow y = -4$

5. Write down the converse of the implication:

   X: Abel is clever $\Rightarrow$ Abel is intelligent

6. Write statements which are equivalent to the following:
a. If you travel by road it will take you 11 hours
b. No student who offer Mathematics will score above 80

7. Use the truth table to prove that: $\sim(\sim p \wedge \sim q) = p \vee q$

8. Determine if the following argument is valid.

   All Asian people are intelligent

   Bill is an Asian

   Bill is intelligent

9. Is the following argument valid?

   White buildings are attractive buildings

   This building is white

   This building is attractive

10. Determine if the argument below is valid.

    All the people in the family are dark in complexion

    Franca is dark in complexion

    Franca is a member of the family

11. Determine if the following argument is valid.

    Joy is a Doctor

    Joy works in a hospital

    All doctors work in hospitals

CHAPTER 9
QUADRATIC EQUATION

An equation such as $x^2 + 7x - 18 = 0$, in which 2 is the highest power of the unknown variable is called quadratic equation. There are four methods of solving quadratic equations. They are:
1. By factorization
2. By Completing the square
3. By use of quadratic equation formula
4. By graphical method

We are going to look at the first three methods.

Factorization of Quadratic Expression

Examples
1. Factorize the following quadratic expressions
a. $3a^2 + 5a$
b. $y^2 - 81$
c. $4m^2 - 49$
d. $18c^2 - 72$

Solutions
a. $3a^2 + 5a$
In this case the common factor (H.C.F) of $3a^2$ and $5a$ is 'a'. Hence we factorize directly as follows:
$3a^2 + 5a = a(3a + 5)$
The terms in the bracket are obtained by dividing each of $3a^2$ and $5a$ by 'a' which is the common factor.

b. $y^2 - 81$
$y^2 - 81$ is a difference (minus) of two squares. So, recall that the difference of two squares such as $a^2 - b^2$ is factorized as follows: $a^2 - b^2 = (a + b)(a - b)$
Therefore, $y^2 - 81 = (y)^2 - (9)^2$
$\qquad\qquad\quad = (y + 9)(y - 9)$

c. $4m^2 - 49$
Each term in this expression is a square. The square root of $4m^2$ is 2m, while the square root of 49 is 7. Hence, $4m^2 - 49$ is a difference (minus) of two squares. So, recall that the difference of two squares such as $a^2 - b^2$ is factorized to obtain: $(a + b)(a - b)$
Therefore, $4m^2 - 49 = (2m)^2 - (7)^2$
$\qquad\qquad\qquad = (2m + 7)(2m - 7)$ [In the same way as $a^2 - b^2 = (a + b)(a - b)$]

d. $18c^2 - 72$
This is not exactly a difference of two squares, but it can be made a difference of two squares by first factorizing with 18 as a common factor. This gives:

$18c^2 - 72 = 18(c^2 - 4)$

The terms in the bracket are now difference of two squares. The square root of c^2 = c, while the square root of 4 = 2. Hence, we now have:

$18(c^2 - 4) = 18[(c)^2 - (2)^2]$

$\qquad\qquad = 18[(c + 2)(c - 2)]$

Therefore, $18c^2 - 72 = 18[(c + 2)(c - 2)]$

In the four examples above, none of the quadratic expressions has a complete three terms of the usual quadratic expression such as $3x^2 + 7x - 4$. Hence it is easy to factorize such expression. However, when we have a complete quadratic expression such as $2x^2 - 11x - 30$, then we have to apply a different method of factorization. This method is explained in the examples below.

2. Factorize $2x^2 - 11x - 30$.

<u>Solution</u>

This is a complete quadratic expression with three terms. In order to factorize this expression, we first multiply the first and last terms. This gives:

$2x^2(-30) = -60x^2$

Then we look for two factors of $-60x^2$ (i.e. two numbers whose product is $-60x^2$) such that their sum will give $-11x$ (i.e. the middle term in the original quadratic expression).

Some simple rules that should be noted when looking for the two factors are as given below:

1. In a case such as $2x^2 + 7x + 5$, where the sign of the constant term (i.e. +5) and the middle term (i.e. +7) are both positive, then the two factors should be positive.

2. In a case such as $2x^2 - 7x + 5$, where the sign of the constant term is positive (i.e. +5) and the sign of the middle term is negative (i.e. −7), then the two factors should be negative.

3. In a case such as $2x^2 - 3x - 5$, where the sign of the constant term is negative (i.e. −5) and the sign of the middle term is negative (i.e. −3), then one factor will be positive and the other will be negative. However, the higher value of the two factors will take the negative sign (i.e. the sign of the middle term).

4. In a case such as $2x^2 + 3x - 5$, where the sign of the constant term is negative (i.e. −5) and the sign of the middle term is positive (i.e. +3), then one factor will be positive and the other will be negative. However, the higher value of the two factors will take the positive sign (i.e. from the middle term).

Let us summarize these four rules as shown below:

S/N	Sign of product	Sign of middle term	Sign of factors
1	+	+	+ and +
2	+	-	- and -
3	-	-	+ and - with the larger value having -
4	-	+	+ and - with the larger value having +

With these four rules in mind, finding the factors in a quadratic equation becomes easy.
Let us continue with the solution to the question above.

The question at hand which is $2x^2 - 11x - 30$, is applicable to rule 3 above. This means that our two factors will have different signs with the higher value having the negative sign.

From the $-60x^2$ obtained above, a simple way of finding the factors of 60 (first ignore the minus sign and x^2) is to break 60 into its prime factors as follows:

2	60
2	30
3	15
5	5
	1

The above division can help us obtain the twelve factors of 60 as follows: 1, 2, 3, 4 (from 2 x 2 in the division above), 5, 6 (from 2 x 3 in the division above), 10 (from 2 x 5 in the division above), 12 (from 2 x 2 x 3 in the division above), 15, 20 (from 2 x 2 x 5 in the division above), 30 and 60.

Note that only 1, 2, 3, 5, 15, 30 and 60 are visible from the division above. The other factors (4, 6, 10, 12 and 20) have to be obtained by multiplying any two or more of the visible factors. Let us rewrite the factors as follows: 1, 2, 3, 4, 5, 6, 10, 12, 15, 20, 30, and 60.

Now that we have our factors, we can pair them up to see which of them will give the required sum of -11. According to rule 3 above, the higher factor should have a negative sign. Let us now write out the pairs and assign a negative sign to the larger factor. Hence, the pair whose product will give us -60 are:

1 and -60 (This pair is out. Their sum cannot give -11)

2 and -30 (This pair is out)

3 and -20 (This pair is out)

5 and -12 (This pair is out)

6 and -10 (This pair is out)

4 and -15 (This is the required pair. Their sum gives -11)

Hence the only possible factors that can give a sum of -11 are 4 and -15. Hence we now include x on them so that their product will give $-60x^2$ and their sum will give $-11x$. Hence the numbers are $-15x$ and $+4x$. We now substitute these two numbers for $-11x$ in the original expression. This gives:

$2x^2 - 11x - 30 = 2x^2 + 4x - 15x - 30$ (Note that any of the two numbers can be written first).

We now factorize each pair of terms in this new expression (i.e. factorization by grouping). This gives:

$2x^2 + 4x - 15x - 30 = 2x(x + 2) - 15(x + 2)$

Note that $-15x - 30 = -15(x + 2)$. The negative signs from the question have change to + in the bracket. This is what happens when a negative number is used for the factorization. A positive sign in the question will also change to a negative sign in the bracket. We now have:

$2x^2 + 4x - 15x - 30 = 2x(x + 2) - 15(x + 2)$ [Here the common factor is $(x + 2)$]

$= (x + 2)(2x - 15)$ (Simply take one of the common brackets i.e. $(x + 2)$, and then the other two terms outside each of the bracket and enclose them in a bracket).

Therefore, $2x^2 - 11x - 30 = (x + 2)(2x - 15)$

3. Factorize $x^2 + 13x - 30$

Here, you will observe that the coefficient of x^2 is 1 (i.e. $1x^2$). This is not like the previous example, i.e. $2x^2 - 11x - 30$, where the coefficient of x^2 is 2. So, a short method of factorizing this expression is to go ahead and look for two numbers whose product is -30 and whose sum is $+13$, without any need of multiplying the first and last terms. This is explained as follows:

$x^2 + 13x - 30 = (x\quad)(x\quad)$ (simply put the variable x into the brackets)

In order to complete the bracket, we find two numbers whose product is -30 and whose sum is $+13$. This means that we have to look for the factors of 30.

The factors of 30 are: 1, 2, 3, 5, 6, 10, 15 and 30.

Now that we have our factors, we can pair them up to see which of them will give a possible sum of $+13$. According to rule 4 above, the higher factor should have a positive sign. Let us now write out the pairs and assign a positive sign to the larger factor. Hence the pair whose product will give us -30 are:

-1 and $+30$ (This is out. Their sum cannot give $+13$)

-3 and $+10$ (This is out. Their sum cannot give $+13$)

-5 and $+6$ (This is out.)

-2 and $+15$ (These are the required factors. Their sum will give $+13$)

Therefore, we now put them into the brackets above as follows:

$x^2 + 13x - 30 = (x + 15)(x - 2)$

4. Factorize $3x^2 + 11x + 6$

Solution

$3x^2 + 11x + 6$

The first thing to do is to multiply the first and last term. This gives:

$3x^2(6) = 18x^2$

We now look for two factors of 18 whose product is 18 and sum is $+11$. This is a case of rule 1 above where the constant term is positive and the middle term is also positive. Hence the two factors will be positive. The factors of 18 are 1, 2, 3 , 6, 9, 18. If we pair them up based on any two of the factors whose product is 18, we have:

(1 and 18), (2 and 9), (3 and 6).

Out of these pairs, only 2 and 9 can give us a sum of 11. Their signs must be positive based on rule 1 above. We now include x on them so that their product will give $18x^2$ and their sum will give $11x$. Hence the numbers are $+9x$ and $+2x$.

We now substitute these two numbers for $11x$ in the original expression. This gives:

$3x^2 + 11x + 6 = 3x^2 + 9x + 2x + 6$ (Note that any of the two numbers can be written first).

We now factorize by grouping. This gives:

$3x^2 + 9x + 2x + 6 = 3x(x + 3) + 2(x + 3)$
$$= (x + 3)(3x + 2)$$

Therefore, $3x^2 + 11x + 6 = (x + 3)(3x + 2)$

5. Factorize $5x^2 - 14x + 8$

<u>Solution</u>

$5x^2 - 14x + 8$

The first thing to do is to multiply the first and last terms. This gives:

$5x^2(8) = 40x^2$

We now find two factors of 40 whose product is 40 and sum is -14 (i.e. the middle term). This is a case of rule 2 above where the constant term is positive and the middle term is negative. Hence the two factors will be negative. The factors of 40 are 1, 2, 4 , 5, 8, 10, 20 and 40. According to rule 2 above, the factors will both have a negative sign. Let us now write out the pairs and assign a negative sign to each factor. Hence the pairs whose product will give us 40 are:

$(-1$ and $-40)$, $(-2$ and $-20)$, $(-4$ and $-10)$, $(-5$ and $-8)$. Note that the product of two negative numbers gives a positive value.

Out of these pairs, only -4 and -10 can give us a sum of -14. Hence we now include x on them so that their product will give $40x^2$ and their sum will give $-14x$. Hence the numbers are $-4x$ and $-10x$.

We now substitute these two numbers for $-14x$ in the original expression. This gives:

$5x^2 - 14x + 8 = 5x^2 - 4x - 10x + 8$ (Note that any of the two numbers can be written first).

We now factorize by grouping. This gives:

$5x^2 - 4x - 10x + 8 = x(5x - 4) - 2(5x - 4)$

Note that in the second pair (i.e. $-10x + 8$) the sign of the first number must be used. Hence the use of -2 to obtain $-2(5x - 4)$.

Therefore with $(5x - 4)$ as a common factor we proceed with the factorization as follows:

$x(5x - 4) - 2(5x - 4) = (5x - 4)(x - 2)$

Therefore, $5x^2 - 14x + 8 = (5x - 4)(x - 2)$

Solving quadratic equations by by Factorization

In solving quadratic equation by factorization, we first factorize the equation and then equate each of the term or bracket to zero to solve the linear equation formed. The general form of a quadratic equation is given by: $ax^2 + bx + c = 0$. Hence if an equation is given by $3x^2 - 11x + 6$, we can compare it with $ax^2 + bx + c = 0$, and obtain:

$a = 3$, $b = -11$ and $c = 6$.

With these values of a, b and c, we can determine if a quadratic expression can be factorized or not. Hence, a quadratic equation can be factorized if:

$b^2 - 4ac$ gives a perfect square.

For example, let us determine which of the following can be factorized:

a. $x^2 + 5x + 6 = 0$

b. $4x^2 - 11x + 5 = 0$

c. $2x^2 + x - 6 = 0$

d. $5x^2 - 6x - 8 = 0$

e. $3x^2 - 13x + 8 = 0$

Now, let us see which of them will give $b^2 - 4ac$ as a perfect square.

a. $x^2 + 5x + 6 = 0$.
   $a = 1$, $b = 5$, $c = 6$

Hence, $b^2 - 4ac = 5^2 - (4 \times 1 \times 6)$
$$= 25 - 24$$
$$= 1$$

Therefore, 1 is a perfect square whose square root is 1. Hence, $x^2 + 5x + 6 = 0$ can be factorized.

b. $4x^2 - 11x + 5 = 0$.
In this case, $a = 4$, $b = -11$, $c = 5$

Hence, $b^2 - 4ac = (-11)^2 - (4 \times 4 \times 5)$
$$= 121 - 80$$
$$= 41$$

Therefore, 41 is not a perfect square. Its square root will not give us a whole number. Hence, $4x^2 - 11x + 5 = 0$ cannot be factorized.

c. $2x^2 + x - 6 = 0$.
   $a = 2$, $b = 1$, $c = -6$

Hence, $b^2 - 4ac = 1^2 - [4 \times 2 \times (-6)]$
$$= 1 - (-48)$$
$$= 1 + 48$$
$$= 49$$

Therefore, 49 is a perfect square whose square root is 7. Hence, $2x^2 + x - 6 = 0$ can be factorized.

d. $5x^2 - 6x - 8 = 0$.
   $a = 5$, $b = -6$, $c = -8$

Hence, $b^2 - 4ac = (-6)^2 - [4 \times 5 \times (-8)]$
$$= 36 - (-160)$$
$$= 36 + 160$$
$$= 196$$

Therefore, 196 is a perfect square whose square root is 14. Hence, $5x^2 - 6x - 8 = 0$ can be factorized.

e. $3x^2 - 13x + 8 = 0$.
   $a = 3, b = -13, c = 8$
Hence, $b^2 - 4ac = (-13)^2 - (4 \times 3 \times 8)$
$$= 169 - 96$$
$$= 73$$

Therefore, 73 is not a perfect square. Its square root will not give us a whole number. Hence, $3x^2 - 13x + 8 = 0$ cannot be factorized.

Examples
1. Solve the following equations by factorization:
a. $2a^2 + 8a = 0$
b. $r^2 - 64 = 0$
c. $9m^2 - 25 = 0$
d. $(3x + 2)^2 = 16$

Solutions
a. $2a^2 + 8a = 0$
Factorizing this quadratic equation by taking 2a as the common term gives:
   $2a(a + 4) = 0$
We now equate each term to zero and solve for 'a'. We equate each term to zero because the product of two or more terms is zero if at least one of the terms is equal to zero. Therefore equating each term above to zero and solving for 'a' gives:
Either,
   $2a = 0$
Dividing both sides by 2 gives:
   $a = \dfrac{0}{2}$
   $a = 0$
Or, $(a + 4) = 0$
   $a + 4 = 0$
   $a = -4$
Therefore, $a = 0$ or $a = -4$

b. $r^2 - 64 = 0$
This can be solved directly as follows:
   $r^2 - 64 = 0$
   $r^2 = 64$
   $r = \sqrt{64}$

$$r = \pm 8$$
Therefore, r = +8 or r = –8

Note that ± is used to obtain two values where one is given a positive value and the other is given a negative value.

Hence, r = 8 or r = –8

c. $9m^2 - 25 = 0$
   $9m^2 = 25$

Divide both sides by 9. This gives:

$$m^2 = \frac{25}{9}$$

$$m = \sqrt{\frac{25}{9}}$$

$$m = \pm \frac{5}{3}$$

Therefore, $m = \frac{5}{3}$ or $m = -\frac{5}{3}$

d. $(3x + 2)^2 = 16$

Let us take the square root of both sides. This gives:

$$\sqrt{(3x + 2)^2} = \sqrt{16}$$
$$(3x + 2) = \pm 4$$
$$3x = \pm 4 - 2$$

$$x = \frac{+4 - 2}{3} \quad \text{or} \quad x = \frac{-4 - 2}{3}$$

$$x = \frac{2}{3} \quad \text{or} \quad x = \frac{-6}{3}$$

$$x = \frac{2}{3} \quad \text{or} \quad x = -2$$

2. Solve the equation $2x^2 - 5x - 3 = 0$

<u>Solution</u>
   $2x^2 - 5x - 3 = 0$

Multiply the first and last term on the left hand side. This gives:
   $2x^2(-3) = -6x^2$

Now look for two numbers in x such that their product is $-6x^2$ and their sum is $-5x$ (i.e. the middle term). For us to obtain the two factors let us recall our rules as shown below:

S/N	Sign of product	Sign of middle term	Sign of factors
1	+	+	+ and +
2	+	-	- and -
3	-	-	+ and - with the larger value having -
4	-	+	+ and - with the larger value having +

Based on the rules, since the product of the terms is negative (i.e. $-6x^2$) and their sum is also negative (i.e. $-5x$), as in rule 3 above, then the two factors of 6 that we need must have different signs, and with the higher of the factors having a negative sign. The pairs of factors of 6 whose product is 6 are (1 and 6) and (2 and 3). The larger factor should have a negative sign. If we assign a negative sign to the larger factors, then the pair can be written as: (+1 and –6), (2 and –3). Out of these two pairs, the one whose sum is –5 is (+1 and –6). Hence, we now include x on them so that their product will give $-6x^2$ and their sum will give $-5x$. Hence the numbers are $+x$ and $-6x$. Note that $+1x$ is also $+x$.

We now substitute these two numbers for $-5x$ in the original expression. This gives:

$$2x^2 - 6x + x - 3 = 0$$

We now factorize by grouping. This gives:

$$2x(x - 3) + 1(x - 3) = 0$$

With $(x - 3)$ as a common factor we proceed as follows:

$$(x - 3)(2x + 1) = 0$$

Now that we have factorized the equation, we equate each of the bracket to zero and solve for x in each of the linear equation formed. This gives:

Either, $x - 3 = 0$

$$x = 3$$

Or, $2x + 1 = 0$

$$2x = -1$$

$$x = -\frac{1}{2} \quad \text{(After dividing both sides by 2)}$$

Therefore, $x = 3$ or $x = -\frac{1}{2}$

3. Solve the quadratic equation $y^2 + y - 12 = 0$

Solution

$$y^2 + y - 12 = 0$$

Since the coefficient of y^2 in this equation is 1, then we can factorize the equation by putting directly the appropriate factors into bracket.

$$y^2 + y - 12 = 0$$

$$(y \quad)(y \quad) = 0$$

Here we look for the factors of 12 whose product is -12 and whose sum is $+1$ (i.e. $+1$ from $+y$ which also means $+1y$). Since the product is negative and the sum is positive, then the two factors must have different sign, and with the positive sign going to the larger factor. The factors of 12 whose product will give -12 and whose sum is +1 are +4 and −3. Note that the larger value has the positive sign (i.e. from +1). We now fill these two values into the bracket above. This gives:

$$y^2 + y - 12 = 0$$
$$(y + 4)(y - 3) = 0$$

Equating each bracket to zero gives:

$$y + 4 = 0$$
$$y = -4$$
$$y - 3 = 0$$
$$y = 3$$

Hence, $y = 3$ or $y = -4$

4. Solve the quadratic equation $m^2 - 9m + 18 = 0$

Solution

$$m^2 - 9m + 18 = 0$$

Since the coefficient of m^2 in this equation is 1, then we can factorize the equation by putting directly the appropriate factors into bracket.

$$m^2 - 9m + 18 = 0$$
$$(m \quad)(m \quad) = 0$$

Look for two factors of 18 whose product is +18 and whose sum is −9. Since the product is positive and the sum is negative, then the two factors must be negative. The factors of 18 whose product is +18 and whose sum is −9 are −6 and −3. We now fill these two values into the bracket above. This gives:

$$m^2 - 9m + 18 = 0$$
$$(m - 6)(m - 3) = 0$$

Equating each bracket to zero gives:

$$m - 6 = 0$$
$$m = 6$$
$$m - 3 = 0$$

$$m = 3$$

Hence, $m = 6$ or $m = 3$

5. Solve the equation $6n^2 = 5 - 13n$

Solution

$$6n^2 = 5 - 13n$$

Rearranging the terms in order to equate the equation to zero (i.e. make the right hand side zero) gives:

$$6n^2 + 13n - 5 = 0$$

Multiply the first and last term on the left hand side. This gives:

$$6n^2(-5) = -30n^2$$

Now look for two numbers in n such that their product is $-30n^2$ and their sum is $+13n$ (i.e. the

95

middle term). Based on the rules given above for obtaining factors, and since the product of the terms is negative (i.e. $-30n^2$) and their sum is positive (i.e. $+13n$), then the two factors of 30 that we need must have different signs, and with the higher of the factors having a positive sign (i.e. from $+13$). The pairs of factors of 30 are: (1 and 30), (2 and 15), (3 and 10), (6 and 5). The larger factor should have a positive sign. If we assign a positive sign to the larger number of each pair of the factors, then the pairs can be written as: (-1 and 30), (-2 and 15), (-3 and 10), (-5 and 6)

Out of these pairs, the one whose sum is $+13$ is (-2 and 15).

Hence, we now include 'n' on them so that their product will give $-30n^2$ and their sum will give $+13n$. Hence the numbers are $-2n$ and $+15n$.

We now substitute these two numbers for $+13n$ in the original expression. This gives:

$$6n^2 + 15n - 2n - 5 = 0$$

We now factorize by grouping. This gives:

$$(6n^2 + 15n) - (2n - 5) = 0$$
$$3n(2n + 5) - 1(2n + 5) = 0$$

With $(2x + 5)$ as a common factor, we proceed as follows:

$$(2n + 5)(3n - 1) = 0$$

Now that we have factorized the equation, we equate each of the bracket to zero and solve for n in each of the linear equation formed. This gives:

Either, $2n + 5 = 0$

$$2n = -5$$
$$n = -\frac{5}{2}$$

Or, $3n - 1 = 0$

$$3n = 1$$
$$n = \frac{1}{3} \quad \text{(After dividing both sides by 3)}$$

Therefore, $n = -\frac{5}{2}$ or $n = \frac{1}{3}$

6. Solve the quadratic equation: $8 - 6x - 5x^2 = 0$

Solution

$$8 - 6x - 5x^2 = 0$$

Multiply the first and last term on the left hand side. This gives:

$$8(-5x^2) = -40x^2$$

Now look for two numbers in x such that their product is $-40x^2$ and their sum is $-6x$ (i.e. the middle term). Based on the rules given above for obtaining factors, and since the product of the terms is negative (i.e. $-40x^2$) and their sum is negative (i.e. $-6x$), then the two factors of 40 that we need must have different signs. The higher of the factors will have a negative sign (i.e. from -6, the middle number). The pairs of factors whose product is 40 are (1 and 40), (2 and 20), (4 and 10), (5 and 8). The larger factor should have a negative sign. If we assign a negative sign to the larger number of each pair of the factors, then the pairs can be written as: (1 and -40), (2 and -20), (4 and -10), (5 and -8)

Out of these pairs, the one whose sum is −6 is (4 and −10), since −10 + 4 = −6.

Hence, we now include 'x' on them so that their product will give −40x^2 and their sum will give −6x. Hence the numbers are +4x and −10x.

We now substitute these two numbers for −6x in the original equation. This gives:
$$8 + 4x - 10x - 5x^2 = 0$$

We now factorize by grouping. This gives:
$$(8 + 4x) - (10x - 5x^2) = 0$$
$$4(2 + x) - 5x(2 + x) = 0$$

With (2 + x) as a common factor, we proceed as follows:
$$(2 + x)(4 - 5x) = 0$$

Now that we have factorized the equation, we equate each of the brackets to zero and solve for x in each of the linear equation formed. This gives:

Either, 2 + x = 0
$$x = -2$$

Or,   4 − 5x = 0
$$4 = 5x$$
$$x = \frac{4}{5} \quad \text{(After dividing both sides by 5)}$$

Therefore, $x = -2$ or $x = \frac{4}{5}$

Construction of Quadratic Equation from Given Roots

The solutions of a quadratic equation are called roots. Since the roots are obtained from the factors, it follows that the factors can also be obtained from the roots. The multiplication of the factors gives the quadratic equation.

Consider the quadratic equation: $ax^2 + bx + c = 0$

If we divide each term by 'a', then we will obtain the quadratic equation in terms of its sum of roots and product of roots as follows:
$$x^2 + \frac{b}{a}x + \frac{c}{a} = 0$$

With this simplification, the sum of roots of the quadratic equation is:
$$-\frac{b}{a}$$

while the product of roots of the quadratic equation is:
$$\frac{c}{a}$$

Therefore, a quadratic equation can be generally expressed as follows:
$$x^2 - (\text{sum of roots})x + (\text{product of roots}) = 0$$

If the roots of a quadratic equation are given as α and β, then the quadratic equation can be represented as follows:
$$x^2 - (\alpha + \beta)x + (\alpha\beta) = 0$$

Let us now apply these principles with the following examples.

Examples

1. Find the quadratic equation whose roots are 4 and −7.

<u>Solution</u>

Method 1

Since the roots/solutions of the quadratic equation are 4 and −7, it follows that:

$x = 4$ or $x = -7$

Equating each solution to zero gives us the factors as follows:

$x - 4 = 0$ or $x + 7 = 0$ (Note that from $x = 4$, we have, $x - 4 = 0$, and from $x = -7$, $x + 7 = 0$)

The multiplication of these two factors gives us the equation as follows:

$(x - 4)(x + 7) = 0$

Expanding the brackets gives:

$x^2 + 7x - 4x - 28 = 0$

$x^2 + 3x - 28 = 0$ (Note that $7x - 4x = 3x$)

The quadratic equation is $x^2 + 3x - 28 = 0$

Method 2

Recall that a quadratic equation is given in term of its roots as follows:

$x^2 -$ (sum of roots)$x +$ (product of roots) $= 0$

Hence, we can now use 4 and −7 to obtain the sum and product of roots as follows:

$x^2 -$ (sum of roots)$x +$ (product of roots) $= 0$

$x^2 - (-7 + 4)x + [-7(4)] = 0$

$x^2 - (-3)x + (-28) = 0$

$x^2 + 3x - 28 = 0$

Hence the quadratic equation is $x^2 + 3x - 28 = 0$.

2. Find the quadratic equation whose roots are $\frac{2}{3}$ and 5.

<u>Solution</u>

Method 1

Since the roots/solutions of the quadratic equation are $\frac{2}{3}$ and 5, it follows that:

$x = \frac{2}{3}$ or $x = 5$

Equating each solution to zero gives us the factors as follows:

$x - \frac{2}{3} = 0$ or $x - 5 = 0$

In order to clear out the fraction in $x - \frac{2}{3} = 0$, we simply multiply each term by 3. This gives:

$3(x) - 3(\frac{2}{3}) = 3(0)$

$3x - 2 = 0$

Hence the two factors are:

$3x - 2 = 0$ and $x - 5 = 0$

The multiplication of these two factors gives us the equation as follows:

$(3x - 2)(x - 5) = 0$

Expanding the brackets gives:

$3x^2 - 15x - 2x + 10 = 0$

$3x^2 - 17x + 10 = 0$

Hence, the quadratic equation is $3x^2 - 17x + 10 = 0$

Method 2

Recall that a quadratic equation is given in term of its roots as follows:

$x^2 - $ (sum of roots)$x + $ (product of roots)$= 0$

Hence, we can now use $\dfrac{2}{3}$ and 5 to obtain the sum and product of roots as follows:

$x^2 - $ (sum of roots)$x + $ (product of roots)$= 0$

$x^2 - (\dfrac{2}{3} + 5)x + [\dfrac{2}{3}(5)] = 0$

$x^2 - (\dfrac{2 + 15}{3})x + (\dfrac{10}{3}) = 0$

$x^2 - (\dfrac{17}{3})x + (\dfrac{10}{3}) = 0$

In order to clear out the fractions, we multiply each term by 3 (i.e. the LCM of the denominators). This gives:

$3(x^2) - 3(\dfrac{17}{3})x + 3(\dfrac{10}{3}) = 0$

The 3 will cancel out each other in the fractions to give:

$3x^2 - 17x + 10 = 0$

Hence the quadratic equation is $3x^2 - 17x + 10 = 0$.

3. Find the quadratic equation whose roots are -2 and $-\dfrac{3}{4}$

Solution

Method 1

Since the roots of the quadratic equation are -2 and $-\dfrac{3}{4}$, it follows that:

$x = -2$ or $x = -\dfrac{3}{4}$

Equating each solution to zero gives us the factors as follows:

$x + 2 = 0$ or $x + \dfrac{3}{4} = 0$

In order to clear out the fraction in $x + \dfrac{3}{4} = 0$, we simply multiply each term by 4. This gives:

$4(x) + 4(\dfrac{3}{4}) = 4(0)$

$4x + 3 = 0$

Hence the two factors are:

$x + 2 = 0$ and $4x + 3 = 0$

The multiplication of these two factors gives us the equation as follows:

$(x + 2)(4x + 3) = 0$

Expanding the brackets gives:

$4x^2 + 3x + 8x + 6 = 0$

$4x^2 + 11x + 6 = 0$

Hence, the quadratic equation is $4x^2 + 11x + 6 = 0$

Method 2

Recall that a quadratic equation is given in term of its roots as follows:

$x^2 - $ (sum of roots)$x + $ (product of roots) $= 0$

Hence, we can now use -2 and $-\dfrac{3}{4}$ to obtain the sum and product of roots as follows:

$x^2 - $ (sum of roots)$x + $ (product of roots) $= 0$

$x^2 - (-2 - \dfrac{3}{4})x + [-2(-\dfrac{3}{4})] = 0$

$x^2 - (\dfrac{-8-3}{4})x + [-2(-\dfrac{3}{4}) = 0$

$x^2 - (-\dfrac{11}{4})x + (\dfrac{6}{4}) = 0$

$x^2 - (-\dfrac{11}{4})x + (\dfrac{3}{2}) = 0$ (Note that $\dfrac{6}{4} = \dfrac{3}{2}$ in its lowest term)

In order to clear out the fractions, we multiply each term by 4 (i.e. the LCM of the denominators). This gives:

$4(x^2) - 4(-\dfrac{11}{4})x + 4(\dfrac{3}{2}) = 0$

This will now give:

$4x^2 + 11x + 6 = 0$ (Note that $-4(-\dfrac{11}{4}) = +11$)

Hence the quadratic equation is $4x^2 + 11x + 6 = 0$.

4. A quadratic equation is given by: $x^2 - 7x - 10 = 0$. Find:

a. the sum of the roots of the equation

b. the product of the roots of the equation

Solution

a. $x^2 - 7x - 10 = 0$

Comparing this equation with $ax^2 + bx + c = 0$, shows that:

a = 1, b = –7, and c = –10   (Note that $x^2 = 1x^2$, hence a = 1)

Therefore, the sum of roots is given by:

$-(\dfrac{b}{a}) = -(-\dfrac{7}{1})$

$= 7$

Hence, the sum of the roots is 7

b. Also, the product of roots is given by:

$\dfrac{c}{a} = \dfrac{-10}{1} = -10$

Hence the product of the roots is −10

5. A quadratic equation is given by: $2x^2 + 13x - 26 = 0$. Find:
a. the sum of the roots of the equation
b. the product of the roots of the equation

Solution
a. $2x^2 + 13x - 26 = 0$
Comparing this equation to $ax^2 + bx + c = 0$, shows that:
   $a = 2, b = 13, c = -26$
Recall that the sum of roots of a quadratic equation is given by $-\dfrac{b}{a}$
Hence, sum of roots of the equation above is:
$$-\dfrac{b}{a} = -\dfrac{13}{2}$$
Therefore sum of roots of the equation is $-\dfrac{13}{2}$

b. Recall that the product of the roots of a quadratic equation is given by: $\dfrac{c}{a}$
Hence product of roots of the equation above is:
$$= -\dfrac{26}{2}$$
$$= -13$$
Therefore, product of roots of the equation is −13.

6. If 3 is a root of the equation $2x^2 + Kx - 21 = 0$, find the value of K. Hence find the other root.

Solution
   $2x^2 + Kx - 21 = 0$
In order to find K, substitute 3 for x in the equation. This gives:
   $2x^2 + Kx - 21 = 0$
   $2(3)^2 + K(3) - 21 = 0$
   $2(9) + 3K - 21 = 0$
   $18 + 3K - 21 = 0$
   $3K = 21 - 18$
   $3K = 3$
   $K = \dfrac{3}{3}$
   $K = 1$
Hence the value of K is 1.
The equation can now be written as $2x^2 + x - 21 = 0$
From this equation, a = 2, b = 1 and c = −21
Hence, product of roots = $\dfrac{c}{a}$

$$= -\frac{21}{2}$$

Let the second root be m.

Since one of the roots is 3, then product of roots which is $-\frac{21}{2}$ is given by:

$3(m) = -\frac{21}{2}$ (Note that the two roots are 3 and m)

$$3m = -\frac{21}{2}$$

$$m = \frac{-\frac{21}{2}}{3}$$

$$= -\frac{21}{2} \times \frac{1}{3}$$

$$m = -\frac{21}{6}$$

$$m = -\frac{7}{2}$$ (In its lowest term)

Therefore the other root is $-\frac{7}{2}$

Note that we can also factorize the equation $2x^2 + x - 21 = 0$, to obtain the two roots, and hence the other root.

7. If 1 is a root of the equation $5x^2 - kx - 3 = 0$, find the value of K. Hence find the other root.

Solution
$$5x^2 - kx - 3 = 0$$
In order to find K, substitute 1 for x in the equation. This gives:
$$5x^2 - kx - 3 = 0$$
$$5(1)^2 - k(1) - 3 = 0$$
$$5(1) - k - 3 = 0$$
$$5 - k - 3 = 0$$
$$5 - 3 = k$$
$$k = 2$$
Hence the value of k is 2.
The equation can now be written as $5x^2 - 2x - 3 = 0$
From this equation, a = 5, b = −2 and c = −3
Hence, product of roots = $\frac{c}{a}$

$$= -\frac{3}{5}$$

Let the second root be y.

Since one of the roots is 1, then product of roots which is $-\frac{3}{5}$ is given by:

$1(y) = -\frac{3}{5}$ (Note that the two roots are 1 and y, hence their product is 1y)

$y = -\frac{3}{5}$

Therefore the other root is $-\frac{3}{5}$

8. 5 and −7 are the roots of the quadratic equation $y^2 - ky + M = 0$. Find the values of k and M.

Solution

$y^2 - ky + M = 0$

From this equation, a = 1, b = −k, c = M

Hence sum of roots $= -\frac{b}{a}$

$\qquad = -(-\frac{k}{1})$ (Since b = −k)

$\qquad = k$

Since the two roots are 5 and −7, and their sum is k, then it follows that:

$\qquad -7 + 5 = k$

$\qquad -2 = k$

$\qquad k = -2$

Also, product of roots $= \frac{c}{a}$

$\qquad = \frac{M}{1}$ (Since c = M)

$\qquad = M$

Since the two roots are 5 and −7, and their product is M, then it follows that:

$\qquad 5(-7) = M$

$\qquad -35 = M$

$\qquad M = -35$

Therefore, k = −2 and M = −35

9. If −3 and 8 are the roots of the quadratic equation $2x^2 + Dx - E = 0$. Find the values of the constants D and E.

Solution

$2x^2 + Dx - E = 0$

From this equation, a = 2, b = D, c = −E

Hence sum of roots $= -\frac{b}{a}$

$$= -\frac{D}{2} \quad \text{(Since b = D)}$$

Since the two roots are –3 and 8, and their sum is $-\frac{D}{2}$, then it follows that:

$$-3 + 8 = -\frac{D}{2}$$

$$5 = -\frac{D}{2}$$

$$-D = 2 \times 5$$
$$-D = 10$$
$$D = -10 \quad \text{(After dividing both sides by –1)}$$

Also, product of roots = $\frac{c}{a}$

$$= -\frac{E}{2} \quad \text{(Since c = –E and a = 2)}$$

Since the two roots are –3 and 8, and their product is $-\frac{E}{2}$, then it follows that:

$$-3(8) = -\frac{E}{2}$$

$$-24 = -\frac{E}{2}$$

$$-E = -24 \times 2$$
$$-E = -48$$
$$E = 48 \quad \text{(After dividing both sides by –1)}$$

Therefore, D = –10 and E = 48

10. If one root of the equation $y^2 - ky + 18 = 0$ is twice the other, find the two possible values of k. Hence, what are the two possible equations?

Solution
Let one root of the equation be m. Therefore, the other root will be 2m since one root is twice the other.
Hence product of roots = m(2m)

$$= 2m^2$$

The equation is $y^2 - ky + 18 = 0$
From this equation, a = 1, b = –k and c = 18
Product of roots = $\frac{c}{a}$

$$= \frac{18}{1} = 18$$

Hence, $2m^2 = 18$ (Note that product of roots is also $2m^2$)

$$m^2 = \frac{18}{2}$$
$$m^2 = 9$$

104

$$m = \sqrt{9}$$
$$m = \pm 3$$
$$m = 3 \text{ or } -3$$

When m = 3, the other root, 2m, is 2(3) = 6   (Since one root is twice the other)

When m = −3, then the other root, 2m, is 2(−3) = −6

Hence, the roots are 3 and 6, or −3 and −6.

Recall that the equation is $y^2 - ky + 18 = 0$, and a = 1, b = −k, and c = 18.

Hence, sum of roots = $-\dfrac{b}{a}$

$$= -(-\dfrac{k}{1}) \qquad \text{(Since b = −k and a = 1)}$$
$$= k$$

Therefore, when the roots are 3 and 6, then sum of roots = 3 + 6 = 9. But sum of roots is also equal to k.

Hence, k = 9.

When the roots are −3 and −6, then sum of roots = −3 − 6 = −9

Hence, k = −9

Therefore, k = 9 or −9

When k = 9, the equation is: $y^2 - ky + 18 = 0$, which gives:

$$y^2 - 9y + 18 = 0$$

When k = −9, the equation is:

$$y^2 - (-9)y + 18 = 0, \text{ which gives:}$$
$$y^2 + 9y + 18 = 0$$

Completing the Square of Quadratic expressions

Another method of solving quadratic equation is by completing the square. In order to make a quadratic expression a perfect square, simply add the square of half the coefficient of x of the quadratic expression.

Hence, completing the square of a quadratic expression means to make the quadratic expression a perfect square.

Examples

1. What must be added to $x^2 + 10x$ to make it a perfect square?

Solution

$$x^2 + 10x$$

Half the coefficient of x is given by: $\dfrac{10}{2} = 5$.

Square this value obtained. This gives:

$$5^2 = 25.$$

Hence, 25 must be added to the quadratic expression to make it a perfect square.

If we add 25 to the quadratic expression it gives:

$$x^2 + 10x + 25$$

This means that $x^2 + 10x + 25$ is a perfect square. Its square root is obtained by taking the square root of the first and last terms and adding them together. The square root of x^2 is x

while the square root of 25 is ± 5, but we take $+5$ (take the sign of the coefficient of x in the expression). Therefore the square root of $x^2 + 10x + 25$ is $(x + 5)$. If we square the square root, it gives the perfect square.

Hence, $x^2 + 10x + 25 = (x + 5)^2$

2. What must be added to $x^2 - 18x$ to make it a perfect square?

Solution

$x^2 - 18x$

The coefficient of x is -18. Divide this coefficient by 2. This gives:

$-\dfrac{18}{2} = -9$

Square this value obtained. This gives:

$-9^2 = 81$.

Hence, 81 must be added to the quadratic expression to make it a perfect square.

If we add 81 to the quadratic expression it gives:

$x^2 - 18x + 81$

This means that $x^2 - 18x + 81$ is a perfect square. Its square root is obtained by taking the square roots of the first and last terms and adding them together. The square root of x^2 is x while the square root of 81 is ± 9, but we take -9 (take the sign of the coefficient of x). Therefore the square root of $x^2 - 18x + 81$ is $(x - 9)$. If we square the square root, it gives the perfect square.

Hence, $x^2 - 18x + 81 = (x - 9)^2$

3. Make $y^2 - 5y$ a perfect square.

Solution

$y^2 - 5y$

The coefficient of y is -5. Divide this coefficient by 2. This gives:

$-\dfrac{5}{2}$

Square this value obtained. This gives:

$(-\dfrac{5}{2})^2 = \dfrac{25}{4}$

Hence, $\dfrac{25}{4}$ must be added to the quadratic expression to make it a perfect square.

If we add $\dfrac{25}{4}$ to the quadratic expression it gives:

$y^2 - 5y + \dfrac{25}{4}$

This means that $y^2 - 5y + \dfrac{25}{4}$ is a perfect square.

Its square root is obtained by taking the square roots of the first and last terms and adding them together. The square root of y^2 is y while the square root of $\dfrac{25}{4}$ is $\pm \dfrac{5}{2}$, but we take $-\dfrac{5}{2}$

since the coefficient of y in the expression is negative.

Therefore the square root of $y^2 - 5y + \frac{25}{4}$ is $(y - \frac{5}{2})$. If we square the root, it gives the perfect square.

Hence, $y^2 - 5y + \frac{25}{4} = (x - \frac{5}{2})^2$

4. Make $y^2 + \frac{9}{2}y$ a perfect square.

Solution

$$y^2 + \frac{9}{2}y$$

The coefficient of y is $\frac{9}{2}$. Divide this coefficient by 2. This gives:

$$\frac{9}{2} \div 2$$

$$= \frac{9}{2} \times \frac{1}{2}$$

$$= \frac{9}{4}$$

Square this value obtained. This gives:

$$(\frac{9}{4})^2 = \frac{81}{16}$$

Hence, $\frac{81}{16}$ must be added to the quadratic expression to make it a perfect square.

If we add $\frac{81}{16}$ to the quadratic expression it gives:

$$y^2 + \frac{9}{2}y + \frac{81}{16}$$

This means that $y^2 + \frac{9}{2}y + \frac{81}{16}$ is a perfect square.

Therefore the square root of $y^2 + \frac{9}{2}y + \frac{81}{16}$ is $(y + \frac{9}{4})$.

Hence, $y^2 + \frac{9}{2}y + \frac{81}{16} = (y + \frac{9}{4})^2$

5. If $x^2 - 6x + (-3)^2$ is a perfect square, find its square root.

Solution

$$x^2 - 6x + (-3)^2$$

Since this is a perfect square, we find its square root by taking each of the squared term and enclosing them in a bracket. The square root of x^2 is x, while the square root of $(-3)^2$ is -3.
Hence, we put these two square root terms in a bracket. This gives: $(x - 3)$
Therefore the square root of $x^2 - 6x + (-3)^2$ is $(x - 3)$
This means that: $x^2 - 6x + (-3)^2 = (x - 3)^2$

6. If $x^2 - \frac{7}{2}x + (-\frac{7}{4})^2$ is a perfect square, find its square root.

Solution

$$x^2 - \frac{7}{2}x + (-\frac{7}{4})^2$$

Since this is a perfect square, we find its square root by taking each of the squared term and enclosing them in a bracket. The square root of x^2 is x, while the square root of $(-\frac{7}{4})^2$ is $-\frac{7}{4}$.

Hence, we put these

two square root terms in a bracket. This gives: $(x - \frac{7}{4})$

Therefore the square root of $x^2 - \frac{7}{2}x + (-\frac{7}{4})^2$ is $(x - \frac{7}{4})$

This means that: $x^2 - \frac{7}{2}x + (-\frac{7}{4})^2 = (x - \frac{7}{4})^2$

7. If $x^2 + 1.5x + (0.75)^2$ is a perfect square, find its square root.

Solution

$$x^2 + 1.5x + (0.75)^2$$

The square root of x^2 is x, while the square root of $(0.75)^2$ is 0.75. Hence, we put these two square root terms in a bracket. This gives: $(x + 0.75)$

Therefore the square root of $x^2 + 1.5x + (0.75)^2$ is $(x + 0.75)$

Solving Quadratic Equation by Completing the Square

Examples

1. Use the method of completing the square to solve the quadratic equation $y^2 + 7y - 30 = 0$

Solution

$$y^2 + 7y - 30 = 0$$

Take the constant term (i.e. −30) to the right hand side of the equation. This gives:

$$y^2 + 7y = 30$$

The next step is to make the left hand side expression a perfect square. In doing this, whatever is added to the left hand side should also be added to the right hand side. Let us do this as follows.

The coefficient of y is 7. Divide this coefficient by 2. This gives:

$$\frac{7}{2}$$

Square the value obtained. This gives:

$$(\frac{7}{2})^2 = \frac{49}{4}$$

Hence, $\frac{49}{4}$ must be added to the left hand expression to make it a perfect square. However, we are not going to write it as $\frac{49}{4}$. We will leave it as $(\frac{7}{2})^2$. We will now add $(\frac{7}{2})^2$ to the left hand side, and also add it to the right hand side. This gives:

$$y^2 + 7y + (\tfrac{7}{2})^2 = 30 + (\tfrac{7}{2})^2$$

While keeping the left hand side as it is, so that we can easily express it as a square of its square root, we go ahead and simplify the right hand side. This gives:

$$y^2 + 7y + (\tfrac{7}{2})^2 = \frac{30}{1} + \frac{49}{4}$$

$$y^2 + 7y + (\tfrac{7}{2})^2 = \frac{120 + 49}{4} \qquad \text{(Note that 4 x 30 = 120)}$$

$$y^2 + 7y + (\tfrac{7}{2})^2 = \frac{169}{4}$$

Take the square root of the left hand side and square it. This is done by taking the terms in square (i.e. y and $\tfrac{7}{2}$), enclosing them in a bracket and squaring the bracket. This gives:

$$(y + \tfrac{7}{2})^2 = \frac{169}{4}$$

Take the square root of both sides in order to remove the square on the left hand side. This gives:

$$\sqrt{(y + \tfrac{7}{2})^2} = \sqrt{\frac{169}{4}}$$

$$(y + \tfrac{7}{2}) = \pm \frac{13}{2}$$

(Note that $\sqrt{x^2} = x$, hence the removal of square from $(y + \tfrac{7}{2})^2$ to get $(y + \tfrac{7}{2})$. Also, remember to add $\pm$ to the square root of a number).

Let us continue with our solution as follows:

$$(y + \tfrac{7}{2}) = \pm \frac{13}{2}$$

Or, $\quad y + \dfrac{7}{2} = \pm \dfrac{13}{2}$

Take the constant term on the left hand side to the right. This gives:

$$y = -\frac{7}{2} \pm \frac{13}{2}$$

We now use $\pm$ to split the values above into two solutions. When we take only the positive sign of $\pm$, our value above gives:

$$y = -\frac{7}{2} + \frac{13}{2}$$

$$y = \frac{-7 + 13}{2}$$

$$y = \frac{6}{2}$$

$$y = 3$$

When we take only the negative sign of $\pm$, our value above gives:

$$y = -\frac{7}{2} - \frac{13}{2}$$

$$y = \frac{-7 - 13}{2}$$

$$y = -\frac{20}{2}$$

$$y = -10$$

∴   y = 3 or y = −10

2. Use the method of completing the square to solve the quadratic equation $x^2 - 11x + 28 = 0$

Solution
$$x^2 - 11x + 28 = 0$$
Take the constant term (i.e. 28) to the right hand side of the equation. This gives:
$$x^2 - 11x = -28$$
The coefficient of x is −11. Divide this coefficient by 2 to obtain:
$$-\frac{11}{2}$$
Square $-\frac{11}{2}$. This gives:
$$(-\frac{11}{2})^2$$
Add $(-\frac{11}{2})^2$ to both sides of the equation. This gives:
$$x^2 - 11x + (-\frac{11}{2})^2 = -28 + (-\frac{11}{2})^2$$
While keeping the left hand side, simplify the right hand side. This gives:

$$x^2 - 11x + (-\frac{11}{2})^2 = -28 + \frac{121}{4}$$

$$x^2 - 11x + (-\frac{11}{2})^2 = \frac{-112 + 121}{4} \qquad \text{(Note that 4 x −28 = −112)}$$

$$x^2 - 11x + (-\frac{11}{2})^2 = \frac{9}{4}$$

Take the square root of the left hand side and square it. This is done by taking the terms in square (i.e. x and $-\frac{11}{2}$), enclosing them in a bracket and squaring the bracket. This gives:
$$(x - \frac{11}{2})^2 = \frac{9}{4}$$
Take the square root of both sides. This gives:

$$\sqrt{(x - \frac{11}{2})^2} = \sqrt{\frac{9}{4}}$$

$$(x - \frac{11}{2}) = \pm\frac{3}{2}$$

Or, $x - \frac{11}{2} = \pm\frac{3}{2}$

$$x = \frac{11}{2} \pm \frac{3}{2}$$

When we take only the positive sign of ±, our value above gives:

$$x = \frac{11}{2} + \frac{3}{2}$$

$$x = \frac{11 + 3}{2}$$

$$x = \frac{14}{2}$$

$$x = 7$$

When we take only the negative sign of ±, our value above gives:

$$x = \frac{11}{2} - \frac{3}{2}$$

$$x = \frac{11 - 3}{2}$$

$$x = \frac{8}{2}$$

$$x = 4$$

∴ $x = 7$ or $x = 4$

3. Solve the quadratic equation $x^2 - 6x - 16 = 0$, by the method of completing the square.

Solution

$$x^2 - 6x - 16 = 0$$

Take the constant term (i.e. −16) to the right hand side of the equation. This gives:

$$x^2 - 6x = 16$$

The coefficient of x is −6. Divide this coefficient by 2 to obtain:

$$-\frac{6}{2} = -3$$

Square the value obtained. This gives:

$$(-3)^2$$

Add $(-3)^2$ to both sides of the equation. This gives:

$$x^2 - 6x + (-3)^2 = 16 + (-3)^2$$
$$x^2 - 6x + (-3)^2 = 16 + 9$$
$$x^2 - 6x + (-3)^2 = 25$$

Take the square root of the left hand side and square it. This gives:

$$(x - 3)^2 = 25$$

Take the square root of both sides. This gives:

$$\sqrt{(x - 3)^2} = \sqrt{25}$$
$$x - 3 = \pm 5$$
$$x = 3 \pm 5$$
$$x = 3 + 5 \quad \text{or} \quad x = 3 - 5$$

∴ $x = 8$ or $x = -2$

4. Solve the quadratic equation $2x^2 - 5x + 3 = 0$, by the method of completing the square.

<u>Solution</u>

$$2x^2 - 5x + 3 = 0$$

In this case, the coefficient of x^2 is 2. It is not 1 like our first three examples above. Therefore our first step will be to make the coefficient of x^2 to be 1. This is done by dividing each term in the equation by the coefficient of x^2 which is 2. This gives:

$$\frac{2x^2}{2} - \frac{5x}{2} + \frac{3}{2} = \frac{0}{2}$$

$$x^2 - \frac{5}{2}x + \frac{3}{2} = 0$$

Now that we have made the coefficient of x^2 to be 1, we can proceed as usual.

Take the constant term (i.e. $\frac{3}{2}$) to the right hand side of the equation. This gives:

$$x^2 - \frac{5}{2}x = -\frac{3}{2}$$

The coefficient of x is $-\frac{5}{2}$. Divide this coefficient by 2 to obtain:

$$-\frac{5}{2} \div 2 = -\frac{5}{2} \times \frac{1}{2}$$

$$= -\frac{5}{4}$$

Square the value obtained. This gives:

$$(-\frac{5}{4})^2$$

Add $(-\frac{5}{4})^2$ to both sides of the equation. This gives:

$$x^2 - \frac{5}{2}x + (-\frac{5}{4})^2 = -\frac{3}{2} + (-\frac{5}{4})^2$$

$$x^2 - \frac{5}{2}x + (-\frac{5}{4})^2 = -\frac{3}{2} + \frac{25}{16}$$

$$x^2 - \frac{5}{2}x + (-\frac{5}{4})^2 = \frac{-24 + 25}{16}$$

$$x^2 - \frac{5}{4}x + (-\frac{5}{4})^2 = \frac{1}{16}$$

Take the square root of the left hand side and square it. This gives:

$$(x - \frac{5}{4})^2 = \frac{1}{16}$$

Take the square root of both sides. This gives:

$$\sqrt{(x - \frac{5}{4})^2} = \sqrt{\frac{1}{16}}$$

$$x - \frac{5}{4} = \pm\frac{1}{4}$$

$$x = \frac{5}{4} \pm \frac{1}{4}$$

$$x = \frac{5}{4} + \frac{1}{4} \quad \text{or} \quad x = \frac{5}{4} - \frac{1}{4}$$

$$x = \frac{5+1}{4} \quad \text{or} \quad x = \frac{5-1}{4}$$

$$x = \frac{6}{4} \quad \text{or} \quad x = \frac{4}{4}$$

$$x = \frac{3}{2} \quad \text{or} \quad x = 1$$

5 Solve the quadratic equation $3y^2 + 9y + 5 = 0$, by the method of completing the square.

<u>Solution</u>

$$3y^2 + 9y + 5 = 0$$

In this case, the coefficient of y^2 is 3. Therefore our first step will be to make the coefficient of y^2 to be 1. This is done by dividing each term in the equation by 3. This gives:

$$\frac{3y^2}{3} + \frac{9y}{3} + \frac{5}{3} = \frac{0}{3}$$

$$y^2 + 3y + \frac{5}{3} = 0$$

Take the constant term to the right hand side of the equation. This gives:

$$y^2 + 3y = -\frac{5}{3}$$

The coefficient of y is 3. Divide this coefficient by 2 to obtain:

$$\frac{3}{2}$$

Square the value obtained. This gives:

$$(\frac{3}{2})^2$$

Add $(\frac{3}{2})^2$ to both sides of the equation. This gives:

$$y^2 + 3y + (\frac{3}{2})^2 = -\frac{5}{3} + (\frac{3}{2})^2$$

$$y^2 + 3y + (\frac{3}{2})^2 = -\frac{5}{3} + \frac{9}{4}$$

$$y^2 + 3y + (\frac{3}{2})^2 = \frac{-20 + 27}{12}$$

$$y^2 + 3y + (\frac{3}{2})^2 = \frac{7}{12}$$

Take the square root of the left hand side and square it. This gives:

$$(y + \frac{3}{2})^2 = \frac{7}{12}$$

Take the square root of both sides. This gives:

$$\sqrt{(y + \frac{3}{2})^2} = \sqrt{0.5833}$$

$$y + \frac{3}{2} = \pm 0.764$$

$$y + 1.5 = \pm 0.764$$

$$y = -1.5 \pm 0.764$$

$$y = -1.5 + 0.764 \quad \text{or} \quad y = -1.5 - 0.764$$

$$y = -0.736 \quad \text{or} \quad y = -2.264$$

6. Solve the equation $5m^2 + 7m - 3 = 0$

<u>Solution</u>

$$5m^2 + 7m - 3 = 0$$

In this case, the coefficient of m^2 is 5. Therefore our first step will be to make the coefficient of m^2 to be 1. This is done by dividing each term in the equation by 5. This gives:

$$\frac{5m^2}{5} + \frac{7m}{5} - \frac{3}{5} = \frac{0}{5}$$

$$m^2 + \frac{7}{5}m - \frac{3}{5} = 0$$

Take the constant term to the right hand side of the equation. This gives:

$$m^2 + \frac{7}{5}m = \frac{3}{5}$$

The coefficient of m is $\frac{7}{5}$. Divide this coefficient by 2 to obtain:

$$\frac{7}{5} \div 2$$

$$= \frac{7}{5} \times \frac{1}{2}$$

$$= \frac{7}{10}$$

Square the value obtained. This gives:

$$(\frac{7}{10})^2$$

Add $(\frac{7}{10})^2$ to both sides of the equation. This gives:

$$m^2 + \frac{7}{5}m + (\frac{7}{10})^2 = \frac{3}{5} + (\frac{7}{10})^2$$

$$m^2 + \frac{7}{5}m + (\frac{7}{10})^2 = \frac{3}{5} + \frac{49}{100}$$

$$m^2 + \frac{7}{5}m + (\frac{7}{10})^2 = \frac{60+49}{100}$$

$$m^2 + \frac{7}{5}m + (\frac{7}{10})^2 = \frac{109}{100}$$

Take the square root of the left hand side and square it. This gives:

$$(m + \frac{7}{10})^2 = \frac{109}{100}$$

Take the square root of both sides. This gives:

114

$$\sqrt{(m + \frac{7}{10})^2} = \sqrt{1.09}$$

$$m + \frac{7}{10} = \pm 1.044$$
$$m + 0.7 = \pm 1.044$$
$$m = -0.7 \pm 1.044$$

Separating the sign $\pm$ into + and − gives the two solutions as follows:

$$m = -0.7 + 1.044 \quad \text{or} \quad m = -0.7 - 1.044$$
$$m = 0.344 \quad \text{or} \quad m = -1.744$$

7. Solve the quadratic equation $2x^2 - 8x + 5 = 0$

<u>Solution</u>
$$2x^2 - 8x + 5 = 0$$
The coefficient of x^2 is 2. Therefore our first step will be to make the coefficient of x^2 to be 1. This is done by dividing each term in the equation by 2. This gives:
$$\frac{2x^2}{2} - \frac{8x}{2} + \frac{5}{2} = \frac{0}{2}$$
$$x^2 - 4x + \frac{5}{2} = 0$$
Take the constant term to the right hand side of the equation. This gives:
$$x^2 - 4x = -\frac{5}{2}$$
The coefficient of x is −4. Divide this coefficient by 2 to obtain:
$$-\frac{4}{2} = -2$$
Square the value obtained. This gives:
$$(-2)^2$$
Add $(-2)^2$ to both sides of the equation. This gives:
$$x^2 - 4x + (-2)^2 = -\frac{5}{2} + (-2)^2$$

$$x^2 - 4x + (-2)^2 = -\frac{5}{2} + 4$$

$$x^2 - 4x + (-2)^2 = \frac{-5 + 8}{2}$$

$$x^2 - 4x + (-2)^2 = \frac{3}{2}$$
Take the square root of the left hand side and square it. This gives:
$$(x - 2)^2 = \frac{3}{2}$$
Take the square root of both sides. This gives:
$$\sqrt{(x - 2)^2} = \sqrt{1.5}$$
$$x - 2 = \pm 1.225$$
$$x = 2 \pm 1.225$$

$$x = 2 + 1.225 \quad \text{or} \quad x = 2 - 1.225$$
$$x = 3.225 \quad \text{or} \quad x = 0.775$$

Quadratic Formula

The general expression of a quadratic equation is given by: $ax^2 + bx + c = 0$, where a, b and c are constants. This equation can be solved by completing the square are follows:

$$ax^2 + bx + c = 0$$

The coefficient of x^2 is a. Therefore we make the coefficient of x^2 to be 1. This is done by dividing each term in the equation by a. This gives:

$$\frac{ax^2}{a} + \frac{bx}{a} + \frac{c}{a} = \frac{0}{a}$$

$$x^2 + \frac{b}{a}x + \frac{c}{a} = 0$$

Take the constant term to the right hand side of the equation. This gives:

$$x^2 + \frac{b}{a}x = -\frac{c}{a}$$

The coefficient of x is $\frac{b}{a}$. Divide this coefficient by 2 to obtain:

$$\frac{b}{a} \div 2$$

$$= \frac{b}{a} \times \frac{1}{2}$$

$$= \frac{b}{2a}$$

Square the value obtained. This gives:

$$\left(\frac{b}{2a}\right)^2$$

Add $\left(\frac{b}{2a}\right)^2$ to both sides of the equation. This gives:

$$x^2 + \frac{b}{a}x + \left(\frac{b}{2a}\right)^2 = -\frac{c}{a} + \left(\frac{b}{2a}\right)^2$$

$$x^2 + \frac{b}{a}x + \left(\frac{b}{2a}\right)^2 = -\frac{c}{a} + \frac{b^2}{4a^2} \qquad \text{(In } \left(\frac{b}{2a}\right)^2, \text{ each term in the bracket is squared)}$$

$$x^2 + \frac{b}{a}x + \left(\frac{b}{2a}\right)^2 = \frac{-4ac + b^2}{4a^2} \qquad \text{(Note that the LCM of } 4a^2 \text{ and a is } 4a^2)$$

$$x^2 + \frac{b}{a}x + \left(\frac{b}{2a}\right)^2 = \frac{b^2 - 4ac}{4a^2}$$

Take the square root of the left hand side and square it. This gives:

$$\left(x + \frac{b}{2a}\right)^2 = \frac{b^2 - 4ac}{4a^2}$$

Take the square root of both sides. This gives:

$$\sqrt{\left(x + \frac{b}{2a}\right)^2} = \frac{b^2 - 4ac}{4a^2}$$

$$x + \frac{b}{2a} = \sqrt{\frac{b^2 - 4ac}{4a^2}}$$ (Note that $\sqrt{\frac{b}{a}}$ can also be expressed as $\frac{\sqrt{b}}{\sqrt{a}}$)

$$x + \frac{b}{2a} = \pm\frac{\sqrt{b^2 - 4ac}}{2a}$$ (Note that the square root of $4a^2$ is $2a$)

$$x = -\frac{b}{2a} \pm \frac{\sqrt{b^2 - 4ac}}{2a}$$

Since their LCM is 2a, we simplify further as follows:

$$x = \frac{-b \pm \sqrt{b^2 - 4ac}}{2a}$$

The above formula is known as quadratic formula. It can be used to solve any quadratic equation.

Examples
1. Solve the following equations by using the quadratic formula:
a. $2x^2 + 3x - 9 = 0$
b. $4x^2 - 8x + 3 = 0$

Solutions
a. $2x^2 + 3x - 9 = 0$
When this equation is compared with $ax^2 + bx + c = 0$, it shows that:
    $a = 2$, $b = 3$ and $c = -9$
We now substitute these values into the quadratic formula as follows:

$$x = \frac{-b \pm \sqrt{b^2 - 4ac}}{2a}$$

$$= \frac{-3 \pm \sqrt{3^2 - [4(2)(-9)]}}{2(2)}$$

$$= \frac{-3 \pm \sqrt{9 - (-72)}}{2 \times 2}$$

$$= \frac{-3 \pm \sqrt{9 + 72}}{4}$$

$$= \frac{-3 \pm \sqrt{81}}{4}$$

$$x = \frac{-3 \pm 9}{4}$$

We now split the sign $\pm$ into + and − to obtain the two solutions as follows:

$$x = \frac{-3 + 9}{4} \quad \text{or} \quad x = \frac{-3 - 9}{4}$$

$$x = \frac{6}{4} \quad \text{or} \quad x = \frac{-12}{4}$$

$$x = \frac{3}{2} \quad \text{or} \quad x = -3$$

b. $4x^2 - 8x + 3 = 0$

When this equation is compared with $ax^2 + bx + c = 0$, it shows that:

$a = 4$, $b = -8$ and $c = 3$

We now substitute these values into the quadratic formula as follows:

$$x = \frac{-b \pm \sqrt{b^2 - 4ac}}{2a}$$

$$= \frac{-(-8) \pm \sqrt{(-8)^2 - [4(4)(3)]}}{2(4)} \qquad \text{(Note that } -b \text{ is } -(-8) \text{ since b is } -8\text{)}$$

$$= \frac{+8 \pm \sqrt{64 - 48}}{8}$$

$$= \frac{8 \pm \sqrt{16}}{8}$$

$$x = \frac{8 \pm 4}{8}$$

We now split the sign $\pm$ into + and − to obtain the two solutions as follows:

$$x = \frac{8 + 4}{8} \quad \text{or} \quad x = \frac{8 - 4}{8}$$

$$x = \frac{12}{8} \quad \text{or} \quad x = \frac{4}{8}$$

$$x = \frac{3}{2} \quad \text{or} \quad x = \frac{1}{2}$$

2. Solve the following equations by using the quadratic formula:

a. $x^2 - 4x - 2 = 0$

b. $4 + x - 5x^2 = 0$

c. $7 - 2x - 3x^2 = 0$

d. $2n^2 - 15n + 18 = 0$

Solutions

a. $x^2 - 4x - 2 = 0$

In this equation, a = 1, b = −4 and c = −2

We now substitute these values into the quadratic formula as follows:

$$x = \frac{-b \pm \sqrt{b^2 - 4ac}}{2a}$$

$$= \frac{-(-4) \pm \sqrt{(-4)^2 - [4(1)(-2)]}}{2(1)}$$

$$= \frac{4 \pm \sqrt{16 - (-8)}}{2}$$

$$= \frac{4 \pm \sqrt{16 + 8}}{2}$$

$$= \frac{4 \pm \sqrt{24}}{2}$$

$$x = \frac{4 \pm 4.90}{2}$$

We now split the sign ± into + and − to obtain the two solutions as follows:

$$x = \frac{4 + 4.90}{2} \quad \text{or} \quad x = \frac{4 - 4.90}{2}$$

$$x = \frac{8.90}{2} \quad \text{or} \quad x = \frac{-0.90}{2}$$

$$x = 4.45 \quad \text{or} \quad x = -0.45$$

b. $4 + x - 5x^2 = 0$

In this equation, a = −5, b = 1 and c = 4

Note that 'a' is the coefficient of x^2, b is the coefficient of x, while c is the constant term. So, do not be confused by the reversed arrangement of the quadratic equation.

We now substitute the values of a, b and c into the quadratic formula as follows:

$$x = \frac{-b \pm \sqrt{b^2 - 4ac}}{2a}$$

$$= \frac{-1 \pm \sqrt{1^2 - [4(-5)(4)]}}{2(-5)}$$

$$= \frac{-1 \pm \sqrt{1 - (-80)}}{-10}$$

$$= \frac{-1 \pm \sqrt{1 + 80}}{-10}$$

119

$$= \frac{-1 \pm \sqrt{81}}{-10}$$

$$x = \frac{-1 \pm 9}{-10}$$

We now split the sign $\pm$ into + and − to obtain the two solutions as follows:

$$x = \frac{-1 + 9}{-10} \quad \text{or} \quad x = \frac{-1 - 9}{-10}$$

$$x = \frac{8}{-10} \quad \text{or} \quad x = \frac{-10}{-10}$$

$$x = -\frac{4}{5} \quad \text{or} \quad x = 1$$

c. $7 - 2x - 3x^2 = 0$

This arrangement is similar to that of question (b) above. However, let us solve this question by changing this arrangement to the usual arrangement of a quadratic equation. In order to do this, we divide each term in the equation by −1 so that the coefficient of x^2 will become a positive value. This gives:

$$\frac{7}{-1} - \frac{2x}{-1} - \frac{3x^2}{-1} = \frac{0}{-1}$$

$$-7 + 2x + 3x^2 = 0$$

Or, $3x^2 + 2x - 7 = 0$

In this new arrangement, a = 3, b = 2 and c = −7

We now substitute the values of a, b and c into the quadratic formula as follows:

$$x = \frac{-b \pm \sqrt{b^2 - 4ac}}{2a}$$

$$= \frac{-2 \pm \sqrt{2^2 - [4(3)(-7)]}}{2(3)}$$

$$= \frac{-2 \pm \sqrt{4 - (-84)}}{6}$$

$$= \frac{-2 \pm \sqrt{4 + 84}}{6}$$

$$= \frac{-2 \pm \sqrt{88}}{6}$$

$$x = \frac{-2 \pm 9.38}{6}$$

$$x = \frac{-2 + 9.38}{6} \quad \text{or} \quad x = \frac{-2 - 9.38}{6}$$

$$x = \frac{7.38}{6} \quad \text{or} \quad x = \frac{-11.38}{6}$$

$$x = 1.23 \qquad \text{or} \quad x = -1.90$$

d. $2n^2 - 15n + 18 = 0$

    $a = 2, b = -15$ and $c = 18$

$$n = \frac{-b \pm \sqrt{b^2 - 4ac}}{2a}$$

$$= \frac{-(-15) \pm \sqrt{(-15)^2 - [4(2)(18)]}}{2(2)}$$

$$= \frac{15 \pm \sqrt{225 - 144}}{4}$$

$$= \frac{15 \pm \sqrt{81}}{4}$$

$$n = \frac{15 \pm 9}{4}$$

$$n = \frac{15 + 9}{4} \quad \text{or} \quad n = \frac{15 - 9}{4}$$

$$n = \frac{24}{4} \quad \text{or} \quad n = \frac{6}{4}$$

$$n = 6 \quad \text{or} \quad n = \frac{3}{2}$$

Exercise 9

1. Factorize the following quadratic expressions
a. $7a^2 + 9a$
b. $m^2 - 16$
c. $9p^2 - 121$
d. $32b^2 - 200$

2. Factorize the following equations:
a. $3x^2 - 4x - 15.$
b. $x^2 + x - 72$
c. $5x^2 + 22x + 8$
d. $10x^2 - 4x - 6$

3. Determine if the following equations can be factorized or not:
a. $x^2 + 14x + 13 = 0$
b. $3x^2 - 11x + 6 = 0$

c. $5x^2 + 2x - 9 = 0$

d. $2x^2 - 7x - 11 = 0$

e. $8x^2 - 21x + 12 = 0$

4. Solve the following equations by factorization:

a. $3a^2 + 12a = 0$

b. $b^2 - 36 = 0$

c. $4y^2 - 49 = 0$

d. $(2x - 5)^2 = 9$

5. Solve the following equations by factorization:

a. $3x^2 - x - 4 = 0$

b. $2y^2 + 2y - 12 = 0$

c. $p^2 - 20p + 36 = 0$

d. $10m^2 = 4 - 3m$

e. $6 - x - 7x^2 = 0$

6. Find the quadratic equation whose roots are:

a. 8 and –5.

b. $\frac{1}{4}$ and -7.

c. –6 and $-\frac{2}{5}$

7. A quadratic equation is given by: $2x^2 - 15x - 8 = 0$. Find:

a. the sum of the roots of the equation

b. the product of the roots of the equation

8. A quadratic equation is given by: $3x^2 + 4x - 11 = 0$. Find:

a. the sum of the roots of the equation

b. the product of the roots of the equation

9. If 2 is a root of the equation $2x^2 + Mx - 18 = 0$, find the value of M. Hence find the other root.

10. If 1 is a root of the equation $3x^2 - kx - 11 = 0$, find the value of K. Hence find the other root.

11. 3 and –9 are the roots of the quadratic equation $b^2 - kb + P = 0$. Find the values of k and P.

12. If –5 and 2 are the roots of the quadratic equation $3x^2 + Ax - B = 0$. Find the values of the constants A and B.

13. If one root of the equation $m^2 + km + 27 = 0$ is thrice the other, find the two possible values of k. Hence, what are the two possible equations?

14. What must be added to $x^2 + 16x$ to make it a perfect square?

15. What must be added to $x^2 - 13x$ to make it a perfect square?

16. Make $a^2 - 6a$ a perfect square.

17. Make $m^2 + \frac{5}{3}m$ a perfect square.

18. If $x^2 - 8x + 16$ is a perfect square, find its square root.

19. If $x^2 + 1.2x + (0.6)^2$ is a perfect square, find its square root.

20. Use the method of completing the square to solve the following quadratic equations:

a. $y^2 + 9y - 22 = 0$

b. $x^2 - 11x + 24 = 0$

c. $2x^2 - 5x - 12 = 0$

d. $3x^2 - 10x + 7 = 0$

e. $5y^2 + 11y + 6 = 0$

f. $2m^2 + 3m - 14 = 0$

g. $x^2 - 7x + 8 = 0$

21. Solve the following equations by using the quadratic formula:

a. $x^2 + x - 12 = 0$

b. $7x^2 - 17x + 10 = 0$

c. $2x^2 - 7x - 4 = 0$

d. $9 + 5x - 3x^2 = 0$

e. $5 - 7x - 5x^2 = 0$

f. $4n^2 - 12n + 9 = 0$

CHAPTER 10
WORD PROBLEMS LEADING TO QUADRATIC EQUATIONS

Examples

1. Find two numbers whose difference is 3 and whose product is 54.

<u>Solution</u>

Let the smaller number be x.

Then the larger number will be $x + 3$ (The larger one is greater than the smaller one by 3)

∴ Their product is $x(x + 3) = 54$

    $x^2 + 3x = 54$

    $x^2 + 3x - 54 = 0$

Solving this quadratic equation by factorization gives:

    $(x + 9)(x - 6) = 0$ (Two numbers whose product is –54 and sum is 3 are 9 and –6)

∴ $x = -9$ or $x = 6$ (When each of the bracket above is equated to zero and solved)

When $x = -9$, then the larger number is:

    $x + 3$ (As stated at the beginning)

    $= -9 + 3$ (Since $x = -9$)

    $= -6$

When $x = 6$, then the larger number is:

    $x + 3$

    $= 6 + 3 = 9$

∴   The two numbers are –9 and –6 or 9 and 6.

2. The difference between two numbers is 12. The sum of their squares is 74. Find the numbers.

<u>Solution</u>

Let the smaller number be 'a'

∴   The larger number is $a + 12$

Their squares are a^2 and $(a + 12)^2$

∴   The sum of their squares is:

    $a^2 + (a + 12)^2 = 74$

    $a^2 + (a + 12)(a + 12) = 74$

    $a^2 + a^2 + 12a + 12a + 144 = 74$

    $2a^2 + 24a + 144 - 74 = 0$

    $2a^2 + 24a + 70 = 0$

Dividing throughout by 2 to simplify it further gives:

    $a^2 + 12a + 35 = 0$

Solving this quadratic equation by factorization gives:

    $(a + 7)(a + 5) = 0$

∴ $a = -7$ or $a = -5$

When $a = -7$, the larger number is:

a + 12

= −7 + 12

= 5

When a = −5, the larger number is:

a + 12

= −5 + 12

= 7

∴ The two numbers are −7 and 5 or −5 and 7.

3. A boy is 5 years older than his brother. The product of their ages is 204. Find their ages.

Solution

Let his brother's age be y years. (Note that his brother's age is smaller since the boy is older)

∴ The boy's age = (y + 5)years

The product of their ages is given by:

$y(y + 5) = 204$

∴ $y^2 + 5y = 204$

$y^2 + 5y − 204 = 0$

Solving this quadratic equation by factorization gives:

$(y − 12)(y + 17) = 0$ (Note: two numbers whose product is −204 and sum is 5 are −12 and 17)

∴ y = 12 or y = −17

Age cannot be negative, therefore, y = 12

∴ The brother's age is 12 years.

The boy's age is y + 5

= 12 + 5

= 17

The boy's age is 17 years.

4. A paper measures 15cm by 10cm. A strip of equal width and having an 'L' shape is cut off from the ends of the paper. If the area of the remaining paper is 84cm^2, find the width of the strip removed.

Solution

Let the width of the strip removed be x cm. Since the L shape affect the length and width, then the new length of the paper is 15 − x, while the new width is 10 − x. This is because a size of x cm has been cut off along a length and along a width of the paper.

∴ The new area (length x width) is given by:

$(15 − x)(10 − x) = 84$

$150 − 15x − 10x + x^2 = 84$ (Note that $−x \times −x = +x^2$

$x^2 − 15x − 10x + 150 − 84 = 0$

$x^2 − 25x + 66 = 0$

Factorizing this quadratic equation gives:

$(x - 22)(x - 3) = 0$

$\therefore$ $x = 22$ or 3

The value 22 is not possible since it is greater than the original dimension of the paper.

$\therefore$ $x = 3$.

$\therefore$ The width of the strip removed is 3cm.

5. A rectangular field is 16m long and 8m wide. A path of equal width runs along one side and one end thereby forming an 'L' shape path. If the total area of the field and path is 180m^2, find the width of the path.

<u>Solution</u>

Let the width of the path be y m.

$\therefore$ The new length of the field including the path is $(16 + y)$m while the new width is $(8 + y)$m.

$\therefore$ The new area (i.e. total area) is:

$(16 + y)(8 + y) = 180$

$128 + 16y + 8y + y^2 = 180$

$y^2 + 16y + 8y + 128 - 180 = 0$

$y^2 + 24y - 52 = 0$

Factorizing this equation gives:

$(y + 26)(y - 2) = 0$

$\therefore$ $y = -26$ or $y = 2$

$y = -26$ is not possible since length/width cannot be negative

$\therefore$ $y = 2$

$\therefore$ The width of the path is 2m.

6. The ages of two brothers are 7 and 2 years. In how many years time will the product of their ages be 234?

<u>Solution</u>

Let the product of their ages be 234 in x years time.

$\therefore$ In x years time, their ages will be $(7 + x)$ years and $(2 + x)$ years

$\therefore$ Product of their ages then will be

$(7 + x)(2 + x) = 234$

$14 + 7x + 2x + x^2 = 234$

$x^2 + 7x + 2x + 14 - 234 = 0$

$x^2 + 9x - 220 = 0$

Factorizing this equation and solving it gives:

$(x + 20)(x - 11) = 0$

$\therefore$ $x = -20$ or $x = 11$

$x = -20$ is not correct since years cannot be negative.

$\therefore$ $x = 11$years.

$\therefore$ In 11 years time, the product of their ages will be 234.

7. Two sisters are 25 years and 18 years old. How many years ago was the product of their ages 144?

<u>Solution</u>

Let the product of their ages be 144, x years ago.

$\therefore$ x years ago, their ages were $(25 - x)$ years and $(18 - x)$ years.

$\therefore$ The product of their ages then was:

$(25 - x)(18 - x) = 144$

$450 - 25x - 18x + x^2 = 144$

$x^2 - 25x - 18x + 450 - 144 = 0$

$x^2 - 43x + 306 = 0$

Factorizing this quadratic equation gives:

$(x - 34)(x - 9) = 0$

$\therefore$ $x = 34$ or $x = 9$

34 is not possible since it is greater than the sisters' ages

$\therefore$ $x = 9$

$\therefore$ 9 years ago, the product of their ages was 144.

8. Find two consecutive odd numbers whose product is 255.

<u>Solution</u>

Let the smaller number be x. Therefore the larger number is $(x + 2)$

Note that any two consecutive odd numbers (or even numbers) differs by 2. For example, 5 and 7.

$\therefore$ Their product is given by:

$x(x + 2) = 255$

$x^2 + 2x - 255 = 0$

Solving the equation by factorizing method gives:

$(x + 17)(x - 15) = 0$

$\therefore$ $x = -17$ or $x = 15$

When $x = -17$, the larger number is $(x + 2)$. This gives:

$-17 + 2 = -15$

When $x = 15$, the larger number is $15 + 2 = 17$.

$\therefore$ The two consecutive odd numbers are -15 and -17 or 15 and 17.

9. Twice a particular whole number is subtracted from 3 times the square of the number. The answer is 133. What is the number?

<u>Solution</u>

Let the number be x.

$\therefore$ Twice the number is $2x$.

The square of the number is x^2.

3 times the square of the number is $3x^2$.

∴ When twice the number is subtracted from 3 times the square of the number to give 133, then this forms an equation as follows:

$$3x^2 - 2x = 133$$
∴ $$3x^2 - 2x - 133 = 0$$

Apart from factorization method, this equation can also be solved by using quadratic equation formula as follows:

$$3x^2 - 2x - 133 = 0$$

Hence , a = 3, b = −2, and c = −133

∴ $$x = \frac{-b \pm \sqrt{b^2 - 4ac}}{2a}$$

$$= \frac{-(-2) \pm \sqrt{(-2)^2 - [4 \times 3 \times (-133)]}}{2 \times 3}$$

$$= \frac{2 \pm \sqrt{4 - (-1596)}}{6}$$

$$= \frac{2 \pm \sqrt{4 + 1596)}}{6}$$

$$= \frac{2 \pm \sqrt{1600}}{6}$$

$$= \frac{2 \pm 40}{6}$$

$$x = \frac{2 + 40}{6} \quad \text{or} \quad x = \frac{2 - 40}{6}$$

$$x = \frac{42}{6} \quad \text{or} \quad x = \frac{-38}{6}$$

$$x = 7 \text{ or } -6.3$$

∴ x = 7 since the number is a whole number

∴ The number is 7.

10 A man is 4 times as old as his daughter. 5 years ago, the product of their ages was 175. Find their present ages.

Solution

Let the daughter's age be y years.

∴ The man's age is 4y years

5 years ago, the daughter's age was (y − 5) years, while the man's age was (4y − 5) years.

∴ The product of their ages then (i.e. 5 years ago) is given by:

$$(4y - 5)(y - 5) = 175$$
$$4y^2 - 20y - 5y + 25 = 175$$
$$4y^2 - 25y + 25 - 175 = 0$$
$$4y^2 - 25y - 150 = 0$$

In order to solve this equation by factorization, multiply the first and last terms on the left hand side. This gives:

$$4y^2 \times (-150) = -600y^2$$

Now look for two numbers in y whose product is $-600y^2$ and sum is −25y (i.e. the middle term in the equation). The two numbers are −40y and +15y. Replace −25y in the original equation by these two numbers. This gives:

$4y^2 - 40y + 15y - 150 = 0$

Factorizing by grouping gives:

$4y(y - 10) + 15(y - 10) = 0$

Take the terms outside each bracket, and then one of the two equal brackets. Note that $(y - 10)$ is common (i.e. the HCF) in the equation above. This gives:

$(y - 10)(4y + 15) = 0$

$\therefore$ y = 10 or $y = -\dfrac{15}{4}$

y = 10   (Ignore $y = -\dfrac{15}{4}$ since age cannot be negative or fraction)

$\therefore$ The daughter's age is 10 years, while the man's age is 40 years (i.e. 4y = 4 x 10 = 40)

Exercise 10

1. Find two numbers whose difference is 14 and whose product is 72.

2. The difference between two numbers is 10. The sum of their squares is 58. Find the numbers.

3. A boy is 7 years older than his brother. The product of their ages is 198. Find their ages.

4. A field measures 21cm by 12cm. A path of equal width and having an 'L' shape is at the ends of the field. If the area of the remaining field is 165cm^2, find the width of the path.

5. A rectangular field is 16m long and 8m wide. A path of equal width runs along one side and one end thereby forming an 'L' shape path. If the total area of the field and path is 180m^2, find the width of the path.

6. The ages of two friends are 13 and 9 years. In how many years time will the product of their ages be 480?

7. Two brothers are 28 years and 10 years old. How many years ago was the product of their ages 144?

8. Find two consecutive even numbers whose product is 288.

9. Thrice a particular whole number is added to 2 times the square of the number. The answer is 230. What is the number?

10 A woman is 3 times as old as his daughter. 5 years ago, the product of their ages was 217. Find their present ages.

CHAPTER 11
VARIATION

Direct Variation

Direct variation involves the relationship between two quantities whereby an increase or decrease in one of them leads to an increase or decrease respectively in the other.

The symbol α means 'varies with' or is 'proportional to'.

Examples

1. If x varies directly as y and $x = 30$ when y=12, find:
a. the formula connecting x and y
b. x when y=10
c. y when $x = 20$

Solution

a. $x \, \alpha \, y$ (This means x varies directly as y)

  $x = Ky$ (Replacing the proportionality sign with the equals sign introduces a constant K)

So, when $x = 30$ and y = 12, the equation above becomes:

    30 = K x 12

    30 = 12K

  $K = \dfrac{30}{12} = \dfrac{5}{2}$

The formula connecting x and y is:

    $x = \dfrac{5}{2} y$ (This is obtained by substituting $\dfrac{5}{2}$ for K in the equation above, i.e. $x = Ky$)

b. When y = 10, x is given by:

    $x = \dfrac{5}{2} y$

    $x = \dfrac{5}{2} \times 10 = \dfrac{50}{2} = 25$

  $\therefore \; x = 25$

c. When $x = 20$, y is given by:

    $x = \dfrac{5}{2} y$

    $20 = \dfrac{5}{2} y$

    5y = 40

    $y = \dfrac{40}{5}$

$y = 8$

2. If m varies as the square of n and m=27 when n=3, find:
a. the relationship between m and n
b. n when m = 48
c. m when $n = 2\frac{1}{3}$

Solution
a. $m \alpha n^2$ (This means m varies as n^2)
 $m = Kn^2$ (Replacing the proportionality sign with the equals sign introduces a constant K)
So, when m = 27 and n = 3, the equation above becomes:
   $27 = K \times 3^2$
   $27 = 9K$
$K = \dfrac{27}{9} = 3$
The relationship between m and n is:
 $m = 3n^2$ (This is obtained by substituting 3 for K in the equation $m = Kn^2$ above)

b. When m = 48, n is given by:
   $m = 3n^2$
   $48 = 3 \times n^2$
   $48 = 3n^2$
   $n^2 = \dfrac{48}{3} = 16$
   $n = \sqrt{16}$
      $n = 4$

c. When $n = 2\frac{1}{3}$, m is given by:
   $m = 3n^2$
   $m = 3 \times (\frac{7}{3})^2$ (Note that $2\frac{1}{3}$ has been converted to $\frac{7}{3}$)
   $m = 3 \times \dfrac{49}{9}$
Canceling out gives:
   $m = \dfrac{49}{3}$

Inverse Variation
In inverse or indirect variation, as one quantity increases the other decreases.

Examples

1. If c varies inversely as d and c=18 when y=4, find:
a. the formula connecting c and d
b. c when d=10
c. d when c=12

<u>Solution</u>

a. $c \alpha \dfrac{1}{d}$ (This means c varies inversely as d)

  $c = \dfrac{K}{d}$ (Replacing the proportionality sign with the equals sign introduces a constant K)

So, when c = 18 and d = 4, the equation above becomes:

  $18 = \dfrac{K}{4}$

  K = 18 x 4

  K = 72

The formula connecting c and d is:

  $c = \dfrac{72}{d}$ (This is obtained by substituting 72 for K in the equation above, i.e. $c = \dfrac{K}{d}$)

b. When d = 10, c is given by:

  $c = \dfrac{72}{d}$

  $c = \dfrac{72}{10}$

  c = 7.2

c. When c = 12, d is given by:

  $c = \dfrac{72}{d}$

  $12 = \dfrac{72}{d}$

  12d = 72

  $d = \dfrac{72}{12}$

  d = 6

2. If r varies inversely as the cube root of t and r=6 when t=64, find:
a. the relationship between r and t

b. t when r = 16

c. r when t = $\dfrac{8}{27}$

Solution

a. $r \alpha \dfrac{1}{\sqrt[3]{t}}$ (This means r varies inversely as the cube root of t)

$r = \dfrac{K}{\sqrt[3]{t}}$ (Replacing the proportionality sign with the equals sign introduces a constant K)

So, when r = 6 and t = 64, the equation above becomes:

$6 = \dfrac{K}{\sqrt[3]{64}}$

$6 = \dfrac{K}{4}$

K = 6 x 4 = 24

The relationship between r and t is:

$r = \dfrac{24}{\sqrt[3]{t}}$ (This is obtained by substituting 24 for K in the equation $r = \dfrac{K}{\sqrt[3]{t}}$)

b. When r = 16, t is given by:

$r = \dfrac{24}{\sqrt[3]{t}}$

$16 = \dfrac{24}{\sqrt[3]{t}}$

$\sqrt[3]{t} = \dfrac{24}{16}$

$\sqrt[3]{t} = \dfrac{3}{2}$

In order to remove the cube root, take the cube of both sides. This gives:

$(\sqrt[3]{t})^3 = (\dfrac{3}{2})^3$

$\therefore t = \dfrac{27}{8}$

c. When $t = \dfrac{8}{27}$, r is given by:

$$r = \dfrac{24}{\sqrt[3]{t}}$$

$$r = \dfrac{24}{\sqrt[3]{8/27}}$$

$$r = \dfrac{24}{2/3}$$

$$r = 24 \times \dfrac{3}{2}$$

After equal division by 2, it gives:

$$r = 12 \times 3$$

$$r = 36$$

Joint Variation

In joint variation, three or more quantities are related directly or inversely or both.

Examples

1. If m varies directly as the square of n and inversely as p, and m=3 when n=2 and p=8, find:
a. the relationship between m, n and p
b. m when n = 3 and p =27
c. p when $m = \dfrac{1}{2}$ and $n = \dfrac{3}{2}$

Solutions

$$m\, \alpha\, \dfrac{n^2}{p} \qquad \text{(This means m varies directly as the square of n and inversely as p)}$$

$$m = \dfrac{Kn^2}{p}$$

So, when m = 3, n = 2, and p = 8, the equation above becomes:

$$3 = \dfrac{K \times 2^2}{8}$$

$$3 = \dfrac{4K}{8}$$

$$4K = 3 \times 8 = 24$$

$$K = \dfrac{24}{4} = 6$$

The relationship between m, n and p is:

$$m = \dfrac{6n^2}{p} \qquad \text{(This is obtained by substituting 6 for K in the equation } m = \dfrac{Kn^2}{p})$$

b. When n = 3 and p = 27, then m is given by:

$$m = \frac{6n^2}{p}$$

$$m = \frac{6 \times 3^2}{27}$$

$$m = \frac{6 \times 9}{27}$$

$$m = \frac{54}{27}$$

$$m = 2$$

c. When $m = \frac{1}{2}$ and $n = \frac{3}{2}$, then p is given by:

$$m = \frac{6n^2}{p}$$

$$\frac{1}{2} = \frac{6 \times \frac{3}{2}}{p}$$

$$\frac{1}{2} = \frac{9}{p}$$

$$p = 9 \times 2$$
$$P = 18$$

2. The weight w of a rod varies jointly as its length L and the square root of its density d. If w = 12 when L = 5 and d = 9, find:

a. L in terms of w and d

b. w when L = 8 and d = 25

c. d when L = 20 and w = 4

Solutions

a. $w \, \alpha \, L\sqrt{d}$ (This means w varies jointly as L and the square root of d)

$w = KL\sqrt{d}$

So, when w = 12, L = 5, and d = 9, the equation above becomes:

$12 = K \times 5 \times \sqrt{9}$

$12 = 15K$

$K = \frac{12}{15} = \frac{4}{5}$

The formula connecting w, L and d is:

$w = \frac{4}{5}L\sqrt{d}$ (This is obtained by substituting $\frac{4}{5}$ for K in the equation $w = KL\sqrt{d}$)

L can now be expressed in terms of w and d as follows:

$$w = \frac{4}{5} L\sqrt{d}$$

$$5w = 4L\sqrt{d}$$

Dividing both sides of the equation by $4\sqrt{d}$ gives:

$$L = \frac{5w}{4\sqrt{d}}$$

b. When L = 8 and d = 25, then w is given by:

$$w = \frac{4}{5} L\sqrt{d}$$

$$w = \frac{4}{5} \times 8 \times \sqrt{25}$$

$$w = \frac{4}{5} \times 8 \times 5$$

Cancelling out the 5 gives:

$$w = 4 \times 8$$

$$w = 32$$

c. When L = 20 and w = 4, then d is given by:

$$w = \frac{4}{5} L\sqrt{d}$$

$$4 = \frac{4}{5} \times 20 \times \sqrt{d}$$

$$4 \times 5 = 4 \times 20 \times \sqrt{d}$$

$$20 = 80\sqrt{d}$$

$$\sqrt{d} = \frac{20}{80} = \frac{1}{4}$$

Taking the square of both sides gives:

$$(\sqrt{d})^2 = (\frac{1}{4})^2$$

$$d = \frac{1}{16}$$

Partial Variation

The fourth type of variation is called partial variation. In partial variation, one quantity is partly constant and partly varies with the other. Two constants are involved in partial variation.

Examples

1. x is partly constant and partly varies as y. When y=2, x=30, and when y=6, x=50.
a. Find the formula which connects x and y.
b. Find x when y=3

Solutions

a. From the first sentence, we have:

$x = C + Ky$ (Let this be equation 1) where C and K are constants.

Substituting y=2 and x=30 in this equation gives:

30 = C + 2K   (Let this be equation 2)

Similarly, when y = 6 and x = 50, we have:

50 = C + 6K   (Let this be equation 3)

Bringing equation 2 and 3 together gives:

$$30 = C + 2K \quad \text{(Equation 2)}$$
$$50 = C + 6K \quad \text{(Equation 3)}$$

Equation 3 – Equation 2:   20 =   4K

Divide both sides by 4. This gives:

$$K = \frac{20}{4} = 5$$

K = 5

Substitute 5 for K in equation 2.

30 = C + 2K

30 = C + (2 x 5)

30 = C + 10

30 – 10 = C

C = 20

We now substitute the values of C and K into equation 1 in order to obtain the formula connecting x and y.

The formula connecting x and y is now given by:

x = 20 + 5y

b. When y = 3, x is obtained by substituting 3 for y in the formula connecting x and y.

x = 20 + 5y

= 20 + (5x3)

= 20 + 15

x = 35

2. m is partly constant and partly varies as n. When n = 4, m = 5, and when n = 12, m = 14.

a. Find the formula which connects m and n.

b. Find m when n = 16

c. Find n when m = 9

Solutions

a. From the first sentence, we have:

m = C + Kn   (Let this be equation 1) where C and K are constants.

Substituting n=4 and m=5 in equation 1 gives:

5 = C + 4K   (Let this be equation 2)

Similarly, when n=12 and m=14, we have:

14 = C + 12K   (Let this be equation 3)

Bringing equation 2 and 3 together gives:

$$5 = C + 4K \quad \text{(Equation 2)}$$

$$\underline{14 = C + 12K} \quad \text{(Equation 3)}$$

Equation 3 – Equation 2: $\quad 9 = \quad 8K$

Divide both sides by 8. This gives:

$$K = \frac{9}{8}$$

Substitute $\frac{9}{8}$ for K in equation 2.

$$5 = C + 4K$$

$$5 = C + (4 \times \frac{9}{8})$$

$$5 = C + \frac{9}{2}$$

$$5 - \frac{9}{2} = C$$

$$C = \frac{1}{2}$$

We now substitute the values of C and K into equation 1 in order to obtain the formula connecting m and n.

The formula connecting m and n is given by:

$$m = \frac{1}{2} + \frac{9}{8} n$$

b. When n = 16, m is obtained by substituting 16 for n in the formula connecting m and n.

$$m = \frac{1}{2} + \frac{9}{8} n$$

$$m = \frac{1}{2} + (\frac{9}{8} \times 16)$$

$$= \frac{1}{2} + 18 = \frac{37}{2}$$

$$m = 18\frac{1}{2}$$

c. When m = 9, n is obtained by substituting 9 for m in the formula connecting m and n.

$$m = \frac{1}{2} + \frac{9}{8} n$$

$$9 = \frac{1}{2} + (\frac{9n}{8})$$

$$9 - \frac{1}{2} = \frac{9n}{8}$$

$$\frac{17}{2} = \frac{9n}{8}$$

$$17 \times 8 = 9n \times 2$$

$$136 = 18n$$

$$n = \frac{136}{18} = \frac{68}{9}$$

$$n = 7\frac{5}{9}$$

Exercise 11

1. If x varies directly as y and $x=10$ when y=8, find:
a. the formula connecting x and y
b. x when y=10
c. y when $x=16$

2. If h varies as the square root of p and h=5 when p=9, find:
a. the relationship between h and p
b. p when h = 20
c. h when $p = 6\frac{1}{4}$

3. If p varies inversely as q and p = 12 when q = 3, find:
a. the formula connecting p and q
b. q when p = 20
c. p when q = 5

4. If m varies inversely as the cube root of n and m=5 when n = 27, find:
a. the relationship between m and n
b. m when n = 8
c. n when $m = \frac{64}{125}$

5. If a varies directly as the square of b and inversely as c, and when a=4 when b=3 and c=6, find:
a. the formula connecting a, b and c
b. a when b = 5 and c = 10
c. b when $a = \frac{1}{2}$ and c = 8

6. The height h of a box varies jointly as its length L and the square of its width w. If h = 20 when L = 4 and w = 3, find:
a. w in terms of h and L
b. w when h = 12 and L = 4
c. L when h = 8 and w = 5

7. x is partly constant and partly varies as y. When y = 4, $x = 14$, and when y = 5, $x = 17$.
a. Find the relationship between x and y.
b. Find x when y = 8

8. E is partly constant and partly varies as F. When F = 2, E = 25, and when F = 5, E = 55.

a. Find the formula which connects E and F.

b. Find E when F = $2\frac{1}{2}$

c. Find F when E = 40

CHAPTER 12
SIMULTANEOUS LINEAR AND QUADRATIC EQUATIONS

Examples

1. Solve the simultaneous equations: $2x^2 - 5y = -8$ and $3x + y = -1$

<u>Solution</u>

$2x^2 - 5y = -8$ ....Equation (1)

$3x + y = -1$ ....Equation (2)

From equation (2), $y = -1 - 3x$ ....Equation (3)

Substitute $-1-3x$ for y in equation (1)

$2x^2 - 5y = -8$ ....Equation (1)

$2x^2 - 5(-1 - 3x) = -8$

$2x^2 + 5 + 15x = -8$

$2x^2 + 15x + 5 + 8 = 0$

$2x^2 + 15x + 13 = 0$

In order to solve this equation by factorization method, we first multiply the first and last terms. This gives:

$2x^2 \times 13 = 26x^2$

Find two numbers in x whose product is $26x^2$ and whose sum is $15x$ (i.e. the middle term in the original equation). The two numbers in x are $13x$ and $2x$ (their product is $26x^2$ and their sum is $15x$). The two numbers are now used to replace the middle term in the original equation. This gives:

$2x^2 + 13x + 2x + 13 = 0$

Factorize this equation by grouping.

$x(2x + 13) + 1(2x + 13) = 0$

$\therefore$ $(2x + 13)(x + 1) = 0.$ (Note that $(2x + 13)$ is the common factor in the expression above)

Equating each bracket above to zero and solving each of the linear equation formed, gives:

$x = -1$ or $x = -\dfrac{13}{2} = -6\dfrac{1}{2}$

Substitute -1 for x in equation (3)

$y = -1 - 3x$ ....Equation (3)

$= -1 - 3(-1)$

$= -1 + 3$

$y = 2$

Substitute $-\dfrac{13}{2}$ for x in equation (3)

$y = -1 - 3x$ ....Equation (3)

$= -1 - 3(-\dfrac{13}{2})$

$= -1 + \dfrac{39}{2}$

$= \dfrac{-2 + 39}{2}$

$y = \dfrac{37}{2}$

$$y = 18\frac{1}{2}$$

Therefore the solutions are $x = -1$ and $y = 2$ or $x = -6\frac{1}{2}$ and $y = 18\frac{1}{2}$. The solution can be given more neatly in x and y coordinates as follows:

$$(-1, 2), (-6\frac{1}{2}, 18\frac{1}{2})$$

2. Solve the simultaneous equations: $x^2 + 3y^2 = 4$
$$x - 2y = 1$$

Solutions

$x^2 + 3y^2 = 4$ ....Equation (1)

$x - 2y = 1$ ....Equation (2)

From equation (2) $x = 1 + 2y$

Substitute $1 + 2y$ for x in equation (1)

$x^2 + 3y^2 = 4$

$(1 + 2y)^2 + 3y^2 = 4$

$(1 + 2y)(1 + 2y) + 3y^2 = 4$

$1 + 2y + 2y + 4y^2 + 3y^2 = 4$

$7y^2 + 4y + 1 - 4 = 0$

$7y^2 + 4y - 3 = 0$

Solving this quadratic equation by factorizing gives:

$7y^2 + 7y - 3y - 3 = 0$

$7y(y + 1) - 3(y + 1) = 0$

$(y + 1)(7y - 3) = 0$

$\therefore$ $y = -1$ or $y = \dfrac{3}{7}$

Substitute -1 for y in equation (2). This gives:

$x - 2y = 1$ ....Equation (2)

$x - 2(-1) = 1$

$x + 2 = 1$

$x = 1 - 2$

$x = -1$

Similarly, substitute $\dfrac{3}{7}$ for y in equation (2).

$x - 2y = 1$ ....Equation (2)

$x - 2(\dfrac{3}{7}) = 1$

$x - \dfrac{6}{7} = 1$

$x = 1 + \dfrac{6}{7}$

$= \dfrac{7 + 6}{7}$

$x = \dfrac{13}{7} = 1\dfrac{6}{7}$

Therefore the solutions are $x = -1$ and $y = -1$, or $x = 1\frac{6}{7}$ and $y = \frac{3}{7}$

The solutions can also be given as: $(-1, -1)$, $(1\frac{6}{7}, \frac{3}{7})$

3. Solve the equations $3x^2 - xy = 0$, $2y - 5x = 1$

Solution

$3x^2 - xy = 0$ ....Equation (1)

$2y - 5x = 1$ ....Equation (2)

From equation (1):

$x(3x - y) = 0$

∴ $x = 0$ or $3x - y = 0$

$x = 0$ or

$y = 3x$ ....Equation (3)

Substitute 0 for x in equation (2). This gives

$2y - 5(0) = 1$ ....Equation (2)

$2y = 1$

$y = \frac{1}{2}$

Similarly, substitute $3x$ for y in equation (2).

$2y - 5x = 1$ ....Equation (2)

$2(3x) - 5x = 1$

$6x - 5x = 1$

$x = 1$

Substitute 1 for x in equation (3)

$y = 3x$ ....Equation (3)

$y = 3(1)$

$y = 3$

∴ $x = 0$ and $y = \frac{1}{2}$ or $x = 1$ and $y = 3$

This can also be given by: $(0, \frac{1}{2})$ or $(1, 3)$

4. Solve simultaneously $3x - 5y = -11$, $xy = -2$

Solution

$3x - 5y = -11$ ....Equation (1)

$xy = -2$ ....Equation (2)

From equation (2), $x = \frac{-2}{y}$ ....Equation (3)

Substitute $\frac{-2}{y}$ for x in equation (1)

$3x - 5y = -11$ ....Equation (1)

$3(\frac{-2}{y}) - 5y = -11$

$$\frac{-6}{y} - 5y = -11$$

Multiply each term by y in order to clear the fraction. This gives:

$$y(\frac{-6}{y}) - y(5y) = y(-11)$$
$$-6 - 5y^2 = -11y$$
$$0 = 5y^2 - 11y + 6$$

Or, $\quad 5y^2 - 11y + 6 = 0$

Solving this equation by factorization gives:

$$5y^2 - 5y - 6y + 6 = 0$$
$$5y(y - 1) - 6(y - 1) = 0 \quad \text{(−6 outside the bracket changes the + sign to − sign in the}$$
bracket)
$$(y - 1)(5y - 6) = 0$$

Equating each bracket to zero and solving each equation gives:

$$y = 1 \text{ or } y = \frac{6}{5} = 1\frac{1}{5}$$

Substitute 1 for y in equation (3)

$$x = \frac{-2}{y} \quad ....\text{Equation (3)}$$
$$= \frac{-2}{1}$$
$$= -2$$

Substitute $\frac{6}{5}$ for y in equation (3)

$$x = \frac{-2}{y} \quad ....\text{Equation (3)}$$
$$x = \frac{-2}{\frac{6}{5}}$$
$$= -2 \times \frac{5}{6}$$
$$= \frac{-10}{6}$$
$$x = \frac{-5}{3} = -1\frac{2}{3}$$

Therefore, the solutions are $x = -2$ and y = 1, or $x = -1\frac{2}{3}$ and y = $1\frac{1}{5}$. These can be given as: $(-2, 1)$, $(-1\frac{2}{3}, 1\frac{1}{5})$.

5. Solve the equations $9y^2 + 8x = 12, \ 2x + 3y = 4$

Solution

$$9y^2 + 8x = 12 \ ....\text{Equation (1)}$$
$$2x + 3y = 4 \ ....\text{Equation (2)}$$

From equation (2):

$$2x = 4 - 3y$$

$$\therefore \quad x = \frac{4 - 3y}{2} \quad \text{.................Equation (3)}$$

Substitute $\frac{4 - 3y}{2}$ for x in equation (1).

$$9y^2 + 8x = 12 \quad \text{.................Equation (1)}$$

$$9y^2 + 8\left(\frac{4 - 3y}{2}\right) = 12$$

$$9y^2 + 4(4 - 3y) = 12 \quad \text{(Note that 8 divided by 2 gives 4 which is outside the bracket)}$$

$$9y^2 + 16 - 12y = 12$$

$$9y^2 - 12y + 16 - 12 = 0$$

$$9y^2 - 12y + 4 = 0$$

Solving this equation by factorization gives:

$$9y^2 - 6y - 6y + 4 = 0$$

$$3y(3y - 2) - 2(3y - 2) = 0$$

$$(3y - 2)(3y - 2) = 0$$

Equation each equation to zero and solving each equation gives:

$$y = \frac{2}{3} \text{ in both cases.}$$

Now substitute $\frac{2}{3}$ for y in equation (3)

$$x = \frac{4 - 3y}{2} \quad \text{.................Equation (3)}$$

$$= \frac{4 - 3\left(\frac{2}{3}\right)}{2}$$

$$= \frac{4 - 2}{2} \quad \text{(Note that } 3 \times \frac{2}{3} = 2\text{)}$$

$$x = \frac{2}{2} = 1$$

Therefore, the solutions are $x = 1$ and $y = \frac{2}{3}$

6. Solve the equations, $x^2 - y^2 = 20$, $x + y = -2$

<u>Solution</u>

$$x^2 - y^2 = 20 \quad \text{.................Equation (1)}$$

$$x + y = -2 \quad \text{.................Equation (2)}$$

Looking at equation (1) carefully shows that it is a difference of two squares, and that one of the factors of equation (1) is what is given in equation (2).

Recall that a difference of two squares such as $a^2 - b^2$ can be factorized as $a^2 - b^2 = (a + b)(a - b)$. Therefore the equations above can be re-written as follows:

$$(x + y)(x - y) = 20 \quad \text{.................Equation (1)} \quad \text{(Note that } x^2 - y^2 \text{ has been factorized)}$$

$$x + y = -2 \quad \text{.................Equation (2)}$$

Substitute -2 for $x + y$ in equation (1)

$$(x + y)(x - y) = 20 \quad \text{.................Equation (1)}$$

$$-2(x - y) = 20 \quad \text{(Since } (x + y) = -2\text{)}$$

Divide both sides by −2 to obtain:

$$x - y = -10 \dots\dots\dots\dots\text{Equation (3)}$$
$$\underline{x + y = -2} \dots\dots\dots\dots\text{Equation (2)}$$

Equation (2) − equation (3) $2y = 8$ (Note that $+y - (-y) = y + y = 2y$ and $-2 - (-10) = -2 + 10 = 8$)

$$y = \frac{8}{2}$$
$$y = 4$$

Substitute 4 for y in equation (2). (Any equation can be used)

$\quad x + y = -2 \dots\dots\dots\dots\text{Equation (2)}$
$\quad x + 4 = -2$
$\quad x = -2 - 4$
$\quad x = -6$

Therefore the solution are $x = -6$ and $y = 4$. This can also be given as $(-6, 4)$.

7. Solve the equations, $4x^2 - 9y^2 = -32$, $2x - 3y = -4$

Solution

$\quad 4x^2 - 9y^2 = -32 \dots\dots\dots\dots\text{Equation (1)}$
$\quad 2x - 3y = -4 \dots\dots\dots\dots\text{Equation (2)}$

Equation (1) is a difference of two squares. It can be expressed as:

$\quad (2x)^2 - (3y)^2 = -32$

Therefore factorizing this difference of two squares gives:

$\quad (2x - 3y)(2x + 3y) = -32 \dots\dots\dots\dots\text{Equation (1)}$
$\quad 2x - 3y = -4 \dots\dots\dots\dots\text{Equation (2)}$

Substitute −4 for $2x - 3y$ in equation (1)

$\quad (2x - 3y)(2x + 3y) = -32 \dots\dots\dots\dots\text{Equation (1)}$
$\quad -4(2x + 3y) = -32$ (Since $(2x - 3y) = -4$)

Divide both sides by −4 to obtain equation (3) as shown below:

$$2x + 3y = 8 \dots\dots\dots\dots\text{Equation (3)}$$
$$\underline{2x - 3y = -4} \dots\dots\dots\dots\text{Equation (2)}$$

Equation (3) − equation (2): $6y = 12$ (Note that $+3y - (-3y) = 6y$ and $8 - (-4) = 8 + 4 = 12$)

$$y = \frac{12}{6}$$
$$y = 2$$

Substitute 2 for y in equation (3). (Note that any equation can be used)

$\quad 2x + 3y = 8 \dots\dots\dots\dots\text{Equation (3)}$
$\quad 2x + 3(2) = 8$
$\quad 2x + 6 = 8$
$\quad 2x = 8 - 6$
$\quad 2x = 2$
$\quad x = \frac{2}{2}$
$\quad x = 1$

Therefore the solution is $x = 1$ and $y = 2$. This can also be represented as (1, 2).

8. A boy is x years old while his father is y years old. The sum of their ages is equal to twice the difference of their ages. The product of their ages is equal to 675. Find their ages.

Solution
The sum of their ages is $x + y$.
The difference of their ages is $y - x$. (The father's age is larger)
Twice the difference of their ages is $2(y - x)$.
The sum of their ages is equal to twice the difference of their ages is given by:

$$x + y = 2(y - x)$$
$$x + y = 2y - 2x$$
$$x + 2x + y - 2y = 0$$
$$3x - y = 0$$

Or $3x = y$ ....Equation (1)

The product of their ages is 675. This is given by:

$$xy = 675$$ ....Equation (2)

Substitute $3x$ for y in equation (2). (Note that $y = 3x$ from equation 1)

$$xy = 675$$ ....Equation (2)
$$x(3x) = 675$$
$$3x^2 = 675$$
$$x^2 = \frac{675}{3}$$
$$x^2 = 225$$
$$x = \sqrt{225}$$
$$x = \pm 15$$
$$x = 15 \quad \text{(Since age cannot be negative)}$$

Substitute 15 for x in equation (1)

$$3x = y$$ ....Equation (1)

Or, $y = 3x$
$$= 3(15)$$
$$y = 45$$

Therefore the boy is 15 years old (i.e. x), while his father is 45 years old (i.e. y)

9. Twice a number added to another number is 19. The first number added to the square of the second number gives 17. Find the two numbers.

Solution
Let the first number be 'a' and the second number be 'b'.
Twice the first number = 2a
Twice the first number added to the second number is 19. This is given by:

$$2a + b = 19$$ ....Equation (1)

The square of the second number = b^2
The first number added to the square of the second number gives 17. This is represented as:

$$a + b^2 = 17$$ ....Equation (2)

147

From equation (1), a = $\dfrac{19-b}{2}$ ....Equation (3)

Substitute $\dfrac{19-b}{2}$ for 'a' in equation (2)

$\qquad a + b^2 = 17$ ....Equation (2)

$\qquad \dfrac{19-b}{2} + b^2 = 17$

Multiply each term by 2 (L.C.M) in order to clear the fraction. This gives:

$\qquad 2(\dfrac{19-b}{2}) + 2(b^2) = 2(17)$

$\qquad 19 - b + 2b^2 = 34$

$\qquad 2b^2 - b + 19 - 34 = 0$

$\therefore \qquad 2b^2 - b - 15 = 0$

Solving this equation by factorization method gives

$\qquad 2b^2 - 6b + 5b - 15 = 0$

$\qquad 2b(b - 3) + 5(b - 3) = 0$

Equating each bracket to zero and solving each equation gives:

$\qquad (b - 3)(2b + 5) = 0$

$\qquad b = 3 \text{ or } b = \dfrac{-5}{2} = -2\dfrac{1}{2}$

From equation (3), when b = 3, we obtain 'a' as follows:

$\qquad a = \dfrac{19-b}{2}$ ....Equation (3)

$\qquad = \dfrac{19-3}{2}$

$\qquad = \dfrac{16}{2}$

$\qquad a = 8$

Similarly, from equation (3), when b = $\dfrac{-5}{2}$, we obtain 'a' as follows:

$\qquad a = \dfrac{19-b}{2}$ ....Equation (3)

$\qquad = \dfrac{19-(\frac{-5}{2})}{2}$

$\qquad = \dfrac{19+\frac{5}{2}}{2}$

$\qquad = \dfrac{\frac{38+5}{2}}{2}$

$\qquad = \dfrac{43}{2} \div 2$

$\qquad = \dfrac{43}{2} \times \dfrac{1}{2}$

$$= \frac{43}{4}$$

$$a = 10\frac{3}{4}$$

Therefore the solutions are a = 8 and b = 3, or a = $10\frac{3}{4}$ and b = $-2\frac{1}{2}$. These can also be given as:
(8, 3) or $(10\frac{3}{4}, -2\frac{1}{2})$.

10. The sum of the squares of the ages of two brothers is 65. If the difference between their ages is 3, find their ages.

<u>Solution</u>
Let the older brother be m years old, and the younger brother be n years old. Therefore the two equations that can be formed from the question are:

$m^2 + n^2 = 65$ ....Equation (1)

$m - n = 3$ ....Equation (2)

From equation (2):

$m = 3 + n$ ....Equation (3)

Substitute 3 + n for m in equation (1). This gives:

$m^2 + n^2 = 65$ ....Equation (1)

$(3 + n)^2 + n^2 = 65$

$(3 + n)(3 + n) + n^2 = 65$

$9 + 3n + 3n + n^2 + n^2 = 65$

$2n^2 + 6n + 9 - 65 = 0$

$2n^2 + 6n - 56 = 0$

Solving this equation by using quadratic equation formula (you can also use factorization method) gives:

$a = 2, b = 6, c = -56$

∴ $n = \dfrac{-b \pm \sqrt{b^2 - 4ac}}{2a}$

$$= \dfrac{-6 \pm \sqrt{6^2 - [4 \times 2 \times (-56)]}}{2 \times 2}$$

$$= \dfrac{-6 \pm \sqrt{36 - (-448)}}{4}$$

$$= \dfrac{-6 \pm \sqrt{36 + 448}}{4}$$

$$= \dfrac{-6 \pm \sqrt{484}}{4}$$

$$= \frac{-6 \pm 22}{4}$$

$$n = \frac{-6 + 22}{4} \text{ or } \frac{-6 - 22}{4}$$

$$= \frac{16}{4} \text{ or } \frac{-28}{4}$$

$$n = 4 \text{ or } -7$$

But, age cannot be negative. Therefore:

$$n = 4$$

Substitute 4 for n in equation (3)

$$m = 3 + n \ldots\ldots\ldots\ldots\ldots\text{Equation (3)}$$

$$= 3 + 4$$

$$m = 7$$

∴   The older brother is 7 years old while the younger brother is 4 years old.

Exercise 12

1. Solve the following simultaneous equations:
a. $x^2 - 2y = -2$
   $5x + y = 13$
b. $x^2 + 2y^2 = 9$
   $2x - y = -4$
c. $5x^2 - 3xy = 26$, $3y - 7x = 17$
d. $6x - 5y = -2$, $2xy = 1$
e. $5y^2 + 2x = 11$, $x - 4y = 7$
f. $x^2 - y^2 = 16$, $x + y = -2$
g. $3x^2 - 2xy + y^2 = 9$, $5x - 2y = 8$

2.   A boy is x years old while his father is y years old. The sum of their ages is equal to thrice the difference of their ages. The product of their ages is equal to 1800. Find their ages.

3.   Four times a number added to twice another number is 34. Twice the first number added to twice the square of the second number gives 108. Find the two numbers.

4.   The sum of the squares of the ages of two brothers is 157. If the difference between their ages is 5, find their ages.

CHAPTER 13
LINEAR INEQUALITY AND INTRODUCTORY LINEAR PROGRAMMING

Linear Inequality

If a statement is expressed as $x = 10$, then this is an equation. However, if x is not exactly equal to 10, but can vary in values from 10 and above or from 10 and below, then we say $x < 10$ or $x > 10$. We can also say $x \leq 10$ or $x \geq 10$. These are called inequalities.

The four symbols of inequalities and their meanings are as follows:

> $<$ means less than
>
> $>$ means greater than
>
> $\leq$ means less than or equal to
>
> $\geq$ means greater than or equal to

Inequalities can be solved like equations. When solving inequalities, the following points should be noted.

1. When both sides of an inequality are divided by a negative number, the inequality sign should be reversed. This is done in order to keep the inequality true. Consider the example below:

$$-3 < 5 \quad \text{(This is true)}$$

Now let us divide both sides by -3. This gives:

$$\frac{-3}{-3} < \frac{5}{-3}$$

$$1 < -\frac{5}{3} \quad \text{(This is not true because the inequality sign has not been reversed)}$$

Therefore the right thing to do is to reverse the inequality sign when dividing by a negative number. This means that the correct division above should be:

$$-3 < 5$$

$$\frac{-3}{-3} > \frac{5}{-3}$$

$$1 > -\frac{5}{3} \quad \text{(This is true because the inequality sign has been reversed)}$$

2. When an inequality is cross multiplied, it is advisable to cross multiply from the lower right hand side to the upper left hand side first, and not the other way round first. Let us examine the example below.

$$\frac{2}{7} < \frac{3}{4}$$

The correct way to cross multiply is as follows:

$$(2 \times 4) < (7 \times 3)$$

$$8 < 21 \quad \text{(This is true)}$$

However, if you cross multiply from the lower left to the upper right first, then it gives:

$$(7 \times 3) < (2 \times 4)$$

$$21 < 8 \quad \text{(This is not true)}$$

In order to make this kind of cross multiplication true, you have to reverse the sign to give $21 > 8$.

Hence, in order to avoid making mistake, it is better to cross multiply from the lower right hand side to the upper left hand side first. In this case your inequality remains true, and you do not need to reverse the inequality sign.

Examples

1. Solve the inequality $5x - 2 > 19 - 2x$. Represent your result on a number line/line graph.

<u>Solution</u>

$$5x - 2 > 19 - 2x$$

We simply solve this problem just like the way we solve an equation. But we must continue to use the inequality sign.

Let us collect terms in x on one side, and constant terms on the other side.

$$5x - 2 > 19 - 2x$$

$5x + 2x > 19 + 2$ (Note that the sign of any term that crosses the inequality sign must change)

$$7x > 21$$

Divide both sides by 7. This gives:

$$\frac{7x}{7} > \frac{21}{7}$$

$$x > 3$$

In representing inequalities on number lines/line graphs, we simply draw a number line and make sure the arrow starts from the solution (i.e. 3) and points to the right when it is >. It points to the left when the inequality sign is <. The beginning of the arrow line should be a small circle. When the sign is ≤ or ≥, the small circle on the number line where the arrow starts should be shaded to show that the value is included in the possible solutions.

∴ $x > 3$ is represented as shown below:

2. Solve the inequality $9x + 5 < 4x - 2(3 - x)$. Represent your result on a number line.

<u>Solution</u>

$$9x + 5 < 4x - 2(3 - x)$$

Let us expand the bracket on the right hand side. This gives:

$$9x + 5 < 4x - 6 + 2x$$

Let us collect terms in x on one side, and constant terms on the other side.

$$9x - 4x - 2x < -6 - 5$$

$$3x < -11$$

Divide both sides by 3. This gives:

$$\frac{3x}{3} < \frac{-11}{3}$$

$$x < -\frac{11}{3}$$

∴ $x < -\dfrac{11}{3}$ is represented on a number line as shown below:

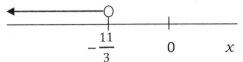

3. Find the range of values of x for which $\dfrac{2x - 1}{3} - \dfrac{1}{2}x \le 8 + x$. Show the result on a line graph.

Solution
$$\frac{2x - 1}{3} - \frac{1}{2}x \le 8 + x$$
Multiply each term by 6 (i.e. the LCM of 3 and 2 which are the denominators). This is done in order to remove the fractions.
$$6(\frac{2x - 1}{3}) - 6(\frac{x}{2}) \le 6(8 + x)$$
$2(2x - 1) - 3(x) \le 48 + 6x$ (Note that 2 and 3 outside the brackets are obtained when 6 outside the brackets above divides 3 and 2 (the denominators) respectively).
$4x - 2 - 3x \le 48 + 6x$
We now collect like terms as follows:
$4x - 3x - 6x \le 48 + 2$
$-5x \le 50$
Divide both sides by –5 and reverse the inequality sign. This gives:
$\dfrac{-5x}{-5} \ge \dfrac{50}{-5}$ (Take note of the reversal of the sign since we are dividing by a negative number)
$x \ge -10$
We now represent this result on a number line as follows

Note that the small circle at the beginning of the arrow is shaded to show that –10 is also part of the solution due to the sign $\ge$. And remember that greater than represents right hand direction of arrow, while less than represents left hand direction of arrow.

4. Find the range of values of x for which $\dfrac{1}{5} - \dfrac{1 - 4x}{2} \ge \dfrac{3}{4} + 3x$. Show the result on a line graph.

Solution
$$\frac{1}{5} - \frac{1 - 4x}{2} \ge \frac{3}{4} + 3x$$
Multiply each term by 20 (i.e. the LCM of 5, 2 and 4 which are the denominators).
$$20(\frac{1}{5}) - 20(\frac{1 - 4x}{2}) \ge 20(\frac{3}{4}) + 20(3x)$$
$4(1) - 10(1 - 4x) \ge 5(3) + 60x$
Note that 20 has been used to divide the denominators of the fractions.

$$4 - 10 + 40x \geq 15 + 60x$$

We now collect like terms as follows:

$$40x - 60x \geq 15 + 10 - 4$$
$$-20x \geq 21$$

Divide both sides by –20 and reverse the sign. This gives:

$$\frac{-20x}{-20} \leq \frac{21}{-20}$$ (Take note of the reversal of the sign since we are dividing by a negative number)

$$x \leq -\frac{21}{20}$$

We now represent this result on a number line as follows

5. Given that m is an integer, find the three greatest values which satisfy the inequality:

$$2(3m + 1) < \frac{1}{2}(2m - 5)$$

Solution

$$2(3m + 1) < \frac{1}{2}(2m - 5)$$

This can also be written as follows:

$$2(3m + 1) < \frac{2m - 5}{2}$$

Cross multiply to obtain:

$$2 \times 2(3m + 1) < 2m - 5$$
$$4(3m + 1) < 2m - 5$$
$$12m + 4 < 2m - 5$$

Collect like terms. This gives:

$$12m - 2m < -5 - 4$$
$$10m < -9$$
$$\therefore \quad m < \frac{-9}{10}$$

An integer is any whole number such as, –1, 7, 0.

Hence, if $m < \frac{-9}{10}$, and m is an integer, then m can be any of the numbers below.

$$m = -1, -2, -3, -4 \ldots\ldots$$

Hence the three greatest values of m are, –1, –2, –3.

Note that the nearest whole number that is just less than $\frac{-9}{10}$ is –1 which is also the greatest whole number for 'm'.

6. Given that x is an integer, find the three lowest values of x which satisfy the following inequality: $2(x - \frac{1}{3}) - \frac{3}{4}(x + 5) > \frac{1}{2}(x - 4)$

Solution

$$2\left(x-\frac{1}{3}\right) - \frac{3}{4}(x+5) > \frac{1}{2}(x-4)$$

Expanding each of the bracket and placing each of the denominator properly gives:

$$2x - \frac{2}{3} - \frac{3x+15}{4} > \frac{x-4}{2}$$

Multiply each term by 12 (i.e. the LCM of the denominators). This gives:

$$12(2x) - 12\left(\frac{2}{3}\right) - 12\left(\frac{3x+15}{4}\right) > 12\left(\frac{x-4}{2}\right)$$

$24x - 4(2) - 3(3x+15) > 6(x-4)$ (Note that 12 has been used to divide each denominator)

$24x - 8 - 9x - 45 > 6x - 24$

Collect like term to give:

$24x - 9x - 6x > 8 + 45 - 24$

$$9x > 29$$

$$x > \frac{29}{9}$$

$$x > 3\frac{2}{9}$$

Since x is an integer (whole number), then x starts from 4 because 4 is the nearest whole number greater than $3\frac{2}{9}$

∴ The three lowest values of x are 4, 5, 6.

7. Given that c is an integer, find the three greatest values which satisfy the inequality:

$$\frac{1}{3}(c-1) > \frac{2}{5}(3c+4)$$

Solution

$$\frac{1}{3}(c-1) > \frac{2}{5}(3c+4)$$

This can also be written as follows:

$$\frac{c-1}{3} > \frac{2(3c+4)}{5}$$

Cross multiply to obtain:

$5(c-1) > 3 \times 2(3c+4)$

$5c - 5 > 6(3c+4)$

$5c - 5 > 18c + 24$

Collect like terms. This gives:

$5c - 18c > 24 + 5$

$-13c > 29$

∴ $\qquad c < \dfrac{29}{-13}$ $\qquad$ (Note that the inequality sign has been reversed due to the division by −13)

Hence, $c < -2\frac{3}{13}$

Hence, if $c < -2\frac{3}{13}$, and c is an integer, then c can be any of the numbers below.

$c = -3, -4, -5, -6 \ldots\ldots$

Hence the three greatest values of c are, −3, −4, −5.

8. Given that r is an integer, find the three lowest values of r which satisfy the following inequality: $\frac{3}{8} - \frac{1}{2}(3r - 1) < \frac{1}{4}(2 - r)$

Solution

$$\frac{3}{8} - \frac{1}{2}(3r - 1) < \frac{1}{4}(2 - r)$$

Multiply each term by 8 (i.e. the LCM of the denominators). This gives:

$$8\left(\frac{3}{8}\right) - \frac{8(3r - 1)}{2} < \frac{8(2 - r)}{4}$$

$3 - 4(3r - 1) < 2(2 - r)$ (Note that 8 has been used to divide each denominator)

$3 - 12r + 4 < 4 - 2r$

Collect like term to give:

$-12r + 2r < 4 - 4 - 3$

$-10r < -3$

$r > \frac{-3}{-10}$ (Take note of the reversal of the inequality sign)

$r > \frac{3}{10}$

Since r is an integer (whole number), then r starts from 1 because 1 is the nearest whole number greater than $\frac{3}{10}$

∴ The three lowest values of r are 1, 2, 3.

9. What is the range of values of y for which $4y - 7 \leq 3y$ and $3y \leq 5y + 8$ are both satisfied? Show your result on a graph.

Solution

We solve each of the inequality to obtain two values of y. This is as shown below.

$4y - 7 \leq 3y$

$4y - 3y \leq 7$

$y \leq 7$

And

$3y \leq 5y + 8$

$3y - 5y \leq 8$

$-2y \leq 8$

$y \geq \frac{8}{-2}$ (Take note of the reversal of the inequality sign due to division by negative number)

∴ $y \geq -4$

Hence, $y \leq 7$ and $y \geq -4$

Combining the two inequalities gives:

$-4 \leq y \leq 7$

When combining the solutions of inequalities, we ensure that the appropriate part of the inequality sign is facing the variable (i.e. y). In the solution, $y \leq 7$, the elbow part of the inequality sign is facing y. This was maintained in the combination above. In the solution y ≥ –4, the open part of the inequality sign is facing y. This was also maintained in the combination above.

Hence, both inequalities are satisfied if: $-4 \leq y \leq 7$

This can be represented on a line graph as shown below.

10. Express 3y – 2 < 10 + y < 2 + 5y in the form a < y < b, where a and b are both integers. Show the range of values on a line graph.

<u>Solution</u>

      3y – 2 < 10 + y < 2 + 5y

We can take out two inequalities from the question above and solve. They are:

      3y – 2 < 10 + y ....Inequality (1)

And, 10 + y < 2 + 5y ....Inequality (2)

 Let us solve each of them.

      3y – 2 < 10 + y ....Inequality (1)

     3y – y < 10 + 2

       2y < 12

        $y < \dfrac{12}{2}$

        y < 6

Now let us solve the second inequality.

      10 + y < 2 + 5y ....Inequality (2)

     y – 5y < 2 – 10

      –4y < –8

       $y > \dfrac{-8}{-4}$

       y > 2

Therefore the two solutions are y < 6 and y > 2

These two solutions show that y ranges from 2 to 6.

Hence the solutions can be combined to give:

     2 < y < 6

This can be represented on a line graph as shown below.

11. Find the range of values of y for which 2y + 1 > 21 or y + 5 ≤ 12 is satisfied. Show your result on a line graph.

<u>Solution</u>

Let us solve the first inequality as follows:

$$2y + 1 > 21$$
$$2y > 21 - 1$$
$$2y > 20$$
$$y > \frac{20}{2}$$
$$y > 10$$

The other inequality is solved as follows:

$$y + 5 \leq 12$$
$$y \leq 12 - 5$$
$$y \leq 7$$

Hence the inequalities are satisfied by either $y > 10$ or $y \leq 7$

Therefore, the line graph is as given below.

12. Express $3m - 5 < 4m + 1 < 13 + m$ in the form $a < y < b$, where a and b are both integers.

<u>Solution</u>

$$3m - 5 < 4m + 1 < 13 + m$$

Two inequalities that can be obtained from the question above are:

$$3m - 5 < 4m + 1 \ldots\ldots\ldots\ldots\text{Inequality (1)}$$

And, $4m + 1 < 13 + m\ldots\ldots\ldots\ldots\text{Inequality (2)}$

Let us solve each of them.

$$3m - 5 < 4m + 1 \ldots\ldots\ldots\ldots\text{Inequality (1)}$$
$$3m - 4m < 1 + 5$$
$$-m < 6$$

Divide both sides by −1 and reverse the sign. This gives:

$$\frac{-m}{-1} > \frac{6}{-1}$$
$$m > -6$$

The second inequality is solved as follows:

$$4m + 1 < 13 + m \ldots\ldots\ldots\ldots\ldots\text{Inequality (2)}$$
$$4m - m < 13 - 1$$
$$3m < 12$$

Hence, $m < \dfrac{12}{3}$

$$m < 4$$

Therefore the two solutions are $m > -6$ and $m < 4$.

These two solutions show that y ranges from −6 to 4.

Hence the solutions can be combined to give:

$$-6 < m < 4$$

This can be represented on a line graph as shown below

$$-6 \qquad\qquad 0 \qquad 4 \quad y$$

Application of Inequalities (Introduction to Linear programming)

Inequality can be applied in real life problems. This process is called linear programming. Linear programming is a mathematical method which is used to determine the best solution of a problem whose requirements have been presented using linear inequalities. A more common and direct method of solving problems involving linear inequalities (linear programming) is by graphical method.

However, we are going to use the method of solving simultaneous equations in solving linear programming.

Note that in problems involving inequalities, we use $\leq$ for 'at most' (which also means not more than), while we use $\geq$ for 'at least' (which also means not less than).

Examples

1. To employ workers in a baking company, a man needs at least 6 mixers and 10 bakers. He does not want to employ more than 25 workers altogether. A mixer needs 2 tables to work on and a baker needs 1 table. There are only 40 tables available.
a. Write down four inequalities that represent the restriction on the man.
b. What is the greatest number of bakers that he can employ
c. Determine the optimum number of mixers and bakers the man should employ.

Solutions

a. Let x represent the number of mixers and y the number of bakers. This gives:

$\quad x \geq 6$ (This mean at least 6 mixers)

$\quad y \geq 10$ (This means at least 10 bakers)

Total number of tables required by mixers = $2x$ (Since each mixer needs 2 tables)

Total number of tables required by bakers = y   (Since each baker needs 1 tables)

Hence, total table is given by:

$\quad 2x + y \leq 40$ (Since total table is not more than 40)

Total number of staff is given by:

$\quad x + y \leq 25$ (Since the man does not want to employ more than 25 workers)

Therefore, the four inequalities that are required are:

$\quad x \geq 6$

$\quad y \geq 10$

$\quad 2x + y \leq 40$

$\quad x + y \leq 25$

b. The greatest number of bakers he can employ will be achieved if the lowest number of mixers is employed. From, $x \geq 6$, the lowest number of mixers he can employ is 6. Recall that the total number of workers is given by:

$x + y \leq 25$

Substitute $x = 6$ into the inequality above in order to obtain y (bakers). This gives:

$x + y \leq 25$

$6 + y \leq 25$

$y \leq 25 - 6$

$y \leq 19$

Therefore the greatest number of bakers he can employ is 19.

c. In order to employ an optimum number of workers, the man must employ a number of workers (up to 25) that will use all or almost all the tables.

The values y = 19 is a maximum number of bakers. So, let us plug in $x = 6$ and y = 19 into $2x + y \leq 40$, and see if the maximum number of table will be used. This gives:

$2x + y \leq 40$

$2(6) + 19 \leq 40$

$12 + 19 \leq 40$

$31 \leq 40$

This is true (because 31 is actually less than 40), but a maximum number of tables have not been used since we still have 9 (40 − 31 = 9) tables left. Hence let us use the other extreme values for number of workers. This is obtained when y = 10 (from y ≥ 10). Hence when y = 10, we obtain x as follows:

$x + y \leq 25$

$x + 10 \leq 25$

$x \leq 25 - 10$

$x \leq 15$

Note that extreme values in this example means one variable is at its lowest possible value while the other is at its highest possible value. Note also that y ≥ 10 means that the lowest possible value of y is 10.

Let us now find out if the values $x = 15$ and y = 10, will make use of a maximum number of table. Hence substitute $x = 15$ and y = 10 in the inequality for number of tables. This gives:

$2x + y \leq 40$

$2(15) + 10 \leq 40$

$30 + 10 \leq 40$

$40 \leq 40$

Hence, this is true, and the maximum number of tables is used. Hence this number of workers gives us the optimum number of workers that the man should employ.

Hence the man should employ 15 mixers and 10 bakers.

A more direct method of obtaining the optimum number of workers is to make the two inequalities to become equations and solve the two equations simultaneously. This is done as follows:

$2x + y = 40$ ....Equation(1)

$x + y = 25$..................Equation (2)

Equation (1) − Equation (2): x = 15   (Note that 40 − 25 = 15)

Hence, $x = 15$

Substitute 15 for x in equation (2)

$x + y = 25$ ....Equation (2)

$15 + y = 25$

$y = 25 - 15$

$y = 10$

Hence, $x = 15$ and $y = 10$

Therefore, the man should employ 15 mixers and 10 bakers.

2. A shop sell large number of laptops, costing $300 each, and television, costing $400 each. The shop owner allows $30,000 to be spent in buying these electronics and needs at least twice as many laptops as televisions. There must be at least 50 laptops and 20 televisions.

a. What is the highest number of electronics that can be bought and in what way?

b. If it is decided to buy as many televisions as conditions allow, how many of each electronic can be bought?

Solutions

a. Let m represent number of laptops and let n represent number of television.

Hence, total cost of laptops is 300m, while total cost of televisions is 400n. Since their combined cost should not exceed $30,000, then we have that:

$300m + 400n \leq 30000$

At least twice as many laptops as televisions can be represented as follows:

$m \geq 2n$

At least 50 laptops and 20 televisions can be represented as follows:

$m \geq 50$ (At least 50 laptops)

$n \geq 20$ (At least 20 televisions)

Therefore, the inequalities formed by this problem are:

$300m + 400n \leq 30000$

$m \geq 2n$

$m \geq 50$

$n \geq 20$

In this problem, the lowest possible value of m is 50 (from $m \geq 50$) and the lowest possible value of n is 20 (from $n \geq 20$).

Hence, when m = 50, we obtain the highest possible value of n from the total cost inequality as follows:

$300m + 400n \leq 30000$

$300(50) + 400n \leq 30000$

$15000 + 400n \leq 30000$

$400n \leq 30000 - 15000$

$400n \leq 15000$

$n \leq \dfrac{15000}{400}$

$n \leq 37.5$

$n \leq 37$ (take the whole number since number of electronics must be whole number)

Hence, when m = 50, n = 37.

Also, when n = 20 (lowest possible value of n), we obtain the highest possible value of m from the total cost inequality as follows:

$$300m + 400n \leq 30000$$
$$300m + 400(20) \leq 30000$$
$$300m + 8000 \leq 30000$$
$$300m \leq 30000 - 8000$$
$$300m \leq 22000$$
$$m \leq \frac{22000}{300}$$
$$m \leq 73.3$$
$$m \leq 73 \quad \text{(take the whole number since number of electronics must be whole number)}$$

Hence, when n = 20, m = 73

Now let us compare the two solutions obtained in order to know which one will give the highest number of electronics.

When m = 50 and n = 37, the total electronics that can be bought is = 50 + 37 = 87

When n = 20 and m = 73, the total electronics that can be bought is = 20 + 73 = 93

Therefore the highest number of electronics that can be bought is 93. This is obtained by buying 73 laptops and 20 televisions.

b. Recall the two extreme points obtained above: When m = 50, n = 37 and when n = 20, m = 73. Let us take the values involving many televisions, i.e. m = 50 and n = 37, and see if it satisfies the two inequalities above. Let us put m = 50 and n = 37 in the inequality: $m \geq 2n$. This gives:

$$m \geq 2n$$
$$50 \geq 2(37)$$
$$50 \geq 74 \quad \text{(This condition is not true)}$$

Hence we have to look for another set of maximum condition for television, i.e. the value of n. A good way of obtaining this set of values is to make the two inequalities above to become equations and solve the two equations simultaneously. This is done as follows:

$$300m + 400n = 30000 \dots\dots\dots\dots\text{Equation (1)}$$
$$m = 2n \dots\dots\dots\dots\text{Equation (2)}$$

Substitute 2n for m in equation (1).

$$300m + 400n = 30000 \dots\dots\dots\dots\text{Equation (1)}$$
$$300(2n) + 400n = 30000$$
$$600n + 400n = 30000$$
$$1000n = 30000$$
$$n = \frac{30000}{1000}$$
$$n = 30$$

Now substitute 30 for n in equation 2. This gives:

$$m = 2n$$
$$m = 2(30)$$
$$m = 60$$

Hence, m = 60 and n = 30.

Therefore, if it is decided to buy as many televisions as condition allows, then 30 televisions should be bought and 60 laptops should be bought.

3. A man needs at least three shoes and three belts. A shoe cost $60 and a belt cost $36, and the man has $360 to spend. The man decides to spend as much as possible of his $360.

a. How many ways can he spend his money, and how many of each item can he buy in each case?

b. Does any of the ways leave him with any extra money? If so, how much?

Solutions

a. Let x represent shoe and let y represent belt. We now form inequalities as follows.

$x \geq 3$ (At least 3 shoes)

$y \geq 3$ (At least 3 belts)

Cost of shoes = $60x$, while cost of belts = 36y. The total cost of the items should not be greater than 360, since the man has $360 to spend. This gives:

$60x + 36y \leq 360$

Now let us substitute $x = 3$, and obtain a value of y. This gives:

$60x + 36y \leq 360$

$60(3) + 36y \leq 360$

$180 + 36y \leq 360$

$36y \leq 360 - 180$

$36y \leq 180$

$y \leq \dfrac{180}{36}$

$y \leq 5$

Hence, when $x = 3$, y = 5

Let us substitute y = 3, and obtain a value of x. This gives:

$60x + 36y \leq 360$

$60x + 36(3) \leq 360$

$60x + 108 \leq 360$

$60x \leq 360 - 108$

$60x \leq 252$

$x \leq \dfrac{252}{60}$

$x \leq 4.2$

$x \leq 4$ (Take only the whole number since x must be a whole number)

Hence, when y = 3, $x = 4$

Therefore the man can spend his money in two ways. They are: $x = 3$ and y = 5, or $x = 4$ and y = 3.

This means that the man can buy 3 shoes and 5 belts, or 4 shoes and 3 belts

b. The condition where $x = 4$ (This is actually 4.2) and y = 3, leaves him with extra money. This is due to the 0.2 left out of 4.2 to obtain 4.

Hence, let us obtain the cost of 4 shoes and 3 belts in order to obtain the extra money. This gives:

$60x + 36y$

$= 60(4) + 36(3)$

$= 240 + 108$

= 348

Hence the extra money that the man will have is: 360 − 348 = 12

Therefore, this condition (4 shoes and 3 belts) leaves the man with extra money of $12.

4. A patient is required to take drugs containing iron and calcium. Two types of drugs available are Fero drugs and Calco drugs. Fero contains 160mg of iron, Calco contains 40mg of iron and the required iron for a day should be at least 840mg. Also, Fero contains 4mg of calcium, Calco contains 3mg of calcium and the required calcium for a day should be at least 31mg. A daily prescription is not up to 10 or more drugs.

a. Write the inequalities to be satisfied by these conditions.

b. What prescriptions should be given for each of the drugs?

Solutions

a. Let y represents the number of Fero drugs and z represents the number of Calco drugs. Hence the total iron from Fero is 160y and the total iron from Calco is 40z. This gives a total iron of 160y + 40z. Since the total daily iron should be at least 840mg, then this gives:

$160y + 40z \geq 840$

Similarly, the total calcium from Fero is 4y and the total calcium from Calco is 3z. This gives a total calcium of 4y + 3z. Since the total daily calcium should be at least 31mg, then this gives:

$4y + 3z \geq 31$

Also, since the total daily prescription is not up to 10 or more drugs, we have:

$y + z < 10$

Therefore, the inequalities to be satisfied by these conditions are:

$160y + 40z \geq 840$

$4y + 3z \geq 31$

$y + z < 10$

b. In order to obtain the right prescriptions that should give optimum number of each of the drug, we have to make the three inequalities above to become equations and solve the right pair simultaneously. However, the pair to solve in this case is the first two equations since they have to be maximized as they contain the $\geq$ sign. This gives:

$160y + 40z = 840$ ....Equation (1)

$4y + 3z = 31$ ....Equation (2)

Solving equation 1 and 2 simultaneously gives:

$160y + 40z = 840$ ....Equation (1)

$4y + 3z = 31$ ....Equation (2)

Divide equation (1) by 40 to obtain a simplified form of the equation. This gives:

$\dfrac{160}{40}y + \dfrac{40}{40}z = \dfrac{840}{40}$ (Note that we have to divide each term by 40)

$4y + z = 21$ ....Equation (4)

Solving equation 4 and 2 simultaneously gives:

$4y + 3z = 31$ ....Equation (2)

$\underline{4y + z = 21}$ ....Equation (4)

Equation (2) – Equation (4): $2z = 10$

$$z = \frac{10}{2}$$

$$z = 5$$

Substitute 5 for z in equation (4). This gives:

$4y + z = 21$ ....Equation (4)

$4y + 5 = 21$

$4y = 21 - 5$

$4y = 16$

$$y = \frac{16}{4}$$

$$y = 4$$

Hence $y = 4$ and $z = 5$

Therefore, 4 Fero and 5 Calco should be prescribed.

5. A tailoring company needs to buy a total of 10 sewing machines. A manual sewing machine cost $250, while an electric sewing machine cost $450. The company must buy at least two of each machine and it is prepared to spend no more than $3,500. The profit on each manual sewing machine is not greater than $35, while that on each electric sewing machine is not greater than $50. Find:

a. the maximum profit the company can make in the business

b. the number of each sewing machine that the company should buy.

Solution

a. This is a case where profit is involved. Hence we have to maximize profit.

Let the number of manual sewing machine be x, and the number of electric sewing machine be y. We now form inequalities using x and y as follows.

$x \geq 2$ (At least two of each machine)

$y \geq 2$ (At least two of each machine)

$x + y = 10$ (The company needs a total of 10 machines)

The cost of buying x manual sewing machines at $250 each is $250x$. The cost of buying y electric sewing machines at $450 each is $450y$. Since the company is not prepared to spend more than $3,500, then we have a total cost inequality as follows:

$250x + 450y \leq 3500$

Also, since the profit on each manual sewing machine is $35, and the profit on each electric sewing machine is $50, then the total profit that the company can make is given by:

$35x + 50y = P$ (Where P is the maximum profit that the company can make)

We are not going to use this profit equation to solve for x and y. We are only going to use it to find out which values of x and y will give us a maximum profit.

Let us bring down our inequalities below. They are:

$x \geq 2$

$y \geq 2$

$x + y = 10$

$250x + 450y \leq 3500$

Let us obtain our extreme values first. The lowest possible value of x is 2 (from $x \geq 2$). When x = 2, we can obtain a value for y from $x + y = 10$. This gives y as follows:

$x + y = 10$

$2 + y = 10$

$y = 10 - 2 = 8$

Hence when $x = 2$, y = 8.

Similarly, the lowest possible value of y is 2. If we substitute y = 2 in $x + y = 10$, we will obtain x = 8.

Hence the two solutions we have obtained so far are:

$x = 2$ and y = 8, or $x = 8$ and y = 2.

Now let us test these solutions in the cost inequality to see if they are feasible.

The cost inequality is: $250x + 450y \leq 3500$

Hence when $x = 2$ and y = 8, we have:

$250(2) + 450(8) \leq 3500$

$500 + 3600 \leq 3500$

$4100 \leq 3500$ (This is not true since 4100 is not less than 3500)

Hence $x = 2$ and y = 8 is not a possible solution.

Let us try the other solution which is $x = 8$ and y = 2. We now substitute these values as follows:

$250(8) + 450(2) \leq 3500$

$2000 + 900 \leq 3500$

$2900 \leq 3500$ (This is true)

Hence $x = 8$ and y = 2 is a possible solution

We can also obtain another solution by solving the following equations simultaneously:

$250x + 450y = 3500$ ....Equation (1)

$x + y = 10$ ....Equation (2)

From equation (2), $x = 10 - y$ ....Equation (3)

Substitute $10 - y$ for x in equation (1). This gives:

$250x + 450y = 3500$ ....Equation (1)

$250(10 - y) + 450y = 3500$

$2500 - 250y + 450y = 3500$

$200y = 3500 - 2500$

$200y = 1000$

$y = \dfrac{1000}{200}$

$y = 5$

From equation (3), we obtain x as follows:

$x = 10 - y$ ....Equation (3)

$x = 10 - 5$

$x = 5$

Hence $x = 5$ and y = 5.

Therefore, the two possible solutions to this problem are:

$x = 8$ and y = 2 or $x = 5$ and y = 5

Now the solution that gives a maximum profit becomes the actual solution. The maximum profit equation is:

$35x + 50y = P$

When $x = 8$ and $y = 2$, the maximum profit is given by:

$\quad$ $35(8) + 50(2) = P$

$\quad$ $280 + 100 = P$

$\quad$ $P = 380$

When $x = 5$ and $y = 5$, the maximum profit is given by:

$\quad$ $35(5) + 50(5) = P$

$\quad$ $175 + 250 = P$

$\quad\quad$ $P = 425$ (This is greater than the first profit)

Hence the maximum profit that the company can make is $425

b. The company should buy 5 of each of the sewing machine.

6. A company plans to produce at most 90 units of two products X and Y. The cost of producing each unit of X is $50, while the cost of producing each unit of Y is $100. After production, the time needed to package each unit of X is 1 hour, while the time needed to package each unit of Y is 3 hours. The profit from each unit of X and Y are $25 and $60 respectively. The company has a total budget of $5,000 and a packaging period of not more than 120 hours.

a. What optimal quantity of X and Y should the company produce?

b. What is the maximum profit that the company will make from this production?

Solution

a. Let us obtain the inequalities for this problem. If X represents the total units of X produced and Y represents the total units of Y produced, then:

$\quad$ $X + Y \leq 90$ $\quad$ (The company plans to produce at most (not more than) 90 units of X and Y)

The cost of producing X at $50 each is 50X. The cost of producing Y at $100 each is 100Y. Since the company's total budget is $5,000, then we have a total cost inequality as follows:

$\quad$ $50X + 100Y \leq 5000$

Also, The packaging time for X at 1 hour each is X hours. The packaging time for Y at 3 hours each is 3Y. Since the company's total packaging time is not more than 120 hours, then we have a total time inequality as follows:

$\quad$ $X + 3Y \leq 120$

The profit on each of X is $25 and the profit on each of Y is $60. Then the total profit that the company can make is given by:

$\quad$ $25X + 60Y = P$ $\quad$ (Where P is the maximum profit that the company can make)

Hence, the three inequalities obtained from this problem are:

$\quad$ $X + Y \leq 90$

$\quad$ $50X + 100Y \leq 5000$

$\quad$ $X + 3Y \leq 120$

In this case, the two inequalities that we have to make to become equations and solve them simultaneously are the ones that involve the production processes. They are:

$\quad$ $50X + 100Y = 5000$ ....Equation (1)

$\quad$ $X + 3Y = 120$ ....Equation (2)

From equation (2), $X = 120 - 3Y$ ....Equation (3)

167

Substitute 120 − 3Y for X in equation (1). This gives:

　　50X + 100Y = 5000 ....Equation (1)
　　50(120 − 3Y) + 100Y = 5000
　　6000 − 150Y + 100Y = 5000
　　6000 − 5000 = 150Y − 100Y
　　1000 = 50Y
　　$Y = \dfrac{1000}{50}$
　　Y = 20

Substitute 20 for Y in equation (3). This gives:

　　X = 120 − 3Y ....Equation (3)
　　X = 120 − 3(20)
　　　= 120 − 60

　　X = 60

Hence, X = 60 and Y = 20.

It is necessary to use the third inequality to test if this solution is true.

This gives: X + Y ≤ 90

　　60 + 20 ≤ 90
　　80 ≤ 90　　(This is a true solution)

Therefore, the company should produce 60 units of X and 20 units of Y.

b. The maximum profit that the company will make from this production will be obtained from the profit equation given above. Let us substitute 60 for X and 20 for Y in the profit equation. This gives:

　　25X + 60Y = P
　　25(60) + 60(20) = P
　　1500 + 1200 = P
　∴　P = 2700

Therefore, the maximum profit that the company can make is $2700

7. Solve the linear program below:

　　Maximize: 7m + 4n
　　Subject to: m + n ≤ 9
　　　　　　　　m − n ≥ 3
　　and   10m + 8n ≤ 66

Given that, m ≥ 0 and n ≥ 0

Solution

We have to first ignore 7m + 4n, which is the function to maximize.

In the three inequalities given in this question, our correct solution will be obtained by solving simultaneously two out of the three equations that we will convert them to. We do not know the two that will give us the solution. Hence, we take two at a time and solve them simultaneously. The solution obtained will be substituted into the third inequality to see if it is the right solution.

Let us bring down the three inequalities as follows.

$m + n \leq 9$ ....Inequality (1)

$m - n \geq 3$ ....Inequality (2)

$10m + 8n \leq 66$ ....Inequality (3)

Converting them to equations gives:

$m + n = 9$ ....Equation (1)

$m - n = 3$ .... Equation (2)

$10m + 8n = 66$ .... Equation (3)

Let us solve equations (1) and (2) simultaneously as follows:

$m + n = 9$ ....Equation (1)

$m - n = 3$ .... Equation (2)

Equation (1) – (2): $2n = 6$ (Note that $n - (-n) = n + n = 2n$, and $9 - 3 = 6$)

$n = \dfrac{6}{2}$

$n = 3$

Substitute 3 for n in equation (1). This gives:

$m + n = 9$ ....Equation (1)

$m + 3 = 9$

$m = 9 - 3$

$m = 6$

Hence, m = 6 and n = 3.

Let us now substitute these values in inequality (3), to see if this solution is correct

$10m + 8n \leq 66$ ....Inequality (3)

$10(6) + 8(3) \leq 66$

$60 + 24 \leq 66$

$84 \leq 66$ (This is not correct)

Hence, m = 6 and n = 3 are not correct solutions to this problem.

Let us solve equation (1) and (3) simultaneously as follows:

$m + n = 9$ ....Equation (1)

$10m + 8n = 66$ .... Equation (3)

From equation (1), $m = 9 - n$.

Substitute $9 - n$ for m in equation (3). This gives:

$10m + 8n = 66$ .... Equation (3)

$10(9 - n) + 8n = 66$

$90 - 10n + 8n = 66$

$90 - 2n = 66$

$90 - 66 = 2n$

$24 = 2n$

$n = \dfrac{24}{2}$

$n = 12$

Substitute 12 for n in equation (1). This gives:

$m + n = 9$ ....Equation (1)

$m + 12 = 9$

169

m = 9 − 12

m = −3

But m cannot be a negative value since m ≥ 0

Hence, m = −3 and n = 12 is not a correct solution.

Finally, let us solve equations (2) and (3) simultaneously as follows:

m − n = 3 .... Equation (2)

10m + 8n = 66 .... Equation (3)

From equation (2), m = 3 + n.

Substitute 3 + n for m in equation (3). This gives:

10m + 8n = 66 .... Equation (3)

10(3 + n) + 8n = 66

30 + 10n + 8n = 66

30 + 18n = 66

18n = 66 − 30

$n = \dfrac{36}{18}$

n = 2

Substitute 2 for n in equation (2). This gives:

m − n = 3 ....Equation (2)

m − 2 = 3

m = 3 + 2

m = 5

Hence m = 5 and n = 2

Let us now substitute these values in inequality (1), to see if this solution is correct

m + n ≤ 9 ....Inequality (1)

5 + 2 ≤ 9

7 ≤ 9   (This is true)

Hence, m = 5 and n = 2 is the correct solution.

Therefore we now maximize 7m + 4n as follows:

7m + 4n = 7(5) + 4(2)

= 35 + 8

= 43

8. Solve the linear program:

Minimize: 8x + 5y

Subject to: x + y ≥ 11

x − y ≤ 5

7x + 12y ≥ 35

x ≥ 0 and y ≥ 0

Solution

We have to first ignore 8x + 5y, which is the function to minimize.

In the three inequalities, our correct solution will be obtained by solving simultaneously two out of the three equations that will convert them to. We do not know the two that will give us the solution. Hence, we take two at a time and solve them simultaneously. The solution obtained will be substituted into the third inequality to see if it is the right solution.

Let us bring down the three inequalities as follows.

$x + y \geq 11$ ....Inequality (1)

$x - y \leq 5$ ....Inequality (2)

$7x + 12y \geq 35$ ....Inequality (3)

Converting them to equations gives:

$x + y = 11$ ....Equation (1)

$x - y = 5$ ....Equation (2)

$7x + 12y = 35$ ....Equation (3)

Let us solve equations (1) and (2) simultaneously as follows:

$x + y = 11$ ....Equation (1)

$\underline{x - y = 5}$ .... Equation (2)

Equation (1) – (2): $2y = 6$

$y = \dfrac{6}{2}$

$y = 3$

Substitute 3 for y in equation (1). This gives:

$x + y = 11$ ....Equation (1)

$x + 3 = 11$

$x = 11 - 3$

$x = 8$

Hence, x = 8 and y = 3.

Let us now substitute these values in inequality (3), to see if this solution is correct

$7x + 12y \geq 35$ ....Inequality (3)

$7(8) + 12(3) \geq 35$

$56 + 36 \geq 35$

$92 \geq 35$ (This is correct)

Hence, x = 8 and y = 3 are the correct solutions to this problem.

Note that if we take other pairs of equations and solve simultaneously, we will not obtain feasible solutions.

Therefore the only correct solution is x = 8 and y = 3.

Hence, we now minimize 8x + 5y as follows:

$8x + 5y = 8(8) + 5(3)$

$= 64 + 15$

$= 79$

Exercise 13

1. Solve the inequality $9x - 12 > 2 + 2x$. Represent your result on a number line/line graph.

2. Solve the inequality $5x + 8 < 5x + 2(3 - x)$. Represent your result on a number line.

3. Find the range of values of x for which $\dfrac{3x+1}{4} - \dfrac{3}{5}x \le 11 - 2x$. Show the result on a line graph.

4. Find the range of values of x for which $\dfrac{3}{4}x - \dfrac{2-3x}{2} \ge \dfrac{1}{2} + \dfrac{21}{8}x$. Show the result on a line

5. Given that m is an integer, find the three greatest values which satisfy the inequality:

$$5(7m+3) < \dfrac{2}{5}(9m-2)$$

6. Given that x is an integer, find the three lowest values of x which satisfy the following inequality: $\quad 3\left(2x-\dfrac{3}{4}\right) - \dfrac{1}{2}(x+3) > \dfrac{1}{3}(5x-2)$

7. Given that c is an integer, find the three greatest values which satisfy the inequality:

$$\dfrac{3}{10}(3c-4) < \dfrac{4}{5}(c+7)$$

8. Given that r is an integer, find the three lowest values of r which satisfy the following inequality:

$$\dfrac{5}{6} - \dfrac{3}{2}(4r-9) < \dfrac{5}{12}(3-r)$$

9. What is the range of values of y for which $5y - 4 \le 3y$ and $9y \le 15y + 6$ are both satisfied? Show your result on a graph.

10. Express $7y - 4 < 11 + 2y < 5y + 17$ in the form $a < y < b$, where a and b are both integers. Show the range of values on a line graph.

11. Find the range of values of y for which $4y - 1 > -5$ or $5 \ge y + 12$ are both satisfied. Show your result on a line graph.

12. Express $5m - 9 < 7m - 17 < 23 - m$ in the form $a < y < b$, where a and b are both integers.

13. To employ workers in a farm, a man needs at least 5 drivers and 2 supervisors. He does not want to employ more than 10 workers altogether. A driver needs 2 cars to drive and a supervisor needs 1 car. There are only 15 cars available.
a. Write down four inequalities that represent the restriction on the man.
b. What is the greatest number of supervisors that he can employ
c. Determine the optimum number of drivers and supervisors the man should employ.

14. A shop sells bicycles, costing $200 each, and tricycles, costing $1300 each. The shop owner allows $21,000 to be spent in buying these items and needs at least three times as many bicycles as tricycles. There must be at least 30 bicycles and 10 tricycles.
a. What is the highest number of items that can be bought and in what way?
b. If it is decided to buy as many tricycles as conditions allow, how many of each item can be bought?

15. A boy needs at least five shirts and three trousers. A shirt cost $120 and a trouser cost $150, and the boy has $1600 to spend. The boy decides to spend as much as possible of his $1600.
a. In what way can he spend the maximum of his money?
b. How much is his balance from (a) above?

16. A paint seller produces two types of tins of paints each containing indigo and cyan. The two types of tins of paints that he makes are Fux tin and Clusion tin. Fux tin contains $320cm^3$ of indigo, Clusion tin contains $80cm^3$ of indigo and the required indigo for a day production should be at least $1680cm^3$. Also, Fux tin contains $8cm^3$ of cyan, Clusion tin contains $6cm^3$ of cyan and the required cyan for a day production should be at least $62cm^3$. A day's production is not more than 20 tins

a. Write the inequalities to be satisfied by these conditions.
b. What production quantity should be made for each of the tin?

17. In a mixing process, there is need to buy a total of 15 mixing machines. A manual mixing machine cost \$8000, while an electric mixing machine cost \$12000. The process must use at least three of each machine and the money to be spent is no more than \$80,000. The profit on each manual mixing machine is not greater than \$600, while that on each electric sewing machine is not greater than \$500. Find:

a. the maximum profit the process can make in the business
b. the number of each mixing machine that should be bought for the process.

18. A company plans to produce at most 120 units of two products A and B. The cost of producing each unit of A is \$60, while the cost of producing each unit of B is \$90. After production, the time needed to package each unit of A is 2 hour, while the time needed to package each unit of B is 5 hours. The profit from each unit of A and B are \$20 and \$40 respectively. The company has a total budget of \$8,700 and a packaging period of not more than 400 hours.

a. What optimal quantity of A and B should the company produce?
b. What is the maximum profit that the company will make from this production?

19. Solve the linear program below:
Maximize: $2a + 3b$
Subject to: $a + b \leq 25$
$\qquad a - b \geq 8$
and $\quad 8a + 6b \leq 180$
Given that, $a \geq 0$ and $b \geq 0$

20. Solve the linear program:
Minimize $9m + 4n$
Subject to: $m + n \geq 14$
$\qquad m - n \leq 1$
$\qquad 11m + 7n \geq 130$
$\qquad m \geq 0$ and $n \geq 0$

CHAPTER 14
QUADRATIC INEQUALITY

If we have a factorized quadratic inequality such as $(x + 2)(x - 1) > 0$, then the product $(x + 2)(x - 1)$ has to be positive (> 0). This means that either $(x + 2)$ and $(x - 1)$ are both positive or both negative.

If $(x + 2)(x - 1) < 0$, then the product $(x + 2)(x - 1)$ has to be negative (< 0). This means that only $(x + 2)$ is negative or only $(x - 1)$ is negative.

Therefore it is necessary to test the solutions of a quadratic inequality in order to know the correct solution.

However, there is a general rule that applies to quadratic inequalities (and even other inequalities having two solutions). If the coefficient of x^2 in a quadratic inequality is a positive value, then of the two values of the solutions obtained, the larger value will take the original sign ($<$ or $>$) of the inequality, while the smaller value will take the reverse sign of the inequality. The following examples illustrate this rule.

Examples

1. Solve the inequality: $x^2 - 2x - 8 < 0$

Solution

$$x^2 - 2x - 8 < 0$$

Let us equate this inequality to zero and solve by factorization.

$$\therefore \quad x^2 - 2x - 8 = 0$$
$$(x - 4)(x + 2) = 0$$
$$\therefore \quad x = 4 \text{ or } x = -2$$

In $x^2 - 2x - 8 < 0$, the coefficient of x^2 is positive. Therefore, between the solutions $x = 4$ and $x = -2$, the larger value i.e. 4 will take the sign of the inequality in the question, i.e. $<$. The lower solution i.e. -2 will take $>$ (i.e. the reverse sign).

Therefore, the solutions is $x < 4$ or $x > -2$. These can be combined to give:

$$-2 < x < 4$$

Hence, $x^2 - 2x - 8 < 0$ is satisfied when $-2 < x < 4$

We can test these solutions to see if our rule and values obtained are correct. When $x < 4$, a possible value is 3 since $3 < 4$. We now substitute $x = 3$ in the original equation as follows:

$$x^2 - 2x - 8 < 0$$
$$(3)^2 - 2(3) - 8 < 0$$
$$9 - 6 - 8 < 0$$
$$-5 < 0 \quad \text{(This is true)}$$

Hence, $x < 4$ is a correct solution to the inequality.

Similarly, when $x > -2$, a possible value is -1 since $-1 > -2$. We now substitute $x = -1$ in the original inequality as follows:

$$x^2 - 2x - 8 < 0$$
$$(-1)^2 - 2(-1) - 8 < 0$$
$$1 + 2 - 8 < 0$$

$-5 < 0$ (This is true)

Hence, $x > -2$ is also a correct solution of the inequality.

2. Solve the inequality $2x^2 + 3x - 5 \geq 0$

Solution

$\qquad 2x^2 + 3x - 5 \geq 0$

Equating to zero and solving by factorization gives:

$\qquad 2x^2 + 3x - 5 = 0$

$\qquad 2x^2 + 5x - 2x - 5 = 0$

$\qquad x(2x + 5) - 1(2x + 5) = 0$

$\therefore \qquad (2x + 5)(x - 1) = 0$

This gives: $x = -\dfrac{5}{2}$ or $x = 1$

Of these two values, the larger solution i.e. 1 will take the inequality sign in the question, while $-\dfrac{5}{2}$ will take the reverse sign. Therefore the solution is $x \geq 1$ or $x \leq -\dfrac{5}{2}$

Hence, $2x^2 + 3x - 5 \geq 0$ is satisfied when $x \geq 1$ or $x \leq -\dfrac{5}{2}$

3. Solve the inequality: $3x^2 - 9x + 4 \leq 0$

Solution

$\qquad 3x^2 - 9x + 4 = 0$ (Equating to zero first)

This equation cannot be factorized. So let's solve it by using quadratic formula as follows:

$\qquad 3x^2 - 9x + 4 = 0$

$\qquad a = 3, b = -9, c = 4$

$\therefore \qquad x = \dfrac{-b \pm \sqrt{b^2 - 4ac}}{2a}$

$\qquad = \dfrac{-(-9) \pm \sqrt{(-9)^2 - (4 \times 3 \times 4)}}{2 \times 3}$

$\qquad = \dfrac{9 \pm \sqrt{81 - 48}}{6}$

$\qquad = \dfrac{9 \pm \sqrt{33}}{6}$

$\qquad x = \dfrac{9 + 5.74}{6}$ or $x = \dfrac{9 - 5.74}{6}$

$\qquad x = \dfrac{14.74}{6}$ or $x = \dfrac{3.26}{6}$

$\therefore \qquad x = 2.46 \quad x = 0.54$

Therefore, $x \leq 2.46$ or $x \geq 0.54$ (Note that the larger value takes the original inequality sign)

Hence, $3x^2 - 9x + 4 \leq 0$ is satisfied when $0.54 \leq x \leq 2.46$

4. Solve the inequality $7 + 6x - x^2 > 0$

<u>Solution</u>

$7 + 6x - x^2 > 0$ (Note that the coefficient of x^2 is negative, i.e. -1)

$7 + 6x - x^2 = 0$ (This can be factorized)

$(7 - x)(1 + x) = 0$

$\therefore \quad x = 7$ or $x = -1$

Since the coefficient of x^2 in this inequality is negative, then the smaller value of the solution i.e. -1 will take the sign of the inequality in the question, while the larger value i.e. 7 will take the reverse sign. This is exactly the opposite of what we do when the coefficient of x^2 is positive like in examples 1 to 3.

$\therefore \quad x > -1$ or $x < 7$

Hence, $7 + 6x - x^2 > 0$ is satisfied when $-1 < x < 7$

5. Solve the inequality $5 + 8x - 2x^2 \geq 0$

<u>Solution</u>

$5 + 8x - 2x^2 \geq 0$

Equating to zero and solving by using quadratic formula gives:

$5 + 8x - 2x^2 = 0$

$a = -2, b = 8, c = 5$

$\therefore \quad x = \dfrac{-b \pm \sqrt{b^2 - 4ac}}{2a}$

$= \dfrac{-8 \pm \sqrt{8^2 - (4 \times -2 \times 5)}}{2(-2)}$

$= \dfrac{-8 \pm \sqrt{64 + 40}}{-4}$

$= \dfrac{-8 \pm \sqrt{104}}{-4}$

$x = \dfrac{-8 + 10.2}{-4}$ or $x = \dfrac{-8 - 10.2}{-4}$

$x = -0.55$ or $x = 4.55$

Since the coefficient of x^2 in this inequality is negative, then the smaller value will take the inequality sign in the question.

$\therefore \quad x \geq -0.55$ or $x \leq 4.55$

Hence $5 + 8x - 2x^2 \geq 0$ is satisfied when $-0.55 \leq x \leq 4.55$

Exercise 14

1. Solve the inequality: $x^2 - x - 30 < 0$
2. Solve the inequality $3x^2 + 10x - 8 \geq 0$
3. Solve the inequality: $5x^2 - 9x + 4 \leq 0$
4. Solve the inequality $9 + 9x - 10x^2 > 0$
5. Solve the inequality $11 + 2x - 2x^2 \geq 0$

6. Solve the inequality: $2x^2 - 5x - 18 < 0$
7. Solve the inequality $7x^2 + x - 6 \geq 0$
8. Solve the inequality: $5x^2 + 17x + 14 \leq 0$
9. Solve the inequality $15 + 2x - x^2 > 0$
10. Solve the inequality $22 + 9x - 4x^2 \geq 0$

CHAPTER 15
INTRODUCTORY VECTOR ALGEBRA

Representation of a Vector

A vector is a quantity which has magnitude and direction. A scalar quantity has only magnitude. A vector can be represented by a line such as AB. This line usually has a corresponding x and y values in the Cartesian plane.

A column vector can be written as $\begin{pmatrix} x \\ y \end{pmatrix}$.

Modulus or Magnitude of a Vector

The modulus or magnitude of a vector can be determined by using Pythagoras theorem as follows:

$|AB| = \sqrt{x^2 + y^2}$ (where AB is the vector).

A zero or null vector represented as $\begin{pmatrix} 0 \\ 0 \end{pmatrix}$ has a magnitude of zero.

Two vectors are parallel if they have the same direction. They are equal if they have the same magnitude.

Examples

1. What is the magnitude of the vector AB = $\begin{pmatrix} 5 \\ 12 \end{pmatrix}$

$$|AB| = \sqrt{5^2 + 12^2}$$
$$= \sqrt{25 + 144}$$
$$= \sqrt{169}$$

$\therefore$ $|AB|$ = 13 units

2. If the magnitude of the vector $\begin{pmatrix} x \\ 8 \end{pmatrix}$ is 17, find the value of x.

Solution

$$\text{Magnitude} = \sqrt{x^2 + 8^2}$$
$$17 = \sqrt{x^2 + 64}$$

Square both sides in order to remove the square root sign. This gives:

$$17^2 = (\sqrt{x^2 + 64})^2$$
$$\therefore \quad 289 = x^2 + 64$$
$$x^2 = 289 - 64$$
$$x^2 = 225$$

Take the square root of both sides. This gives:

$$\sqrt{x^2} = \sqrt{225}$$
$$x = 15$$

Vector Algebra

Examples

(1) Given that $p = \begin{pmatrix} -3 \\ 4 \end{pmatrix}$, $q = \begin{pmatrix} -2 \\ -5 \end{pmatrix}$, $r = \begin{pmatrix} 4 \\ 1 \end{pmatrix}$, find:

a. $p - q + r$

b. $2p + q - 3r$

c. $\frac{1}{5}(p + q) - \frac{1}{2}r$

d. $|3p + q - 2r|$

<u>Solution</u>

a. $p - q + r = \begin{pmatrix} -3 \\ 4 \end{pmatrix} - \begin{pmatrix} -2 \\ -5 \end{pmatrix} + \begin{pmatrix} 4 \\ 1 \end{pmatrix}$

$\qquad = \begin{pmatrix} -3 - (-2) + 4 \\ 4 - (-5) + 1 \end{pmatrix}$

$\qquad = \begin{pmatrix} -3 + 2 + 4 \\ 4 + 5 + 1 \end{pmatrix}$

$\qquad = \begin{pmatrix} 3 \\ 10 \end{pmatrix}$

In solving this problem, simply combine the upper numbers separately and the lower numbers separately. Make sure the signs are well organized. For example, where $4 - (-5)$ should be written, should not be written as $4 - 5$.

`

b. $2p + q - 3r = 2\begin{pmatrix} -3 \\ 4 \end{pmatrix} + \begin{pmatrix} -2 \\ -5 \end{pmatrix} - 3\begin{pmatrix} 4 \\ 1 \end{pmatrix}$

$\qquad = \begin{pmatrix} -6 \\ 8 \end{pmatrix} + \begin{pmatrix} -2 \\ -5 \end{pmatrix} - \begin{pmatrix} 12 \\ 3 \end{pmatrix}$

$\qquad = \begin{pmatrix} -6 - 2 - 12 \\ 8 - 5 - 3 \end{pmatrix}$

$\qquad = \begin{pmatrix} -20 \\ 0 \end{pmatrix}$

Note that a number multiplying a bracket is used to multiply each term in the bracket.

c. $\frac{1}{5}(p + q) - \frac{1}{2}r = \frac{1}{5}\left[\begin{pmatrix} -3 \\ 4 \end{pmatrix} + \begin{pmatrix} -2 \\ -5 \end{pmatrix}\right] - \frac{1}{2}\begin{pmatrix} 4 \\ 1 \end{pmatrix}$

$\qquad = \frac{1}{5}\begin{pmatrix} -3 - 2 \\ 4 - 5 \end{pmatrix} - \begin{pmatrix} 2 \\ \frac{1}{2} \end{pmatrix}$

$\qquad = \frac{1}{5}\begin{pmatrix} -5 \\ -1 \end{pmatrix} - \begin{pmatrix} 2 \\ \frac{1}{2} \end{pmatrix}$

$\qquad = \begin{pmatrix} -1 \\ -\frac{1}{5} \end{pmatrix} - \begin{pmatrix} 2 \\ \frac{1}{2} \end{pmatrix}$

$\qquad = \begin{pmatrix} -1 - 2 \\ -\frac{1}{5} - \frac{1}{2} \end{pmatrix}$

$\qquad = \begin{pmatrix} -3 \\ -\frac{7}{10} \end{pmatrix}$ (Note that $-\frac{1}{5} - \frac{1}{2} = -\frac{7}{10}$)

d. In order to determine $|3p + q - 2r|$, we have to first determine $3p + q - 2r$

$$\therefore \quad 3p + q - 2r = 3\binom{-3}{4} + \binom{-2}{-5} - 2\binom{4}{1}$$
$$= \binom{-9}{12} + \binom{-2}{-5} - \binom{8}{2}$$
$$= \binom{-9 - 2 - 8}{12 - 5 - 2}$$
$$3p + q - 2r = \binom{-19}{5}$$

$$\therefore \quad |3p + q - 2r| = \sqrt{(-19)^2 + 5^2} \quad \text{(Note that we are now calculating the magnitude of } \binom{-19}{5})$$
$$= \sqrt{361 + 25}$$
$$= \sqrt{386}$$
$$= 19.6 \text{ units}$$

2. If $a = \binom{3}{1}$, $b = \binom{2}{-1}$, $c = \binom{8}{6}$, find the values of:

a. m and n such that $ma + nb = c$

b. $|d|$ if $d = 2b - a$

<u>Solution</u>

(a) $ma + nb = c$, means that:

$$m\binom{3}{1} + n\binom{2}{-1} = \binom{8}{6}$$
$$\binom{3m}{m} + \binom{2n}{-n} = \binom{8}{6}$$
$$\binom{3m + 2n}{m - n} = \binom{8}{6}$$

By comparing the upper part of the vectors, and then the lower part, two equations can be obtained as shown below:

$3m + 2n = 8$ ....Equation (1)

$m - n = 6$ ....Equation (2)

Equation (2) x 2 gives:

$2m - 2n = 12$ ....Equation (3)

Bringing equations (1) and (3) together in order to solve them simultaneously gives:

$3m + 2n = 8$ ....Equation (1)

$2m - 2n = 12$ ....Equation (3)

Equation (1) + Equation (3): $5m \quad = 20$ (n has been eliminated, since $+2n + -2n = 0$

$$m = \frac{20}{5}$$
$$m = 4$$

Substitute 4 for m in equation (2)

$m - n = 6$ ....Equation (2)

$4 - n = 6$

$4 - 6 = n$

$n = -2$

$\therefore \quad n = -2$ and $m = 4$

b.	$d = 2b - a$

$$= 2\binom{2}{-1} - \binom{3}{1}$$

$$= \binom{4}{-2} - \binom{3}{1}$$

$$= \binom{4 - 3}{-2 - 1}$$

$$d = \binom{1}{-3}$$

∴	$|d| = \sqrt{1^2 + (-3)^2}$

$$= \sqrt{1 + 9}$$

$$|d| = \sqrt{10} \text{ units}$$

Direction of a Vector

If the position vector of a point A is $(x_1 \; y_1)$, and that of point B is $(x_2 \; y_2)$, then the component of the vector AB is given by:

$$AB = (x_2 - x_1, \; y_2 - y_1)$$

This can be written in column vector as:

$$AB = \binom{x_2 - x_1}{y_2 - y_1}$$

If a vector is expressed as $CD = \binom{x}{y}$, then the angle that CD makes with the x-axis (horizontal axis) is given by:

$$\text{Tan } \theta = \frac{y}{x}$$

∴	$\theta = \tan^{-1}\frac{y}{x}$

Examples

1. If $u = \binom{3}{2}$ and $v = \binom{-2}{5}$

a.	Find $u - 2v$

b.	Express u in the form $(k, \theta°)$ where k is magnitude of u and θ is the angle u makes with the x-axis.

Solution

a.	$u - 2v = \binom{3}{2} - 2\binom{-2}{5}$

$$= \binom{3}{2} - \binom{-4}{10}$$

$$= \binom{3 - -4}{2 - 10}$$

$$= \binom{3 + 4}{2 - 10}$$

$$= \binom{7}{-8}$$

b.	$u = \binom{3}{2}$

The angle that u makes with the x axis is given by:

$$\text{Tan } \theta = \frac{y}{x}$$

$$Tan\ \theta = \frac{2}{3}$$ (Note that the upper number of the vector is x while the lower number is y)

$Tan\ \theta = 0.6667$

$\therefore$ $\theta = tan^{-1}\ 0.6667$

$\theta = 33.7^0$ (From the use of calculator)

Also, $|u| = \sqrt{3^2 + 2^2}$ [Note that u = $\binom{3}{2}$]

$= \sqrt{9 + 4}$

$= \sqrt{13}$ units

$\therefore$ u in the form (k, θ) is given by: $(\sqrt{13}, 33.7^0)$

2. If b = $\binom{-2}{5}$, find the angle θ that b makes with the horizontal axis. Hence express b in the form (u, θ^0), where u is the magnitude of b.

Solution

b = $\binom{-2}{5}$,

The angle that b makes with the horizontal is given by:

$Tan\ \theta = \frac{y}{x}$

$Tan\ \theta = \frac{5}{-2}$ (Note that in the vector$\binom{-2}{5}$, −2 represents x, while 5 represents y)

$Tan\ \theta = \frac{5}{2}$ (Ignore the negative sign of −2)

$\theta = tan^{-1}\ 2.5$

$\therefore$ $\theta = 68.2^0$

Also, $|b| = u = \sqrt{(-2)^2 + 5^2}$ [Since b = $\binom{-2}{5}$]

$= \sqrt{4 + 25}$

$u = \sqrt{29}$

$\therefore$ Expressing b in the form (u, θ^0), gives: $(\sqrt{29}, 68.2^0)$

3. The vertices of a quadrilateral WXYZ are W(2, −1), X(−1, 3), Y(−2, 1), Z(−3, −2). Determine the components and lengths of:
a. WX b. WZ c. −2XY d. YW

Solutions
a. With respect to the origin, each of the points can be expressed as:

W = $\binom{2}{-1}$, X = $\binom{-1}{3}$, Y = $\binom{-2}{1}$, Z = $\binom{-3}{-2}$

$\therefore$ WX = X − W   (Take note of the arrangement of the letters, as the second letter comes first)

$= \binom{-1}{3} - \binom{2}{-1}$

$$= \begin{pmatrix} -1-2 \\ 3--1 \end{pmatrix}$$

$$= \begin{pmatrix} -1-2 \\ 3+1 \end{pmatrix}$$

$\therefore$ WX = $\begin{pmatrix} -3 \\ 4 \end{pmatrix}$

Also, the length of WX is the magnitude of WX. It is given by:

$|WX| = \sqrt{(-3)^2 + 4^2}$
$\quad\quad = \sqrt{9+16}$
$\quad\quad = \sqrt{25}$
$\quad\quad = 5$

$\therefore \quad |WX| = 5$ units

b. Similarly, WZ = Z – W   (Take note of the arrangement of the letters, as the second letter comes first in the subtraction)

$WZ = \begin{pmatrix} -3 \\ -2 \end{pmatrix} - \begin{pmatrix} 2 \\ -1 \end{pmatrix}$

$\quad\quad = \begin{pmatrix} -3-2 \\ -2--1 \end{pmatrix}$

$\quad\quad = \begin{pmatrix} -3-2 \\ -2+1 \end{pmatrix}$

$\therefore$ WZ = $\begin{pmatrix} -5 \\ -1 \end{pmatrix}$

Also, the length of WZ is the magnitude of WZ. It is given by:

$|WZ| = \sqrt{(-5)^2 + (-1)^2}$
$\quad\quad = \sqrt{25+1}$
$|WZ| = \sqrt{26}$ units

c. $-2XY = -2(Y - X)$
$\quad\quad\quad = -2[\begin{pmatrix} -2 \\ 1 \end{pmatrix} - \begin{pmatrix} -1 \\ 3 \end{pmatrix}]$
$\quad\quad\quad = -2[\begin{pmatrix} -2--1 \\ 1-3 \end{pmatrix}]$
$\quad\quad\quad = -2\begin{pmatrix} -2+1 \\ -2 \end{pmatrix}$
$\quad\quad\quad = -2\begin{pmatrix} -1 \\ -2 \end{pmatrix}$
$\quad\quad\quad = \begin{pmatrix} 2 \\ 4 \end{pmatrix}$ (Take note of the change in sign due to the multiplication by –2)

$\therefore \quad -2XY = \begin{pmatrix} 2 \\ 4 \end{pmatrix}$

Also, the length of –2XY is its magnitude. It is given by:

$\therefore \ |-2XY| = \sqrt{2^2 + 4^2}$
$\quad\quad\quad\quad = \sqrt{4+16}$
$\quad\quad\quad\quad = \sqrt{20}$

$\therefore$ $|-2XY| = \sqrt{20}$ units

d. YW = W − Y

$$YW = \begin{pmatrix} 2 \\ -1 \end{pmatrix} - \begin{pmatrix} -2 \\ 1 \end{pmatrix}$$

$$= \begin{pmatrix} 2 - -2 \\ -1 - 1 \end{pmatrix}$$

$$= \begin{pmatrix} 2 + 2 \\ -1 - 1 \end{pmatrix}$$

$\therefore$ $YW = \begin{pmatrix} 4 \\ -2 \end{pmatrix}$

Also, $|YW| = \sqrt{4^2 + (-2)^2}$

$$= \sqrt{16 + 4}$$

$\therefore$ $|YW| = \sqrt{20}$ units

4. A(4, 7) is the vertex of a triangle ABC. BA = $\begin{pmatrix} 5 \\ 3 \end{pmatrix}$ and AC = $\begin{pmatrix} 4 \\ -3 \end{pmatrix}$.

a. Find the coordinates of B and C

b. If M is the midpoint of the line BC, find AM.

<u>Solution</u>

a. With respect to the origin, A = $\begin{pmatrix} 4 \\ 7 \end{pmatrix}$

Let the column vector B be, B = $\begin{pmatrix} d \\ e \end{pmatrix}$

$\therefore$ The component, BA = A − B.   Substitute each value into this equation. This gives:

BA = A − B

$$\begin{pmatrix} 5 \\ 3 \end{pmatrix} = \begin{pmatrix} 4 \\ 7 \end{pmatrix} - \begin{pmatrix} d \\ e \end{pmatrix}$$

$$\begin{pmatrix} 5 \\ 3 \end{pmatrix} = \begin{pmatrix} 4 - d \\ 7 - e \end{pmatrix}$$

$\therefore$ 5 = 4 − d   (By equating the upper parts)

d = 4 − 5

d = −1

Similarly, 3 = 7 − e   (By equating the lower parts)

e = 7 − 3

e = 4

Hence, B = $\begin{pmatrix} -1 \\ 4 \end{pmatrix}$

$\therefore$ The coordinates of B are (−1, 4)

Also, let the column vector C be, C = $\begin{pmatrix} f \\ g \end{pmatrix}$

$\therefore$ The component:

AC = C − A

$$\begin{pmatrix} 4 \\ -3 \end{pmatrix} = \begin{pmatrix} f \\ g \end{pmatrix} - \begin{pmatrix} 4 \\ 7 \end{pmatrix}$$

$$\begin{pmatrix} 4 \\ -3 \end{pmatrix} = \begin{pmatrix} f - 4 \\ g - 7 \end{pmatrix}$$

$\therefore$ 4 = f − 4   (By equating the upper parts)

$$4 + 4 = f$$
$$f = 8$$

Similarly, $-3 = g - 7$ (By equating the lower parts)
$$-3 + 7 = g$$
$$g = 4$$

Hence, $C = \begin{pmatrix} 8 \\ 4 \end{pmatrix}$

∴ The coordinates of C are (8, 4)

b. The midpoint of BC is M = ½(B + C)

$$M = \frac{1}{2}\left[\begin{pmatrix} -1 \\ 4 \end{pmatrix} + \begin{pmatrix} 8 \\ 4 \end{pmatrix}\right]$$

$$= \frac{1}{2}\left[\begin{pmatrix} -1 + 8 \\ 4 + 4 \end{pmatrix}\right]$$

$$= \frac{1}{2}\begin{pmatrix} 7 \\ 8 \end{pmatrix}$$

$$M = \begin{pmatrix} \frac{7}{2} \\ 4 \end{pmatrix}$$

∴ AM = M − A

$$= \begin{pmatrix} \frac{7}{2} \\ 4 \end{pmatrix} - \begin{pmatrix} 4 \\ 7 \end{pmatrix}$$

$$= \begin{pmatrix} \frac{7}{2} - 4 \\ 4 - 7 \end{pmatrix}$$

$$AM = \begin{pmatrix} -\frac{1}{2} \\ -3 \end{pmatrix}$$

Exercise 15

1. What is the magnitude of the vector AB = $\begin{pmatrix} 8 \\ 6 \end{pmatrix}$

2. If the magnitude of the vector $\begin{pmatrix} x \\ 5 \end{pmatrix}$ is 13, find the value of x.

3. Given that $p = \begin{pmatrix} -6 \\ 10 \end{pmatrix}$, $q = \begin{pmatrix} -3 \\ 1 \end{pmatrix}$, $r = \begin{pmatrix} 2 \\ -8 \end{pmatrix}$, find:

a. $p + q - 2r$

b. $p - q + r$

c. $\frac{2}{3}(p - q) - \frac{1}{4}r$

d. $|2p + 5q - 3r|$

4. If $a = \begin{pmatrix} 5 \\ 2 \end{pmatrix}$, $b = \begin{pmatrix} 3 \\ -8 \end{pmatrix}$, $c = \begin{pmatrix} -4 \\ -3 \end{pmatrix}$, find the values of:

a. p and q such that pa − qb = 2c

b. $|m|$ if m = 5a − 3b

5. If $u = \begin{pmatrix} 1 \\ 7 \end{pmatrix}$ and $v = \begin{pmatrix} 3 \\ -2 \end{pmatrix}$

a. Find $4u - 5v$

b. Express u in the form (K, θ°) where k is magnitude of u and θ is the angle u makes with the x– axis.

6. If $n = \begin{pmatrix} 6 \\ -4 \end{pmatrix}$, find the angle θ that n makes with the horizontal axis. Hence express n in the form (u, θ°), where u is the magnitude of n.

7. The vertices of a quadrilateral ABCD are A(3, –2), B(1, 2), C(–1, 3), D(–3, –4). Determine the components and lengths of:

a. AB b. AD c. –3BC d. BD

8. P(2, 5) is the vertex of a triangle PQR. $QP = \begin{pmatrix} 3 \\ -1 \end{pmatrix}$ and $PR = \begin{pmatrix} 6 \\ 4 \end{pmatrix}$.

a. Find the coordinates of Q and R

b. If M is the midpoint of the line QR, find PM.

9. What is the magnitude of the vector $AB = \begin{pmatrix} 24 \\ 17 \end{pmatrix}$

10. If the magnitude of the vector $\begin{pmatrix} x \\ 14 \end{pmatrix}$ is 20, find the value of x.

11. Given that $p = \begin{pmatrix} -1 \\ 5 \end{pmatrix}$, $q = \begin{pmatrix} -4 \\ 7 \end{pmatrix}$, $r = \begin{pmatrix} 2 \\ 3 \end{pmatrix}$, find:

a. $2p - 3q + r$

b. $p + q - r$

c. $\frac{1}{2}(p + 2q) - \frac{1}{4}r$

d. $|5p + 3q - 4r|$

12. If $a = \begin{pmatrix} 2 \\ 5 \end{pmatrix}$, $b = \begin{pmatrix} 3 \\ -3 \end{pmatrix}$, $c = \begin{pmatrix} 5 \\ 6 \end{pmatrix}$, find the values of:

a. p and q such that $pa + qb = 2c$

b. $|d|$ if $d = 2a - 3b$

13. The vertices of a quadrilateral PQRS are P(3, –2), Q(–5, 1), R(–1, 3), S(–2, –4). Determine the components and lengths of:

a. PR b. RQ c. –5QS d. 2PQ

CHAPTER 16
LIMIT OF A FUNCTION

The concept of limits is very important in differential calculus. A function has a limiting value when its variable approaches a certain value. For example, if the limiting value of f(x) as x approaches 4 is 16, it is written as:

$$\lim_{x \to 4} f(x) = 16$$

In evaluating limit, some problems are simply solved by just putting in the values of the variables, while some problems are solved by applying certain rules.

Some important limits

1. $\lim\limits_{x \to 0} \dfrac{\sin x}{x} = 1$ and $\lim\limits_{x \to 0} \dfrac{\tan x}{x} = 1$ (Or $\lim\limits_{x \to 0} \dfrac{\sin ax}{ax} = 1$ and $\lim\limits_{x \to 0} \dfrac{\tan ax}{ax} = 1$)

2. $\lim\limits_{x \to 0} \dfrac{1 - \cos x}{x} = 0$ and $\lim\limits_{x \to 0} \dfrac{1 - \cos ax}{x} = 0$

3. $\lim\limits_{x \to \infty} \left(1 + \dfrac{1}{x}\right)^x = e$

4. $\lim\limits_{x \to 0} \dfrac{a^x - 1}{a^x} = 1$

5. $\lim\limits_{x \to 0} \dfrac{e^x - 1}{x} = 1$

6. $\lim\limits_{x \to 0} \dfrac{1 + x}{x} = 1$

7. $\lim\limits_{x \to a} \dfrac{x^n - a^n}{x - a} = na^{n-1}$

8. $\lim\limits_{x \to 0} \dfrac{a^x - 1}{x} = \ln a$

9. $\lim\limits_{x \to \infty} \dfrac{\ln x}{x} = 0$

10. $\lim\limits_{x \to \infty} x^{\frac{1}{x}} = 0$

11. $\lim\limits_{x \to 0} (1 + x)^{\frac{1}{x}} = e$

12. $\lim\limits_{x \to 0} (1 + \sin x)^{\frac{1}{x}} = e$

13. $\lim\limits_{x \to a} c = c$

Examples

1. Evaluate $\lim_{x \to 0} 3x^3 - 5x^2 + 2x + 6$

<u>Solution</u>

$\lim_{x \to 0} 3x^3 - 5x^2 + 2x + 6$

Simply substitute zero for x in the given expression. This gives:

$= 3(0^3) - 5(0^2) + 2(0) + 6$

$= 0 - 0 + 0 + 6$

$= 6$

2. Evaluate $\lim_{x \to 0} \dfrac{x^2 + 3x - 11}{4x^2 - 5x - 5}$

<u>Solution</u>

$\lim_{x \to 0} \dfrac{x^2 + 3x - 11}{4x^2 - 5x - 5}$

Substituting 0 for x into the expression gives:

$= \dfrac{0^2 + 3(0) - 11}{4(0)^2 - 5(0) - 5}$

$= \dfrac{0 + 0 - 11}{0 - 0 - 5}$

$= \dfrac{-11}{-5}$

$= \dfrac{11}{5}$

$= 2\dfrac{1}{5}$

3. Evaluate $\lim_{x \to 4} \dfrac{x^2 - 2x - 8}{x - 4}$

<u>Solution</u>

A close look at the expression shows that if 4 is substituted for x, it will give $\dfrac{0}{0}$. This has no value as it is indeterminate. Therefore, in order to solve a limit such as this, we have to factorize the denominator and then simplify the expression. This is done as follows:

$\lim_{x \to 4} \dfrac{x^2 - 2x - 8}{x - 4} = \lim_{x \to 4} \dfrac{(x+2)(x-4)}{x - 4}$

Cancelling out $(x - 4)$ gives:

$\lim_{x \to 4} (x + 2)$

We now substitute 4 for x to obtain our final answer as follows:

$\lim_{x \to 4} (x + 2) = 4 + 2$

$= 6$

Therefore, $\lim_{x \to 4} \dfrac{x^2 + x - 8}{x - 4} = 6$

4. Evaluate $\lim\limits_{x \to 2} \dfrac{x^2 - 4}{x - 2}$

Solution

This is similar to example 3 above since substituting 2 for x in the expression will $\dfrac{0}{0}$ which has no value. Hence we factorize the numerator and simplify as follows:

$\lim\limits_{x \to 2} \dfrac{x^2 - 4}{x - 2} = \lim\limits_{x \to 2} \dfrac{(x-2)(x+2)}{x - 2}$ [Note that $x^2 - 4 = x^2 - 2^2$ and recall that $a^2 - b^2 = (a + b)(a - b)$]

Cancelling out $(x - 2)$ gives:

$\lim\limits_{x \to 2} (x + 2)$

We now substitute 2 for x to obtain our final answer as follows:

$\lim\limits_{x \to 2} (x + 2) = 2 + 2$

$\qquad\qquad\qquad = 4$

Therefore, $\lim\limits_{x \to 2} \dfrac{x^2 - 4}{x - 2} = 4$

5. Evaluate $\lim\limits_{x \to 0} (2x - 5)(x + 2)(3x - 1)$

Solution

$\lim\limits_{x \to 0} (2x - 5)(x + 2)(3x - 1)$

Substituting 0 for x in the expression gives:

$\lim\limits_{x \to 0} (2x - 5)(x + 2)(3x - 1) = [2(0) - 5][(0) + 2][3(0) - 1]$

$\qquad\qquad\qquad\qquad\qquad\qquad = (-5)(2)(-1)$

$\qquad\qquad\qquad\qquad\qquad\qquad = 10$

6. Evaluate $\lim\limits_{x \to \infty} \dfrac{x^3 + 2x^2 + 5x + 3}{x^3 + x^2 + 3x + 7}$

Solution

In this case of x approaching infinity, we evaluate it by first dividing each term in the numerator and denominator by the variable having the highest exponent (power). Hence we divide each term by x^3. This is done as follows:

$\lim\limits_{x \to \infty} \dfrac{x^3 + 2x^2 + 5x + 3}{x^3 + x^2 + 3x + 7} = \lim\limits_{x \to \infty} \dfrac{\frac{x^3}{x^3} + \frac{2x^2}{x^3} + \frac{5x}{x^3} + \frac{3}{x^3}}{\frac{x^3}{x^3} + \frac{x^2}{x^3} + \frac{3x}{x^3} + \frac{7}{x^3}}$

$\qquad\qquad\qquad\qquad\qquad = \lim\limits_{x \to \infty} \dfrac{1 + \frac{2}{x} + \frac{5}{x^2} + \frac{3}{x^3}}{1 + \frac{1}{x} + \frac{3}{x^2} + \frac{7}{x^3}}$

$\qquad\qquad\qquad\qquad\qquad = \dfrac{1 + \frac{2}{\infty} + \frac{5}{\infty^2} + \frac{3}{\infty^3}}{1 + \frac{1}{\infty} + \frac{3}{\infty^2} + \frac{7}{\infty^3}}$

$$= \frac{1+0+0+0}{1+0+0+0} \qquad \text{(Note that a number divided by } \infty \text{ gives 0)}$$

$$= \frac{1}{1}$$

$$= 1$$

7. Evaluate $\displaystyle \lim_{x \to \infty} \frac{2x^4 - 3x^2 + 8x - 1}{x^4 - 5x + 3}$

Solution

The variable in its highest exponent (power) is x^4. Hence we divide each term by x^4 as follows:

$$\lim_{x \to \infty} \frac{2x^4 - 3x^2 + 8x - 1}{x^4 - 5x + 3} = \lim_{x \to \infty} \frac{\frac{2x^4}{x^4} - \frac{3x^2}{x^4} + \frac{8x}{x^4} - \frac{1}{x^4}}{\frac{x^4}{x^4} - \frac{5x}{x^4} + \frac{3}{x^4}}$$

$$= \lim_{x \to \infty} \frac{2 - \frac{3}{x^2} + \frac{8}{x^3} - \frac{1}{x^4}}{1 - \frac{5}{x^3} + \frac{3}{x^4}}$$

$$= \frac{2 - \frac{3}{\infty} + \frac{8}{\infty} - \frac{1}{\infty}}{1 - \frac{5}{\infty} + \frac{3}{\infty}}$$

$$= \frac{2 - 0 + 0 - 0}{1 - 0 + 0} \qquad \text{(Note that a number divided by } \infty \text{ gives 0)}$$

$$= \frac{2}{1}$$

$$= 2$$

8. Evaluate $\displaystyle \lim_{x \to 4} \frac{2x^2 - 7x - 4}{x^2 - 3x - 4}$

Solution

If 4 is substituted for x in the function above, it will give $\frac{0}{0}$ which is an indeterminate value.

Hence we factorize the expression and simplify as follows:

$$\lim_{x \to 4} \frac{2x^2 - 7x - 4}{x^2 - 3x - 4} = \lim_{x \to 4} \frac{(2x+1)(x-4)}{(x+1)(x-4)}$$

Cancelling out $(x - 4)$ gives:

$$\lim_{x \to 4} \frac{2x + 1}{x + 1}$$

We now substitute 4 for x to obtain our final answer as follows:

$$\lim_{x \to 4} \frac{2x + 1}{x + 1} = \frac{2(4) + 1}{(4) + 1}$$

$$= \frac{8+1}{4+1}$$

$$= \frac{9}{5}$$

$$= 1\frac{4}{5}$$

Therefore, $\displaystyle \lim_{x \to 4} \frac{2x^2 - 7x - 4}{x^2 - 3x - 4} = 1\frac{4}{5}$

9. Evaluate $\displaystyle \lim_{x \to 3} \frac{x^2 - 9}{x^2 + 6x + 9}$

Solution

Substituting 3 for x in the function can give us the final answer as follows:

$$\lim_{x \to 3} \frac{x^2 - 9}{x^2 + 6x + 9} = \frac{3^2 - 9}{3^2 + 6(3) + 9}$$

$$= \frac{9 - 9}{9 + 18 + 9}$$

$$= \frac{0}{36}$$

$$= 0$$

Hence, $\displaystyle \lim_{x \to 3} \frac{x^2 - 9}{x^2 + 6x + 9} = 0$

Note that even if this expression had been factorized and simplified, the final answer would still give zero.

10. Determine the limiting value of $\dfrac{x + 2}{x^2 + 6x + 8}$ as x tends to -2 .

Solution

The question can also be written as $\displaystyle \lim_{x \to -2} \frac{x + 2}{x^2 + 6x + 8}$

If -2 is substituted for x in the function above, it will give $\dfrac{0}{0}$. Hence we factorize the expression and simplify as follows:

$$\lim_{x \to -2} \frac{x + 2}{x^2 + 6x + 8} = \lim_{x \to -2} \frac{x + 2}{(x + 2)(x + 4)}$$

Cancelling out $(x + 2)$ gives:

$$\lim_{x \to -2} \frac{1}{x + 4}$$

We now substitute -2 for x as follows:

$$\lim_{x \to -2} \frac{1}{x + 4} = \frac{1}{-2 + 4}$$

$$= \frac{1}{2}$$

11. Find the limit of $\dfrac{x^3 - 125}{x - 5}$ as $x \to 5$

Solution

The question can also be written as $\displaystyle\lim_{x \to 5} \dfrac{x^3 - 125}{x - 5}$

We factorize the expression and simplify as follows:

$$\lim_{x \to 5} \dfrac{x^3 - 125}{x - 5} = \lim_{x \to 5} \dfrac{x^3 - 5^3}{x - 5}$$

$$= \lim_{x \to 5} \dfrac{(x - 5)(x^2 + 5x + 5^2)}{x - 5} \quad \text{[Recall the identity } a^3 - b^3 = (a - b)(a^2 + ab + b^2)]$$

Cancelling out $(x - 5)$ gives:

$$\lim_{x \to 5} (x^2 + 5x + 5^2)$$

We now substitute 5 for x as follows:

$$\lim_{x \to 5} (x^2 + 5x + 5^2) = (5^2 + 5(5) + 25)$$

$$= 25 + 25 + 25$$

$$= 75$$

12. Find the limit of $\dfrac{x^3 + 64}{x + 4}$ as $x \to -4$

Solution

The question can also be written as $\displaystyle\lim_{x \to -4} \dfrac{x^3 + 4^3}{x + 4}$

We factorize the expression and simplify as follows:

$$\lim_{x \to -4} \dfrac{x^3 + 64}{x + 4} = \lim_{x \to -4} \dfrac{x^3 + 4^3}{x + 4}$$

$$= \lim_{x \to -4} \dfrac{(x + 4)(x^2 - 4x + 4^2)}{x + 4} \quad \text{[Recall the identity } a^3 + b^3 = (a + b)(a^2 - ab + b^2)]$$

Cancelling out $(x + 4)$ gives:

$$\lim_{x \to -4} (x^2 - 4x + 4^2)$$

We now substitute −4 for x as follows:

$$\lim_{x \to -4} (x^2 - 4x + 4^2) = (-4^2 - 4(-4) + 16)$$

$$= 16 + 16 + 16$$

$$= 48$$

13. Evaluate $\displaystyle\lim_{x \to 9} \dfrac{3 - \sqrt{x}}{9 - x}$

Solution

If 9 is substituted for x in the expression above, it will give $\dfrac{0}{0}$ which is not the desired answer.

Hence a way to go around this problem is to multiply the top and bottom by the conjugate of the surd at the top. The conjugate of $3 - \sqrt{x}$ is $3 + \sqrt{x}$ (only a difference in their middle signs).

Hence we multiply top and bottom by $3 + \sqrt{x}$ and simplify as follows:

$$\lim_{x \to 9} \frac{3 - \sqrt{x}}{9 - x} = \lim_{x \to 9} \frac{(3 - \sqrt{x})(3 + \sqrt{x})}{(9 - x)(3 + \sqrt{x})}$$

$$= \lim_{x \to 9} \frac{3^2 - (\sqrt{x})^2}{(9 - x)(3 + \sqrt{x})} \quad \text{(Note that the top was simplified by using } (a - b)(a + b) = a^2 - b^2)$$

$$= \lim_{x \to 9} \frac{9 - x}{(9 - x)(3 + \sqrt{x})}$$

$$= \lim_{x \to 9} \frac{1}{(3 + \sqrt{x})} \quad \text{(Since } 9 - x \text{ cancels out)}$$

We now substitute 9 for x as follows:

$$\lim_{x \to 9} \frac{1}{(3 + \sqrt{x})} = \frac{1}{(3 + \sqrt{9})}$$

$$= \frac{1}{3 + 3}$$

$$= \frac{1}{6}$$

14. Evaluate $\lim_{x \to 1} \dfrac{x^2 - 1}{x^5 - 1}$

Solution

If 2 is substituted for x in the expression above, it will give $\dfrac{0}{0}$ which is not a good answer. Hence a way to go around this problem is to divide the top and bottom by $x - 1$ (i.e. the difference between the variable and 1). This is done as follows:

$$\lim_{x \to 1} \frac{x^2 - 1}{x^5 - 1} = \lim_{x \to 1} \frac{\dfrac{x^2 - 1}{x - 1}}{\dfrac{x^5 - 1}{x - 1}} \quad \text{(Note that this has not changed the original fraction)}$$

Note that $\lim_{x \to a} \dfrac{x^n - a^n}{x - a} = na^{n-1}$. With this rule we now simplify the expression above as follows:

$$\lim_{x \to 1} \frac{\dfrac{x^2 - 1}{x - 1}}{\dfrac{x^5 - 1}{x - 1}} = \frac{2 \times 1^{(2 - 1)}}{5 \times 1^{(5 - 1)}}$$

$$= \frac{2 \times 1^1}{5 \times 1^5}$$

$$= \frac{2}{5}$$

Continuity of a function

A function is said to be continuous if the three conditions below are satisfied.

1. f(a) exists

2. $\lim_{x \to a} f(x)$ exists

3. $\lim_{x \to a} f(x) = f(a)$ (i.e. if the values of conditions 1 and 2 above are equal)

Examples

1. Determine if the function $f(x) = 2x^2 - 8x + 5$ is continuous at the point $x = -1$

Solution

When $x = -1$, the f(−1) is obtained as follows:

$$f(x) = 2x^2 - 8x + 5$$
$$f(-1) = 2(-1)^2 - 8(-1) + 5$$
$$= 2 + 8 + 5$$
$$= 15$$

Since f(−1) = 15, it means that f(−1) exists.

The next step is to find the value of $\lim_{x \to -1} f(x)$ as follows:

$$\lim_{x \to -1} 2x^2 - 8x + 5 = 2(-1)^2 - 8(-1) + 5$$
$$= 2 + 8 + 5$$
$$= 15$$

Hence $f(-1) = \lim_{x \to -1} f(x) = 15$

Therefore, the function is continuous.

2. Determine if the function $f(x) = \dfrac{2x + 1}{x^2 - 3x + 7}$ is continuous at $x = 2$

Solution

When $x = 2$, the f(2) is obtained as follows:

$$f(x) = \frac{2x + 1}{x^2 - 3x + 7}$$
$$f(2) = \frac{2(2) + 1}{2^2 - 3(2) + 7}$$
$$= \frac{4 + 1}{4 - 6 + 7}$$
$$= \frac{5}{5}$$
$$= 1$$

Since f(2) = 1, it means that f(2) exists.

The next step is to find the value of $\lim_{x \to 2} f(x)$ as follows:

$$\lim_{x \to 2} \frac{2x + 1}{x^2 - 3x + 7} = \frac{2(2) + 1}{2^2 - 3(2) + 7}$$
$$= \frac{4 + 1}{4 - 6 + 7}$$
$$= \frac{5}{5}$$

$$= 1$$

Hence f(2) = $\underset{x \to 2}{lim}$ f(x) = 1

Therefore, the function is continuous.

3. Determine if the function f(x) = $\dfrac{2x^2 - 8}{x + 4}$ is continuous at $x = -4$.

When $x = -4$, the f(−4) is obtained as follows:

$$f(x) = \dfrac{2x^2 - 8}{x + 4}$$

$$f(-4) = \dfrac{2(-4^2) - 8}{-4 + 4}$$

$$= \dfrac{2(16) - 8}{0}$$

$$= \dfrac{24}{0} \quad \text{(Undefined)}$$

Hence f(−4) is undefined, and does not exist. This shows that the function is not continuous. It is discontinuous as $x = -4$.

4. Determine if the function f(x) = $\dfrac{x^2 - 9}{x - 3}$ is continuous at $x = 3$.

When $x = 3$, the f(3) is obtained as follows:

$$f(x) = \dfrac{x^2 - 9}{x - 3}$$

$$f(3) = \dfrac{3^2 - 9}{3 - 3}$$

$$= \dfrac{0}{0} \quad \text{(This has no value)}$$

Hence f(3) does not exist.

Let us determine the value of $\underset{x \to 3}{lim}$f(x) as follows:

$$\underset{x \to 3}{lim} \dfrac{x^2 - 9}{x - 3} = \underset{x \to 3}{lim} \dfrac{(x - 3)(x + 3)}{x - 3}$$

$$= \underset{x \to 3}{lim} (x + 3) \quad \text{(Since } x - 3 \text{ cancels out)}$$

$$= 3 + 3 \quad \text{(When 3 is substituted for } x)$$

$$= 6$$

Hence, f(3) has no value while $\underset{x \to 3}{lim}$ f(x) = 6

This shows that f(3) $\neq$ $\underset{x \to 3}{lim}$ f(x)

Therefore the function discontinuous.

5. Test for the continuity of the function: f(x) = $\begin{cases} -5x + 7 & \text{if } x > 1 \\ x^2 + 2 & \text{if } x \leq 1 \end{cases}$

Solution

$$f(x) = \begin{cases} -5x + 7 & \text{if } x > 1 \\ x^2 + 2 & \text{if } x \leq 1 \end{cases}$$

195

When $x = 1$, we use the expression $x^2 + 2$ to evaluate f(1). Note that $x \leq 1$ means $x < 1$ or $x = 1$.

Hence, $f(x) = x^2 + 2$ (for $x = 1$)

$\qquad f(1) = 1^2 + 2$

$\qquad\qquad = 3$

Since f(1) = 3, it means that f(1) exist.

Let us now determine the limiting value of f(x).

If x approaches 1 from the left, which is $x \leq 1$ (left hand elbow) we write it as $\lim\limits_{x \to 1-} f(x)$, and we pick the expression given by:

$\qquad f(x) = x^2 + 2$

Hence, $\lim\limits_{x \to 1} f(x) = 1^2 + 3$

$\qquad\qquad = 3$

If x approaches 1 from the right, which is $x > 1$ (right hand elbow), we write it as $\lim\limits_{x \to 1+} f(x)$, and we pick the expression given by:

$\qquad f(x) = -5x + 7$

Hence, $\lim\limits_{x \to 1} f(x) = -5(1) + 7$

$\qquad\qquad = 2$

Hence, $\lim\limits_{x \to 1-} f(x) = 3$ while $\lim\limits_{x \to 1+} f(x) = 2$. They are not equal. This shows that $\lim\limits_{x \to 1} f(x)$ does not exist. Therefore, the function is not continuous.

6. Determine the continuity of the function: $f(x) = \begin{cases} x^2 & \text{if } x \geq -2 \\ x+6 & \text{if } x < -2 \end{cases}$

Solution

$\qquad f(x) = \begin{cases} x^2 & \text{if } x \geq -2 \\ x+6 & \text{if } x < -2 \end{cases}$

A direct way of determining the continuity of functions such as these is to find the limits of the two expressions in the function as x tends to the values given in the question.

Therefore, when x tends to -2 in the function $f(x) = x^2$, we have:

$\qquad f(x) = x^2$

Hence, $\lim\limits_{x \to -2} f(x) = (-2)^2$

$\qquad\qquad = 4$

Also, when x tends to -2 in the function $f(x) = x + 6$, we have:

$\qquad f(x) = x + 6$

Hence, $\lim\limits_{x \to -2} f(x) = -2 + 6$

$\qquad\qquad = 4$

Hence, $\lim\limits_{x \to -2} x^2 = \lim\limits_{x \to -2} x + 6 = 4$. Their limits give the same value. This shows that $\lim\limits_{x \to -2} f(x)$ exist.

Therefore, the function is continuous.

7. Determine if the function below is continuous or not.

$$f(x) = \begin{cases} \dfrac{x^2-16}{x+4} & \text{if } x \geq -4 \\ \dfrac{x+4}{2x-3} & \text{if } x < -4 \end{cases}$$

Solution

As x tends to -4 in the function $f(x) = \dfrac{x^2-16}{x+4}$ we simplify as follows:

$$f(x) = \frac{(x+4)(x-4)}{x+4}$$

Hence, $\underset{x \to -4}{lim} f(x) = (x-4)$ (Note that $(x+4)$ cancels out)

$$= -4 - 4$$
$$= -8$$

Also, when x tends to -4 in the function $f(x) = \dfrac{x+4}{2x-3}$, we have:

$$\underset{x \to -4}{lim} f(x) = \frac{-4+4}{2(-4)-3}$$
$$= \frac{-4+4}{-8-3}$$
$$= \frac{0}{-11}$$
$$= 0$$

Their limits give different values. This shows that $\underset{x \to -4}{lim} f(x)$ does not exist. Therefore, the function is not continuous.

8. If $f(x) = \begin{cases} \dfrac{x^2-4}{x-2} \\ 4 \text{ if } x = 2 \end{cases}$ test for the continuity of the function at $x = 2$.

Solution

$$f(x) = \begin{cases} \dfrac{x^2-4}{x-2} \\ 4 \text{ if } x = 2 \end{cases}$$

Let us find the limits of the two expressions in the function as x tends 2.
From the first expression, as x tends to 2 we have:

$$f(x) = \frac{x^2-4}{x-2}$$

Hence, $\underset{x \to 2}{lim} f(x) = \dfrac{(x-2)(x+2)}{x-2}$ (When the numerator is factorized)

$$\underset{x \to 2}{lim} f(x) = (x+2) \quad (x-2 \text{ has cancelled out})$$
$$\underset{x \to 2}{lim} f(x) = 2+2$$
$$= 4$$

Also, when x tends to 2 in the function $f(x) = 4$, we have:

$$f(x) = 4$$

Hence, $\lim\limits_{x \to 2} 4 = 4$ (Since $\lim\limits_{x \to a} c = c$)

The question also tells us that when $x = 2$, f(x) = 4

Hence, $\lim\limits_{x \to 2} 4 = \lim\limits_{x \to 2} \dfrac{x^2 - 4}{x - 2} = 4$. Their limits give the same value of 4. This shows that f(x) exist, and that f(2) = $\lim\limits_{x \to 2}$ f(x). Therefore, the function is continuous.

Examples on Limits of Trigonometric Functions

1. Evaluate $\lim\limits_{x \to 0} \dfrac{\sin 2x}{5x}$

Solution

$$\lim\limits_{x \to 0} \dfrac{\sin 2x}{5x}$$

Substituting zero directly into the expression above will give $\dfrac{0}{0}$. Hence we have to apply the appropriate rule of limit. Let us first make an adjustment to the expression as follows.

Multiply the expression by $\dfrac{2x}{5x}$ and change the denominator to $2x$ just like the numerator, as follows:

$$\lim\limits_{x \to 0} \dfrac{\sin 2x}{5x} = \lim\limits_{x \to 0} \left(\dfrac{\sin 2x}{2x} \times \dfrac{2x}{5x}\right)$$

Note that when $2x$ cancels out, the original expression remains the same.

Recall that $\lim\limits_{x \to 0} \dfrac{\sin ax}{ax} = 1$

Hence, $\lim\limits_{x \to 0} \dfrac{\sin 2x}{2x} = 1$

Substituting 1 for $\lim\limits_{x \to 0} \dfrac{\sin 2x}{2x}$ in the simplification given above to gives:

$$\lim\limits_{x \to 0} \left(\dfrac{\sin 2x}{2x} \times \dfrac{2x}{5x}\right) = \lim\limits_{x \to 0} \dfrac{\sin 2x}{2x} \times \lim\limits_{x \to 0} \dfrac{2x}{5x}$$

$$= 1 \times \lim\limits_{x \to 0} \dfrac{2}{5} \quad (x \text{ cancels out})$$

$$= 1 \times \dfrac{2}{5} \quad (\text{Since } \lim\limits_{x \to a} c = c)$$

$$= \dfrac{2}{5}$$

2. 1. Evaluate $\lim\limits_{x \to 0} \dfrac{\sin 5x}{7x}$

Solution

$$\lim\limits_{x \to 0} \dfrac{\sin 5x}{7x}$$

Multiply the expression by $\dfrac{5x}{7x}$ and change the denominator to $5x$ just like the numerator. This gives:

$$\lim_{x \to 0} \frac{\sin 5x}{7x} = \lim_{x \to 0} \left(\frac{\sin 5x}{5x} \times \frac{5x}{7x}\right)$$

Hence, $\lim_{x \to 0} \frac{\sin 5x}{5x} = 1$

Substituting 1 for $\lim_{x \to 0} \frac{\sin 5x}{5x}$ in the simplification given above to gives:

$$\lim_{x \to 0} \left(\frac{\sin 5x}{5x} \times \frac{5x}{7x}\right) = \lim_{x \to 0} \frac{\sin 5x}{5x} \times \lim_{x \to 0} \frac{5x}{7x}$$

$$= 1 \times \lim_{x \to 0} \frac{5}{7} \quad (x \text{ cancels out})$$

$$= 1 \times \frac{5}{7} \quad (\text{Since } \lim_{x \to a} c = c)$$

$$= \frac{5}{7}$$

The two examples above show that $\lim_{x \to 0} \frac{\sin a\, x}{x} = a$ or $\lim_{x \to 0} \frac{\sin a\, x}{bx} = \frac{a}{b}$. This rule also applies to tangent as $\lim_{x \to 0} \frac{\tan a\, x}{x} = a$ or $\lim_{x \to 0} \frac{\tan a\, x}{bx} = \frac{a}{b}$.

3. Find the value of $\lim_{x \to \frac{\pi}{2}} \frac{\cos 2x}{\sin 3x}$

Solution

$$\lim_{x \to \frac{\pi}{2}} \frac{\cos 2x}{\sin 3x}$$

We can simply substitute in the value of x and obtain our answer.

$$\lim_{x \to \frac{\pi}{2}} \frac{\cos 2x}{\sin 3x} = \frac{\cos 2\left(\frac{\pi}{2}\right)}{\sin 3\left(\frac{\pi}{2}\right)}$$

$$= \frac{\cos \pi}{\sin \frac{3\pi}{2}}$$

$$= \frac{\cos 180}{\sin 270} \quad (\text{Note that } \pi \text{ radians} = 180^\circ)$$

$$= \frac{-1}{-1}$$

$$= 1$$

This problem can also be solved by using the angles directly in radians, but I prefer to work in degrees. Note that cos π in radians = –1, and $\sin\frac{3\pi}{2}$ in radians = –1 as obtained above.

4. Evaluate $\lim_{x \to 0} \frac{\cos x}{\sin x - 3}$

Solution

$$\lim_{x \to 0} \frac{\cos x}{\sin x - 3} = \frac{\cos 0}{\sin 0 - 3}$$

$$= \frac{1}{0-3}$$

$$= -\frac{1}{3}$$

5. Evaluate $\displaystyle\lim_{x \to 0} \frac{\sin 3x}{\sin 2x}$

Solution

Multiply by $\dfrac{x}{\sin 2x}$ and change the denominator to x as follows.

$\displaystyle\lim_{x \to 0} \frac{\sin 3x}{\sin 2x} = \lim_{x \to 0} \left(\frac{\sin 3x}{x} \ X \ \frac{x}{\sin 2x} \right)$ (When x cancels out it gives the original expression)

$\quad = \displaystyle\lim_{x \to 0} \left(\frac{\sin 3x}{x} \ X \ \left(\frac{\sin 2x}{x} \right)^{-1} \right)$ (Note that the inverse of $\dfrac{x}{\sin 2x}$ has been taken)

$\quad = \displaystyle\lim_{x \to 0} \frac{\sin 3x}{x} \ X \ \left(\lim_{x \to 0} \frac{\sin 2x}{x} \right)^{-1}$

$\quad = 3 \times 2^{-1}$ (Note that $\displaystyle\lim_{x \to 0} \frac{\sin 3x}{x} = 3$ and $\displaystyle\lim_{x \to 0} \frac{\sin 2x}{x} = 2$)

$\quad = 3 \times \dfrac{1}{2}$

$\quad = \dfrac{3}{2}$

This example shows that: $\displaystyle\lim_{x \to 0} \frac{\sin a x}{\sin b x} = \frac{a}{b}$

6. Evaluate $\displaystyle\lim_{x \to 0} \frac{\sin 2x}{\sin 5x}$

Solution

$\displaystyle\lim_{x \to 0} \frac{\sin 2x}{\sin 5x}$

From the rule established in example 5 above, we can see that the solution to this problem is $\dfrac{2}{5}$.

Hence, $\displaystyle\lim_{x \to 0} \frac{\sin 2x}{\sin 5x} = \frac{2}{5}$ (Since $\displaystyle\lim_{x \to 0} \frac{\sin a x}{\sin b x} = \frac{a}{b}$)

Exercise 16

1. Evaluate $\displaystyle\lim_{x \to 0} 2x^3 - 4x^2 + x + 9$

2. Evaluate $\displaystyle\lim_{x \to 0} \frac{5x^2 + x - 8}{x^2 - 2x + 5}$

3. Evaluate $\displaystyle\lim_{x \to 5} \frac{x^2 - x - 20}{x - 5}$

4. Evaluate $\displaystyle\lim_{x \to 7} \frac{x^2 - 49}{x - 7}$

5. Evaluate $\lim\limits_{x\to 0}$ $(5x-1)(3x+7)(3x+2)$

6. Evaluate $\lim\limits_{x\to\infty}$ $\dfrac{2x^3+x^2-9x-5}{5x^3+2x^2-7x+2}$

7. Evaluate $\lim\limits_{x\to\infty}$ $\dfrac{x^4-5x^2-2x-6}{3x^4+2x+3}$

8. Evaluate $\lim\limits_{x\to 4}$ $\dfrac{x^2-25}{x^2+2x-15}$

9. Evaluate $\lim\limits_{x\to 8}$ $\dfrac{2x^2-17x+8}{8-x}$

10. Evaluate $\lim\limits_{m\to 0}$ $\dfrac{(5+m)^2-25}{m}$

11. Determine the limiting value of $\dfrac{\sqrt{x}-2}{x-4}$ as x tends to 4 .

12. Find the limit of $\dfrac{x^3-27}{x-3}$ as $x\to 3$

13. Find the limit of $\dfrac{x^3+8}{x+2}$ as $x\to -2$

14. Evaluate $\lim\limits_{x\to 25}$ $\dfrac{5-\sqrt{x}}{25-x}$

15. Evaluate $\lim\limits_{x\to 1}$ $\dfrac{x^3-1}{x^7-1}$

16. Determine if the function $f(x)=x^2+3x-1$ is continuous at the point $x=2$

17. Determine if the function $f(x)=\dfrac{5x-2}{3x^2-x-2}$ is continuous at $x=-1$

18. Determine if the function $f(x)=\dfrac{2x^2-18}{x-3}$ is continuous at $x=3$.

19. Determine if the function $f(x)=\dfrac{x^3+64}{x+4}$ is continuous at $x=-4$.

20. Test for the continuity of the function: $f(x)=\begin{cases} -x+5 & \text{if } x>2 \\ x^3+10 & \text{if } x\leq 2 \end{cases}$

21. Determine the continuity of the function: $f(x)=\begin{cases} 3x^2 & \text{if } x\geq -3 \\ x+30 & \text{if } x<-3 \end{cases}$

22. Determine if the function below is continuous or not.

$$f(x)=\begin{cases} \dfrac{x^2-25}{x+5} & \text{if } x\geq -5 \\ \dfrac{x+5}{2x+10} & \text{if } x<-5 \end{cases}$$

23. If $f(x)=\begin{cases} \dfrac{x^3-27}{x-3} \\ 1 \ \text{if } x=3 \end{cases}$ test for the continuity of the function at $x=3$.

24. If $f(x)=\begin{cases} 2-3x & x<1 \\ x^3+4 & x\geq 1 \end{cases}$ evaluate the following limits, if they exist.

(a) $\lim\limits_{x\to -2} f(x)$

(b) $\lim\limits_{x\to 1} f(x)$

25. Evaluate $\lim\limits_{x \to 2} (5 + |x - 2|)$ if it exists.

26. Evaluate $\lim\limits_{x \to 0} \dfrac{\sin 7x}{2x}$

27. Evaluate $\lim\limits_{x \to 0} \dfrac{\sin x}{3x}$

28. Find the value of $\lim\limits_{x \to \pi} \dfrac{\cos 2x}{\sin \frac{x}{2}}$

29. Evaluate $\lim\limits_{x \to 0} \dfrac{\cos 5x}{\sin 3x - 2}$

30. Evaluate $\lim\limits_{x \to 0} \dfrac{\sin 9x}{\sin 4x}$

CHAPTER 17
DIFFERENTIATION FROM FIRST PRINCIPLE

If $y = f(x)$, then the gradient function of y or $f(x)$ is given by:

$$g(x) = \frac{f(x + \Delta x) - f(x)}{\Delta x}$$

where Δx is the increment in x.

Applying limits has shown that as Δx tends to zero,

$$f'(x) = \lim_{\Delta x \to 0} \frac{f(x + \Delta x) - f(x)}{\Delta x}$$

$$\frac{dy}{dx} = \lim_{\Delta x \to 0} \frac{f(x + \Delta x) - f(x)}{\Delta x}$$

$\frac{dy}{dx}$ reads dee y dee x.

Note that $f'(x)$ and $\frac{dy}{dx}$ can be interpreted as instantaneous rate of change of y with respect to x.

The technique of finding the derivative of a function by considering the limiting value is called differentiation from first principle.

Examples

1. Find the derivative of x from first principle.

Solution

Let, $y = x$

Increasing x by Δx will give an increase in y by Δy. This gives:

$y + \Delta y = x + \Delta x$

$\Delta y = x + \Delta x - y$

$\Delta y = x + \Delta x - x$ (substitute x for y since $y = x$ as given above from the question)

$\Delta y = \Delta x$ (x cancels out)

Dividing both sides by Δx in order to get the derivative gives:

$$\frac{\Delta y}{\Delta x} = \frac{\Delta x}{\Delta x}$$

$$\frac{\Delta y}{\Delta x} = 1$$

Taking limits as Δx tends to zero gives:

$\lim_{\Delta x \to 0} \frac{\Delta y}{\Delta x} = 1$ [Recall that $\lim_{x \to 0} c = c$, hence $\lim_{\Delta x \to 0} f(\Delta x) = 1$ gives 1, since $\frac{\Delta y}{\Delta x}$ is regarded as $f(\Delta x)$]

We now write $\lim_{\Delta x \to 0} \frac{\Delta y}{\Delta x}$ as $\frac{dy}{dx}$. This gives:

$$\frac{dy}{dx} = 1$$

2. Find from first principle, the derivative of $2x^3$.

<u>Solution</u>

Let, $y = 2x^3$

Increasing x by Δx will give an increase in y by Δy. This gives:

$$y + \Delta y = 2(x + \Delta x)^3$$

$$y + \Delta y = 2[x^3 + 3x^2\Delta x + 3x(\Delta x)^2 + (\Delta x)^3]$$

$$= 2x^3 + 6x^2\Delta x + 6x(\Delta x)^2 + 2(\Delta x)^3$$

$$\Delta y = 2x^3 + 6x^2\Delta x + 6x(\Delta x)^2 + 2(\Delta x)^3 - y$$

$$= 2x^3 + 6x^2\Delta x + 6x(\Delta x)^2 + 2(\Delta x)^3 - 2x^3 \quad \text{(Since } y = 2x^3\text{)}$$

$$\Delta y = 6x^2\Delta x + 6x(\Delta x)^2 + 2(\Delta x)^3 \quad \text{(Since } 2x^3 \text{ cancels out)}$$

Dividing both sides by Δx in order to get the derivative gives:

$$\frac{\Delta y}{\Delta x} = \frac{6x^2\Delta x}{\Delta x} + \frac{6x(\Delta x)^2}{\Delta x} + \frac{2(\Delta x)^3}{\Delta x}$$

$$\frac{\Delta y}{\Delta x} = 6x^2 + 6x\Delta x + 2(\Delta x)^2$$

Taking limits as Δx tends to zero gives:

$$\lim_{\Delta x \to 0} \frac{\Delta y}{\Delta x} = 6x^2 + 6x(0) + 2(0)^2 \quad \text{(Note that zero replaces } \Delta x\text{)}$$

$$\lim_{\Delta x \to 0} \frac{\Delta y}{\Delta x} = 6x^2$$

We now write $\lim_{\Delta x \to 0} \frac{\Delta y}{\Delta x}$ as $\frac{dy}{dx}$. This gives:

$$\frac{dy}{dx} = 6x^2$$

3. Find the derivative of $f(x) = \dfrac{1}{x^2}$ from first principle.

<u>Solution</u>

$$f(x) = \Delta x$$

Increasing x by Δx gives:

$$f(x + \Delta x) = \frac{1}{(x + \Delta x)^2}$$

$$f(x + \Delta x) = \frac{1}{x^2 + 2x\Delta x + (\Delta x)^2}$$

Subtract f(x) from both sides of the equation

$$f(x + \Delta x) - f(x) = \frac{1}{x^2 + 2x\Delta x + (\Delta x)^2} - f(x)$$

$$f(x + \Delta x) - f(x) = \frac{1}{x^2 + 2x\Delta x + (\Delta x)^2} - \frac{1}{x^2} \quad \text{(Note that } f(x) = \frac{1}{x^2}\text{)}$$

$$= \frac{x^2 - (x^2 + 2x\Delta x + (\Delta x)^2)}{x^2(x^2 + 2x\Delta x + (\Delta x)^2)}$$

$$= \frac{x^2 - x^2 - 2x\Delta x - (\Delta x)^2}{x^2(x^2 + 2x\Delta x + (\Delta x)^2)}$$

$$= \frac{-2x\Delta x - (\Delta x)^2}{x^2(x^2 + 2x\Delta x + (\Delta x)^2)}$$

$$f(x + \Delta x) - f(x) = \frac{\Delta x(-2x - \Delta x)}{x^2(x^2 + 2x\Delta x + (\Delta x)^2)}$$

Dividing both sides by Δx gives:

$$\frac{f(x + \Delta x) - f(x)}{\Delta x} = \frac{-2x - \Delta x}{x^2(x^2 + 2x\Delta x + (\Delta x)^2)} \qquad (\Delta x \text{ has cancelled out from the right hand side})$$

Taking limits as Δx tends to zero gives:

$$\lim_{\Delta x \to 0} \frac{f(x + \Delta x) - f(x)}{\Delta x} = \frac{-2x - 0}{x^2(x^2 + 2x(0) + (0)^2)}$$

Replacing $\lim\limits_{\Delta x \to 0} \dfrac{f(x + \Delta x) - f(x)}{\Delta x}$ with f'(x) gives the derivative of f(x) as:

$$f'(x) = \frac{-2x}{x^2(x^2)}$$

$$f'(x) = \frac{-2}{x^3}$$

4. If $f(x) = 3x^2$

(a) write down and simplify the expression $\dfrac{f(x + h) - f(x)}{h}$ $(h \neq 0)$

(b) find $\lim\limits_{\Delta x \to 0} \dfrac{f(x + h) - f(x)}{h}$

Solution

(a) Let, $f(x) = 3x^2$

Increasing x by h gives:

$$f(x + h) = 3(x + h)^2$$
$$= 3[x^2 + 2xh + h^2]$$
$$= 3x^2 + 6xh + 3h^2$$

Subtracting f(x) from both sides gives:

$$f(x + h) - f(x) = 3x^2 + 6xh + 3h^2 - 3x^2 \qquad (\text{Note that } f(x) = 3x^2 \text{ from the question})$$
$$= 6xh + 3h^2$$

Dividing both side by h gives:

$$\frac{f(x + h) - f(x)}{h} = \frac{6xh}{h} + \frac{3h^2}{h}$$

$$\frac{f(x + h) - f(x)}{h} = 6x + 3h$$

(b) $\dfrac{f(x + h) - f(x)}{h} = 6x + 3h$

Taking limits as h tends to zero gives:

$$\lim_{h \to 0} \frac{f(x + h) - f(x)}{h} = 6x + 3(0)$$

$$\lim_{h \to 0} \frac{f(x + h) - f(x)}{h} = 6x$$

5. If $f(x) = 4x^3 - 5$,

(a) evaluate $\dfrac{f(x+h) - f(x)}{h}$, where $h \neq 0$

(b) From your result in (5a) above, find the derivatives of $f(x)$ with respect to x.

Solution

(a) Let, $f(x) = 4x^3 - 5$

Increasing x by h gives:

$$f(x + h) = 4(x + h)^3 - 5$$
$$= 4[x^3 + 3x^2h + 3xh^2 + h^3] - 5$$
$$= 4x^3 + 12x^2h + 12xh^2 + 4h^3 - 5$$

Subtracting $f(x)$ from both sides gives:

$$f(x + h) - f(x) = 4x^3 + 12x^2h + 12xh^2 + 4h^3 - 5 - (4x^3 - 5) \quad (f(x) = 4x^3 - 5 \text{ from the question})$$
$$= 4x^3 + 12x^2h + 12xh^2 + 4h^3 - 5 - 4x^3 + 5$$
$$= 12x^2h + 12xh^2 + 4h^3$$

Dividing both side by h gives:

$$\frac{f(x+h) - f(x)}{h} = 12x^2 + 12xh + 4h^2$$

(b) $\dfrac{f(x+h) - f(x)}{h} = 12x^2 + 12xh + 4h^2$

Taking limits as h tends to zero gives:

$$\lim_{h \to 0} \frac{f(x+h) - f(x)}{h} = 12x^2 + 12x(0)\ 4(0)^2$$

$$\lim_{h \to 0} \frac{f(x+h) - f(x)}{h} = 12x^2$$

6. If $f(x) = \dfrac{x^2 + 1}{2x}$

(a) write down and simplify the expression $\dfrac{f(x + \Delta x) - f(\Delta x)}{\Delta x}$, where $\Delta x \neq 0$

(b) Find the derivatives of $y = \dfrac{x^2 + 1}{2x}$

Solution

$$f(x) = \frac{x^2 + 1}{2x}$$

$$f(x + \Delta x) = \frac{(x + \Delta x)^2 + 1}{2(x + \Delta x)}$$

$$f(x + \Delta x) = \frac{x^2 + 2x\Delta x + (\Delta x)^2 + 1}{2x + 2\Delta x}$$

Subtract $f(x)$ from both sides of the equation

$$f(x + \Delta x) - f(x) = \frac{x^2 + 2x\Delta x + (\Delta x)^2 + 1}{2x + 2\Delta x} - \frac{x^2 + 1}{2x}$$

206

$$f(x + \Delta x) - f(x) = \frac{2x[x^2 + 2x\Delta x + (\Delta x)^2 + 1] - (x^2 + 1)(2x + 2\Delta x)}{2x(2x + 2\Delta x)}$$

$$= \frac{2x^3 + 4x^2\Delta x + 2x(\Delta x)^2 + 2x - (2x^3 + 2x^2\Delta x + 2x + 2\Delta x)}{2x(2x + 2\Delta x)}$$

$$= \frac{2x^3 + 4x^2\Delta x + 2x(\Delta x)^2 + 2x - 2x^3 - 2x^2\Delta x - 2x - 2\Delta x)}{2x(2x + 2\Delta x)}$$

$$= \frac{2x^2\Delta x + 2x(\Delta x)^2 - 2\Delta x)}{2x(2x + 2\Delta x)}$$

$$f(x + \Delta x) - f(x) = \frac{\Delta x(2x^2 + 2x\Delta x - 2)}{2x(2x + 2\Delta x)} \qquad \text{(After factorizing the numerator)}$$

Dividing both sides by Δx gives:

$$\frac{f(x + \Delta x) - f(x)}{\Delta x} = \frac{2x^2 + 2x\Delta x - 2}{2x(2x + 2\Delta x)}$$

(b)Taking limits as Δx tends to zero gives:

$$\lim_{\Delta x \to 0} \frac{f(x + \Delta x) - f(x)}{\Delta x} = \frac{2x^2 + 2x(0) - 2}{2x(2x + 2(0))}$$

$$= \frac{2x^2 - 2}{2x(2x)}$$

$$= \frac{2(x^2 - 1)}{4x^2}$$

Replacing $\displaystyle\lim_{\Delta x \to 0} \frac{f(x + \Delta x) - f(x)}{\Delta x}$ with f'(x) gives the derivative of f(x) as:

$$f'(x) = \frac{x^2 - 1}{2x^2}$$

Separating into fractions gives:

$$f'(x) = \frac{x^2}{2x^2} - \frac{1}{2x^2}$$

$$f'(x) = \frac{1}{2} - \frac{1}{2x^2}$$

7. Find from first principle the derivatives with respect to x of $y = 5x^3 - x + 7$

<u>Solution</u>

$$y = 5x^3 - x + 7$$
$$y + \Delta y = 5(x + \Delta x)^3 - (x + \Delta x) + 7$$
$$y + \Delta y = 5[x^3 + 3x^2\Delta x + 3x(\Delta x)^2 + (\Delta x)^3] - x - \Delta x + 7$$
$$= 5x^3 + 15x^2\Delta x + 15x(\Delta x)^2 + 5(\Delta x)^3 - x - \Delta x + 7$$
$$\Delta y = 5x^3 + 15x^2\Delta x + 15x(\Delta x)^2 + 5(\Delta x)^3 - x - \Delta x + 7 - y$$
$$= 5x^3 + 15x^2\Delta x + 15x(\Delta x)^2 + 5(\Delta x)^3 - x - \Delta x + 7 - (5x^3 - x + 7) \quad \text{(Since } y = 5x^3 - x + 7)$$
$$\Delta y = 5x^3 + 15x^2\Delta x + 15x(\Delta x)^2 + 5(\Delta x)^3 - x - \Delta x + 7 - 5x^3 + x - 7$$
$$= 15x^2\Delta x + 15x(\Delta x)^2 + 5(\Delta x)^3 - \Delta x$$

Dividing both sides by Δx gives:

$$\frac{\Delta y}{\Delta x} = \frac{15x^2 \Delta x}{\Delta x} + \frac{15x(\Delta x)^2}{\Delta x} + \frac{5(\Delta x)^3}{\Delta x} - \frac{\Delta x}{\Delta x}$$

$$\frac{\Delta y}{\Delta x} = 15x^2 + 15x\Delta x + 5(\Delta x)^2 - 1$$

Taking limits as Δx tends to zero gives:

$$\lim_{\Delta x \to 0} \frac{\Delta y}{\Delta x} = 15x^2 + 15x(0) + 5(0)^2 - 1$$

$$\lim_{\Delta x \to 0} \frac{\Delta y}{\Delta x} = 15x^2 - 1$$

We now replace $\lim_{\Delta x \to 0} \frac{\Delta y}{\Delta x}$ with $\frac{dy}{dx}$. This gives:

$$\frac{dy}{dx} = 15x^2 - 1$$

8. Differentiate $3x - \dfrac{1}{2x^2}$ from first principle.

<u>Solution</u>

$$y = 3x - \frac{1}{x^2}$$

$$y + \Delta y = 3(x + \Delta x) - \frac{1}{2(x + \Delta x)^2}$$

$$= 3x + 3\Delta x - \frac{1}{2[x^2 + 2x\Delta x + (\Delta x)^2]}$$

Subtract y from both sides of the equation

$$\Delta y = 3x + 3\Delta x - \frac{1}{2x^2 + 4x\Delta x + 2(\Delta x)^2} - y$$

$$= 3x + 3\Delta x - \frac{1}{2x^2 + 4x\Delta x + 2(\Delta x)^2} - \left(3x - \frac{1}{2x^2}\right)$$

$$= 3x + 3\Delta x - \frac{1}{2x^2 + 4x\Delta x + 2(\Delta x)^2} - 3x + \frac{1}{2x^2}$$

$$= 3\Delta x - \frac{1}{2x^2 + 4x\Delta x + 2(\Delta x)^2} + \frac{1}{2x^2}$$

Combining them into one fraction gives:

$$\Delta y = \frac{3\Delta x[2x^2(2x^2 + 4x\Delta x + 2(\Delta x)^2] - 2x^2 + 2x^2 + 4x\Delta x + 2(\Delta x)^2}{2x^2[2x^2 + 4x\Delta x + 2(\Delta x)^2]}$$

$$= \frac{3\Delta x[4x^4 + 8x^3\Delta x + 4x^2(\Delta x)^2] + 4x\Delta x + 2(\Delta x)^2}{2x^2[2x^2 + 4x\Delta x + 2(\Delta x)^2]} \quad (-2x^2 + 2x^2 \text{ has cancelled out})$$

$$= \frac{12x^4\Delta x + 24x^3(\Delta x)^2 + 12x^2(\Delta x)^3 + 4x\Delta x + 2(\Delta x)^2}{2x^2[2x^2 + 4x\Delta x + 2(\Delta x)^2]}$$

Factorizing the numerator gives:

$$= \frac{\Delta x[12x^4 + 24x^3\Delta x + 12x^2(\Delta x)^2 + 4x + 2\Delta x]}{2x^2[2x^2 + 4x\Delta x + 2(\Delta x)^2]}$$

Dividing both sides by Δx gives:

$$\frac{\Delta y}{\Delta x} = \frac{12x^4 + 24x^3\Delta x + 12x^2(\Delta x)^2 + 4x + 2\Delta x}{2x^2[2x^2 + 4x\Delta x + 2(\Delta x)^2]} \quad (\Delta x \text{ that is outside the bracket cancels out})$$

Taking limits as Δx tends to zero gives:

$$\lim_{\Delta x \to 0} \frac{\Delta y}{\Delta x} = \frac{12x^4 + 24x^3(0) + 12x^2(0)^2 + 4x + 2(0)}{2x^2[2x^2 + 4x(0) + 2(0)^2]}$$

$$= \frac{12x^4 + 4x}{2x^2[2x^2]}$$

$$\lim_{\Delta x \to 0} \frac{\Delta y}{\Delta x} = \frac{12x^4 + 4x}{4x^4}$$

$$= \frac{12x^4}{4x^4} + \frac{4x}{4x^4}$$

$$= 3 + \frac{1}{x^3}$$

Replacing $\lim\limits_{\Delta x \to 0} \dfrac{\Delta y}{\Delta x}$ with $\dfrac{dy}{dx}$ gives the derivative as:

$$\frac{dy}{dx} = 3 + \frac{1}{x^3}$$

Exercise 17

1. Find the derivative of $2x$ from first principle.

2. Find from first principle, the derivative of x^2.

3. Find the derivative of $f(x) = \dfrac{1}{x^3}$ from first principle.

4. If $f(x) = 5x^2$

(a) write down and simplify the expression $\dfrac{f(x+h) - f(x)}{h}$ $\quad (h \neq 0)$

(b) find $\lim\limits_{h \to 0} \dfrac{f(x+h) - f(x)}{h}$

5. If $f(x) = 9x^3$,

(a) evaluate $\dfrac{f(x+h) - f(x)}{h}$, where $h \neq 0$

(b) From your result above , find the derivatives of $f(x)$ with respect to x.

6. If $f(x) = \dfrac{x^3 - 2}{x}$

(a) write down and simplify the expression $\dfrac{f(x + \Delta x) - f(\Delta x)}{\Delta x}$, where $\Delta x \neq 0$

(b) Find the derivatives of $y = \dfrac{x^3 - 2}{x}$

7. Find from first principle the derivatives with respect to x of $y = 3x^2 - 10x$

8. Differentiate $x + \dfrac{3}{x}$ from first principle.

9. From first principle, find the derivatives with respect to x of $y = 5x - 3x^2$

10. Differentiate $2x + \dfrac{x}{5}$ from first principle.

CHAPTER 18
GENERAL RULE OF DIFFERENTIATION AND COMPOSITE FUNCTIONS

The general rule for the derivative/differentiation of a function is as given below.

If $y = x^n$

then, $\dfrac{dy}{dx} = nx^{n-1}$

The rule for differentiating a composite function (or function of a function) is given as follows:

If $y = (2x + 5)^4$

then we write, $u = 2x + 5$

and express y as:

$y = u^4$

Therefore, $\dfrac{dy}{dx} = \dfrac{dy}{du} \times \dfrac{du}{dx}$

This rule is called the chain rule.

Examples

1. Find the derivatives of the following:

(a) $y = 2x^7$

(b) $y = \dfrac{3}{4}x^8$

(c) $y = 3\sqrt{x}$

(d) $y = \dfrac{10}{\sqrt[5]{x^3}}$

(e) $y = \dfrac{1}{x^{\frac{1}{4}}}$

Solution

(a) $y = 2x^7$

Solutions

(a) $y = 2x^7$

$\dfrac{dy}{dx} = 7 \times 2x^{7-1}$

This is done by multiplying the exponent (power) by the term and subtracting 1 from the exponent (power). Hence, the answer is:

$\dfrac{dy}{dx} = 14x^6$

(b) $y = \dfrac{3}{4}x^8$

Multiply the term by the exponent (power) (i.e. 8) and subtract 1 from the exponent. This gives:

$\dfrac{dy}{dx} = 8 \times \dfrac{3}{4}x^{8-1}$

$$= \frac{24}{4}x^7$$

$$\frac{dy}{dx} = 6x^7$$

(c) $y = 3\sqrt{x}$

Since $\sqrt{x} = x^{\frac{1}{2}}$, we can rewrite the expression as:

$$y = 3x^{\frac{1}{2}}$$

$$\frac{dy}{dx} = \frac{1}{2} \times 3x^{\frac{1}{2}-1}$$

$$= \frac{3}{2}x^{-\frac{1}{2}}$$

$$= \frac{3}{2} \times \frac{1}{x^{\frac{1}{2}}} \qquad \text{(Note that } x^{-\frac{1}{2}} = \frac{1}{x^{\frac{1}{2}}} \text{ from indices)}$$

$$\frac{dy}{dx} = \frac{3}{2x^{\frac{1}{2}}}$$

$$\frac{dy}{dx} = \frac{3}{2\sqrt{x}} \qquad \text{(Since } x^{\frac{1}{2}} = \sqrt{x}\text{)}$$

(d) $y = \dfrac{10}{\sqrt[5]{x^3}}$

Expressing the root in fractional form gives:

$$y = \frac{10}{(x^3)^{\frac{1}{5}}}$$

$$y = \frac{10}{x^{\frac{3}{5}}} \qquad \text{(The two exponents (powers) 3 and } \frac{1}{5}\text{ have been multiplied)}$$

Taking the denominator to the numerator changes the sign of its exponent as follows:

$$y = 10x^{-\frac{3}{5}}$$

Hence, $\dfrac{dy}{dx} = -\dfrac{3}{5} \times 10x^{-\frac{3}{5}-1}$

$$= -\frac{30}{5}x^{-\frac{8}{5}}$$

$$= -6x^{-\frac{8}{5}}$$

$$= -6 \times \frac{1}{x^{\frac{8}{5}}} \qquad \text{(Note that the inverse of a term changes the sign of its exponent)}$$

$$\frac{dy}{dx} = \frac{-6}{x^{\frac{8}{5}}}$$

Or, $\dfrac{dy}{dx} = \dfrac{-6}{\sqrt[5]{x^8}}$

(e) $y = \dfrac{1}{x^{\frac{1}{4}}}$

This can be expressed as:

$y = x^{-\frac{1}{4}}$

$\dfrac{dy}{dx} = -\dfrac{1}{4} x^{-\frac{1}{4}-1}$

$\qquad = -\dfrac{1}{4} x^{-\frac{5}{4}}$

$\qquad = -\dfrac{1}{4} \times \dfrac{1}{x^{\frac{5}{4}}}$

$\dfrac{dy}{dx} = -\dfrac{1}{4x^{\frac{5}{4}}}$ \qquad (Note that the inverse of a term changes the sign of its exponent)

Or, $\dfrac{dy}{dx} = \dfrac{1}{4\sqrt[4]{x^5}}$

2. Find the derivative of each of the following:

(a) $5x^3 - 7x^2 - 3x + 8$

(b) $\dfrac{3}{5}x^5 + 2x^3 - x$

(c) $\dfrac{2x^4 - 5x^3 - 4x^2 + 3}{x^2}$

(d) $\sqrt{x} + \dfrac{1}{\sqrt{x}}$

Solutions

(a) Let the expression be $y = 5x^3 - 7x^2 - 3x + 8$

Hence, $\dfrac{dy}{dx} = \dfrac{d(5x^3)}{dx} - \dfrac{d(7x^2)}{dx} - \dfrac{d(3x)}{dx} + \dfrac{d(8)}{dx}$

This means that each part should be differentiated separately.

Hence, $\dfrac{dy}{dx} = (3 \times 5x^{3-1}) - (2 \times 7x^{2-1}) - (1 \times 3x^{1-1}) + 0$

$\dfrac{dy}{dx} = 15x^2 - 14x - 3$ \qquad (Note that $x^0 = 1$)

Note that the derivative of a constant is zero as shown by the derivative of 8

213

(b) $y = \dfrac{3}{5}x^5 + 2x^3 - x$

$\quad \dfrac{dy}{dx} = \left(5 \times \dfrac{3}{5}x^{5-1}\right) + \left(3 \times 2x^{3-1}\right) - \left(1 \times x^{1-1}\right)$

$\qquad = 3x^4 + 6x^2 - 1$

(c) $y = \dfrac{2x^4 - 5x^3 - 4x^2 + 3}{x^2}$

Dividing each term in the numerator by the denominator in order to separate the expression into it different fractions gives:

$\quad y = \dfrac{2x^4}{x^2} - \dfrac{5x^3}{x^2} - \dfrac{4x^2}{x^2} + \dfrac{3}{x^2}$

$\qquad = 2x^2 - 5x - 4 + \dfrac{3}{x^2}$

$\quad y = 2x^2 - 5x - 4 + 3x^{-2}$

$\quad \dfrac{dy}{dx} = (2 \times 2x) - 5 - 0 + \left(-2 \times 3x^{-2-1}\right)$

$\qquad = 4x - 5 - 6x^{-3}$

$\quad \dfrac{dy}{dx} = 4x - 5 - \dfrac{6}{x^2}$

(d) $y = \sqrt{x} + \dfrac{1}{\sqrt{x}}$

$\qquad = x^{\frac{1}{2}} + \dfrac{1}{x^{\frac{1}{2}}}$

$\quad y = x^{\frac{1}{2}} + x^{-\frac{1}{2}}$

$\quad \dfrac{dy}{dx} = \dfrac{1}{2}x^{-\frac{1}{2}} - \dfrac{1}{2}x^{-\frac{3}{2}} \qquad$ (Note that $-\dfrac{1}{2} - 1 = -\dfrac{3}{2}$)

$\qquad = \dfrac{1}{2} \times \dfrac{1}{x^{\frac{1}{2}}} - \dfrac{1}{2} \times \dfrac{1}{x^{\frac{3}{2}}}$

$\qquad = \dfrac{1}{2x^{\frac{1}{2}}} - \dfrac{1}{2x^{\frac{3}{2}}}$

$\quad \dfrac{dy}{dx} = \dfrac{1}{2\sqrt{x}} - \dfrac{1}{2\sqrt{x^3}}$

3. If $y = (5x - 2)^3$, find $\dfrac{dy}{dx}$

Solution

$\quad y = (5x - 2)^3 \quad$ (This is a composite function)

214

Let us take u = 5x – 2

If 5x – 2 is replaced with U, then the question (i.e. y = $(5x – 2)^3$) becomes:

$$y = u^3$$

Hence, $\dfrac{dy}{du} = 3u^2$

Since u = 5x – 2

then $\dfrac{du}{dx} = 5$

Therefore, $\dfrac{dy}{dx} = \dfrac{dy}{du} \times \dfrac{du}{dx}$ (Chain rule)

$$= 3u^2 \times 5$$

$$= 15u^2$$

Now substitute 5x – 2 for u to obtain $\dfrac{dy}{dx}$ as follows:

$$\dfrac{dy}{dx} = 15(5x – 2)^2$$

4. If y = $\dfrac{1}{(5x^2 – 1)^4}$ find $\dfrac{dy}{dx}$

Solution

$$y = \dfrac{1}{(5x^2 – 1)^4}$$ (This is a composite function)

It can also be represented as follows:

$$y = (5x^2 – 1)^{-4}$$ (Its inverse changes the sign of its exponent)

Now, let us take u = $5x^2 – 1$

Hence, $y = u^{-4}$ (When $5x^2 – 1$ is replaced with u in the original question)

Therefore, $\dfrac{dy}{du} = -4u^{-5}$

Since u = $5x^2 – 1$

then, $\dfrac{du}{dx} = 10x$

Therefore, $\dfrac{dy}{dx} = \dfrac{dy}{du} \times \dfrac{du}{dx}$ (Chain rule)

$$= -4u^{-5} \times 10x$$

$$\dfrac{dy}{dx} = -40xu^{-5}$$

Now substitute $5x^2 – 1$ for u. This gives:

$$\dfrac{dy}{dx} = -40x(5x^2 – 1)^{-5}$$

Or, $\dfrac{dy}{dx} = \dfrac{-40x}{(5x^2 – 1)^5}$ (Note the change in the sign of the exponent as it becomes

denominator)

5. If $y = (2x^3 + 7x)^{\frac{1}{2}}$ find $\dfrac{dy}{dx}$

<u>Solution</u>

$$y = (2x^3 + 7x)^{\frac{1}{2}}$$

Let $u = 2x^3 + 7x$

Hence, $y = u^{\frac{1}{2}}$

$$\frac{dy}{du} = \frac{1}{2}u^{-\frac{1}{2}}$$

Also, $u = 2x^3 + 7x$

$$\frac{du}{dx} = 6x^2 + 7$$

Therefore, $\dfrac{dy}{dx} = \dfrac{dy}{du} \times \dfrac{du}{dx}$

$$= \frac{1}{2}u^{-\frac{1}{2}} \times 6x^2 + 7$$

$$= \frac{6x^2 + 7}{2}u^{-\frac{1}{2}}$$

$$= \frac{6x^2 + 7}{2} \times \frac{1}{u^{1/2}}$$

$$\frac{dy}{dx} = \frac{6x^2 + 7}{2u^{1/2}}$$

Now, replace u with $2x^3 + 7x$. This gives:

$$\frac{dy}{dx} = \frac{6x^2 + 7}{2(2x^3 + 7x)^{1/2}}$$

Or, $\dfrac{dy}{dx} = \dfrac{6x^2 + 7}{2\sqrt{2x^3 + 7x}}$ $\qquad$ (Note that $(2x^3 + 7x)^{\frac{1}{2}} = \sqrt{2x^3 + 7x}$)

6. Find the derivative of $3x^2 - x + 9)^4$

<u>Solution</u>

$$y = (3x^2 - x + 9)^4$$

Let $u = 3x^2 - x + 9$

Hence, $y = u^4$

$$\frac{dy}{du} = 4u^3$$

$$\frac{du}{dx} = 6x - 1$$

Therefore, $\dfrac{dy}{dx} = \dfrac{dy}{du} \times \dfrac{du}{dx}$

$$= 4u^3 \times 6x - 1$$

$$= 4(6x - 1)u^3$$

$$= (24x - 4)u^3$$

$$\frac{dy}{dx} = (24x - 4)(3x^2 - x + 9)^3 \qquad \text{(When u is replaced with } 3x^2 - x + 9)$$

7. Find the derivative of $\left(x - \dfrac{5}{x}\right)^4$

Solution

$$y = \left(x - \frac{5}{x}\right)^4$$

Let $u = x - \dfrac{5}{x}$

Therefore, $y = u^4$

$$\frac{dy}{du} = 4u^3$$

$$\frac{du}{dx} = \frac{d(x)}{dx} - \frac{d(5x^{-1})}{dx} \qquad \text{(Note that } = \frac{5}{x} = 5x^{-1})$$

$$= 1 - (-1)5x^{-2}$$

$$= 1 + 5x^{-2}$$

$$\frac{du}{dx} = 1 + \frac{5}{x^2}$$

Therefore, $\dfrac{dy}{dx} = \dfrac{dy}{du} \times \dfrac{du}{dx}$

$$= 4u^3 \times \left(1 + \frac{5}{x^2}\right)$$

$$= \left(4 + \frac{20}{x^2}\right)u^3$$

$$\frac{dy}{dx} = \left(4 + \frac{20}{x^2}\right)\left(x - \frac{5}{x}\right)^3 \qquad \text{(When u is replaced with } \left(x - \frac{5}{x}\right))$$

8. Differentiate with respect to x: $\dfrac{1}{2x^5 - 3x + 1}$

Solution

Let $y = \dfrac{1}{2x^5 - 3x + 1}$

Or, $y = (2x^5 - 3x + 1)^{-1} \qquad$ (Recall from indices that $\dfrac{1}{a} = a^{-1}$)

Let us take $u = 2x^5 - 3x + 1$

Therefore, $y = u^{-1}$

$$\frac{dy}{du} = -1u^{-2}$$

$$\frac{du}{dx} = 10x^4 - 3$$

Therefore, $\dfrac{dy}{dx} = \dfrac{dy}{du} \times \dfrac{du}{dx}$

$$= -1u^{-2} \times 10x^4 - 3$$

$$= -1(10x^4 - 3)u^{-2}$$

$$= (-10x^4 + 3)u^{-2}$$

$$= \frac{3 - 10x^4}{u^2}$$

$$\frac{dy}{dx} = \frac{3 - 10x^4}{(2x^5 - 3x + 1)^2}$$

9. Differentiate with respect to x: $\sqrt{7 - 5x^3}$

<u>Solution</u>

Let $y = \sqrt{7 - 5x^3}$

Or, $y = (7 - 5x^3)^{\frac{1}{2}}$ (Recall from indices that $\sqrt{a} = a^{\frac{1}{2}}$)

Let $u = 7 - 5x^3$

Hence, $y = u^{\frac{1}{2}}$

$$\frac{dy}{du} = \frac{1}{2} u^{-\frac{1}{2}}$$

$$\frac{du}{dx} = -15x^2$$

Therefore, $\dfrac{dy}{dx} = \dfrac{dy}{du} \times \dfrac{du}{dx}$

$$= \frac{1}{2} u^{-\frac{1}{2}} \times (-15x^2)$$

$$= -\frac{15}{2} x^2 (u^{-\frac{1}{2}})$$

$$= -\frac{15x^2}{2u^{\frac{1}{2}}}$$

$$= -\frac{15x^2}{2\sqrt{u}}$$

$$\frac{dy}{dx} = -\frac{15x^2}{2\sqrt{7 - 5x^3}}$$

10. Find $\dfrac{dy}{dx}$ if $y = \dfrac{1}{\sqrt{2x^3 - 5}}$

<u>Solution</u>

$$y = \frac{1}{\sqrt{2x^3 - 5}}$$

Let $u = 2x^3 - 5$

hence, $y = \dfrac{1}{\sqrt{u}}$

$$y = \frac{1}{u^{\frac{1}{2}}}$$

$$y = u^{-\frac{1}{2}}$$

$$\frac{dy}{du} = -\frac{1}{2} u^{-\frac{3}{2}}$$ (Note that $-\frac{1}{2} - 1 = -\frac{3}{2}$)

$$\frac{du}{dx} = 6x^2$$

Therefore, $\dfrac{dy}{dx} = \dfrac{dy}{du} \times \dfrac{du}{dx}$

$$= -\frac{1}{2}u^{-\frac{3}{2}} \times (6x^2)$$

$$= -\frac{6}{2}x^2\,(u^{-\frac{3}{2}})$$

$$= -3x^2(u^{-\frac{3}{2}})$$

$$= -\frac{3x^2}{u^{\frac{3}{2}}}$$

$$= -\frac{3x^2}{\sqrt{u^3}}$$

$$\frac{dy}{dx} = -\frac{3x^2}{\sqrt{(2x^3-5)^3}}$$

Exercise 18

1. Find the derivatives of the following:

(a) $y = 8x^5$

(b) $y = \dfrac{2}{5}x^5$

(c) $y = \sqrt[3]{x}$

(d) $y = 7\sqrt[7]{x}$

(e) $y = \dfrac{1}{\sqrt[8]{x^5}}$

(f) $y = \dfrac{2}{x^{\frac{5}{2}}}$

2. Find the derivative of each of the following:

(a) $2x^5 - 3x^4 - 4x^3 + 5x^2 - 6x + 7$

(b) $x^7 + 2x^4 - \dfrac{3}{x}$

(c) $\dfrac{3x^9 - x^7 - 5x^4 + 2x^2 - 1}{x^3}$

(d) $5(\sqrt[4]{x}) + \dfrac{5}{\sqrt[3]{2x}}$

3. If $y = (2x - 5)^4$, find $\dfrac{dy}{dx}$

4. If $y = \dfrac{3}{(x^3 - 7)^2}$ find $\dfrac{dy}{dx}$

5. If $y = (2x^3 + 7x)^{\frac{1}{2}}$ find $\dfrac{dy}{dx}$

6. Find the derivative of $(7x^3 - x^2 + 3)^5$

7. Find the derivative of $\left(3x - \dfrac{2}{3x}\right)^3$

8. Differentiate with respect to x: $-\dfrac{9}{3x^2 - x - 10}$

9. Differentiate with respect to x: $\sqrt{1 - 2x^4}$

10. Find $\dfrac{dy}{dx}$ if $y = \dfrac{1}{\sqrt[3]{5x^3 - 1}}$

11. Find the derivative of $(x^5 - 3)^9$

12. Find the derivative of $\left(x - \dfrac{1}{5x}\right)^2$

13. Differentiate with respect to x: $\dfrac{2}{x^3 - x - \frac{1}{x}}$

14. Differentiate with respect to x: $\sqrt{5x - x^2}$

15. Find $\dfrac{dy}{dx}$ if $y = \dfrac{1}{\sqrt[5]{x^3 + 2}}$

CHAPTER 19
PRODUCT RULE OF DERIVATIVE

If $y = uv$ where u and v are functions of x, then:

$$\frac{dy}{dx} = u\frac{dv}{dx} + v\frac{du}{dx}$$

This is called the product rule of differentiation.

Similarly, If $y = uvw$ where u, v and w are functions of x, then:

$$\frac{dy}{dx} = \frac{du}{dx}vw + \frac{dv}{dx}uw + \frac{dw}{dx}uv$$

Examples

1. If $f(x) = (x - 3)(x + 4)$, find $f'(x)$

Solution

$$f(x) = (x - 3)(x + 4)$$

This is a product of two functions of x.

Let $u = x - 3$

and $v = x + 4$

Hence, $\dfrac{du}{dx} = 1$

$\dfrac{dv}{dx} = 1$

Therefore, the derivative of $f(x)$ is:

$$f'(x) = u\frac{dv}{dx} + v\frac{du}{dx}$$

$$= (x - 3) \times 1 + (x + 4) \times 1$$

$$= x - 3 + x + 4$$

$$f'(x) = 2x + 1$$

Note that another way to differentiate the function in example 1 above is to expand the bracket and differentiate directly.

2. Find the derivative of $y = (4x^2 + 1)(x^2 - 3)$

Solution

$$y = (4x^2 + 1)(x^2 - 3)$$

Let $u = 4x^2 + 1$

and $v = (x^2 - 3)$

Hence, $\dfrac{du}{dx} = 8x$

$\dfrac{dv}{dx} = 2x$

$$\frac{dy}{dx} = u\frac{dv}{dx} + v\frac{du}{dx}$$
$$= (4x^2 + 1)2x + (x^2 - 3)8x$$
$$= 8x^3 + 2x + 8x^3 - 24x$$
$$\frac{dy}{dx} = 16x^3 - 22x$$

3. If $y = x^2(1 + 2x)^{\frac{1}{2}}$, find $\frac{dy}{dx}$

Solution

$$y = x^2(1 + 2x)^{\frac{1}{2}}$$
$$u = x^2$$

and $v = (1 + 2x)^{\frac{1}{2}}$

Hence, $\frac{du}{dx} = 2x$

$$\frac{dv}{dx} = \frac{1}{2} \times 2 \times (1 + 2x)^{\frac{1}{2} - 1} \qquad \text{(Use of chain rule)}$$
$$= (1 + 2x)^{-\frac{1}{2}}$$
$$\frac{dv}{dx} = \frac{1}{(1+2x)^{\frac{1}{2}}}$$

Hence, $\frac{dy}{dx} = u\frac{dv}{dx} + v\frac{du}{dx}$

$$= x^2\frac{1}{(1+2x)^{\frac{1}{2}}} + (1 + 2x)^{\frac{1}{2}} \times 2x$$
$$= \frac{x^2}{(1+2x)^{\frac{1}{2}}} + 2x(1 + 2x)^{\frac{1}{2}}$$

let us now simplify by using $(1 + 2x)^{\frac{1}{2}}$ as the LCM as follows:

$$= \frac{x^2 + 2x(1+2x)^{\frac{1}{2}}(1+2x)^{\frac{1}{2}}}{(1+2x)^{\frac{1}{2}}}$$
$$= \frac{x^2 + 2x(1+2x)}{(1+2x)^{\frac{1}{2}}} \qquad \text{(Note that } (1 + 2x)^{\frac{1}{2}} \times (1 + 2x)^{\frac{1}{2}} = (1 + 2x)^{\frac{1}{2} + \frac{1}{2}} = 1 + 2x\text{)}$$
$$= \frac{x^2 + 2x + 4x^2}{(1+2x)^{\frac{1}{2}}}$$
$$\frac{dy}{dx} = \frac{5x^2 + 2x}{\sqrt{1+2x}}$$

4. Find the derivative of $(2x + 3)^3(4x^2 - 1)^2$

Solution

$$y = (2x + 3)^3(4x^2 - 1)^2$$

Let $u = (2x + 3)^3$

and $v = (4x^2 - 1)^2$

$\dfrac{du}{dx} = 3(2x + 3)^{3-1} \times 2$ (Use of chain rule. Also note that 2 is from the derivative of $2x + 3$)

$\quad = 6(2x + 3)^2$

$\dfrac{dv}{dx} = 2(4x^2 - 1)^{2-1} \times 8x$ (Note that $8x$ is from the derivative of $4x^2 - 1$)

$\quad = 16x(4x^2 - 1)$

Hence, $\dfrac{dy}{dx} = u\dfrac{dv}{dx} + v\dfrac{du}{dx}$

$\quad = (2x + 3)^3 \times 16x(4x^2 - 1) + (4x^2 - 1)^2 \times 6(2x + 3)^2$

Let us factorize the expression by taking out $(2x + 3)^2$ and $(4x^2 - 1)$ which are the common terms as follows:

$\dfrac{dy}{dx} = (2x + 3)^2(4x^2 - 1)[(2x + 3)16x + (4x^2 - 1)6]$

$\quad = (2x + 3)^2(4x^2 - 1)(32x^2 + 48x + 24x^2 - 6)$

$\quad = (2x + 3)^2(4x^2 - 1)(56x^2 + 48x - 6)$

$\quad = (2x + 3)^2(4x^2 - 1)\,2(28x^2 + 24x - 3)$

$\dfrac{dy}{dx} = 2(2x + 3)^2(4x^2 - 1)(28x^2 + 24x - 3)$

5. Differentiate: $y = x(x + 1)(x^2 - 4)$

Solution

$y = x(x + 1)(x^2 - 4)$

Expanding the first two brackets gives:

$y = (x^2 + x)(x^2 - 4)$

Hence, $u = (x^2 + x)$

and $v = (x^2 - 4)$

Hence, $\dfrac{du}{dx} = 2x + 1$

$\dfrac{dv}{dx} = 2x$

Therefore, $\dfrac{dy}{dx} = u\dfrac{dv}{dx} + v\dfrac{du}{dx}$

$\quad = (x^2 + x)2x + (x^2 - 4)(2x + 1)$

$\quad = 2x^3 + 2x^2 + 2x^3 + x^2 - 8x - 4$

$\dfrac{dy}{dx} = 4x^3 + 3x^2 - 8x - 4$

6. Find the derivative of $(x^2 + 3x - 2)^2 \sqrt{x}$

Solution

$y = (x^2 + 3x - 2)^2 \sqrt{x}$

223

$$u = (x^2 + 3x - 2)^2$$

$$v = \sqrt{x}$$

Or, $v = x^{\frac{1}{2}}$

$\dfrac{du}{dx} = 2(x^2 + 3x - 2)^{2-1} \times 2x + 3$ (Note that $2x + 3$ is from the derivative of $x^2 + 3x - 2$)

$\quad = (4x + 6)(x^2 + 3x - 2)$

$\dfrac{dv}{dx} = \dfrac{1}{2} x^{-\frac{1}{2}}$ (Note that $8x$ is from the derivative of $4x^2 - 1$)

$\quad = \dfrac{1}{2x^{\frac{1}{2}}}$

$\dfrac{dv}{dx} = \dfrac{1}{2\sqrt{x}}$

Hence, $\dfrac{dy}{dx} = u\dfrac{dv}{dx} + v\dfrac{du}{dx}$

$\quad = (x^2 + 3x - 2)^2 \dfrac{1}{2\sqrt{x}} + \sqrt{x}\,(4x + 6)(x^2 + 3x - 2)$

$\quad = (x^2 + 3x - 2)^2 \dfrac{1}{2\sqrt{x}} + 2\sqrt{x}\,(2x + 3)(x^2 + 3x - 2)$

Factorize the expression by taking out $(x^2 + 3x - 2)$ which is the common factor. This gives:

$\dfrac{dy}{dx} = (x^2 + 3x - 2)\left[\dfrac{x^2 + 3x - 2}{2\sqrt{x}} + 2\sqrt{x}\,(2x + 3)\right]$

Simplifying the part in the bracket by taking $2\sqrt{x}$ as the LCM gives:

$\dfrac{dy}{dx} = (x^2 + 3x - 2)\left[\dfrac{x^2 + 3x - 2 + 4x(2x + 3)}{2\sqrt{x}}\right]$ (Note that $2\sqrt{x} \times 2\sqrt{x} = 4x$)

$\quad = (x^2 + 3x - 2)\left[\dfrac{x^2 + 3x - 2 + 8x^2 + 12x}{2\sqrt{x}}\right]$

$\dfrac{dy}{dx} = \dfrac{(x^2 + 3x - 2)(9x^2 + 15x - 2)}{2\sqrt{x}}$

7. Find the derivative of $(1 + x)(5x - 2)^{\frac{3}{2}}$

<u>Solution</u>

$y = (1 + x)(5x - 2)^{\frac{3}{2}}$

$u = (1 + x)$

$v = (5x - 2)^{\frac{3}{2}}$

$\dfrac{du}{dx} = 1$

$\dfrac{dv}{dx} = \dfrac{3}{2}(5x - 2)^{\frac{3}{2} - 1} \times 5$ (Note that the derivative of $5x - 2$ is 5)

$\quad = \dfrac{15}{2}(5x - 2)^{\frac{1}{2}}$

Hence, $\dfrac{dy}{dx} = u\dfrac{dv}{dx} + v\dfrac{du}{dx}$

$$= (1 + x) \times \frac{15}{2}(5x - 2)^{\frac{1}{2}} + (5x - 2)^{\frac{3}{2}} \times 1$$

$$= \frac{15}{2}(1 + x)(5x - 2)^{\frac{1}{2}} + (5x - 2)^{\frac{3}{2}}$$

Factorize by taking out $(5x - 2)^{\frac{1}{2}}$ (i.e. the lower exponent) which is the common factor gives:

$$\frac{dy}{dx} = (5x - 2)^{\frac{1}{2}}\left[\frac{15}{2}(1 + x) + 5x - 2\right] \quad \text{(Note that } \frac{(5x-2)^{\frac{3}{2}}}{(5x-2)^{\frac{1}{2}}} = (5x - 2)^{\frac{3}{2} - \frac{1}{2}} = 5x - 2\text{)}$$

$$= (5x - 2)^{\frac{1}{2}}\left[\frac{15}{2} + \frac{15x}{2} + 5x - 2\right]$$

$$= (5x - 2)^{\frac{1}{2}}\left[\frac{25x}{2} + \frac{11}{2}\right]$$

$$\frac{dy}{dx} = \sqrt{5x - 2}\left[\frac{25x + 11}{2}\right]$$

8. If y = (1 + x)(2 − 3x)(2x − 1), find $\frac{dy}{dx}$ by using product rule.

Solution

y = (1 + x)(2 − 3x)(2x − 1)

This is a product of three expressions, u, v and w.

Hence, u = (1 + x)

v = (2 − 3x)

and   w = (2x − 1)

Therefore, $\frac{du}{dx} = 1$

$\frac{dv}{dx} = -3$

$\frac{dw}{dx} = 2$

Hence the formula for product rule of three terms is given by:

$$\frac{dy}{dx} = \frac{du}{dx}vw + \frac{dv}{dx}uw + \frac{dw}{dx}uv$$

$$= 1(2 - 3x)(2x - 1) + (-3)(1 + x)(2x - 1) + 2(1 + x)(2 - 3x)$$

$$= 4x - 2 - 6x^2 + 3x + (-3 - 3x)(2x - 1) + (2 + 2x)(2 - 3x)$$

$$= 7x - 2 - 6x^2 - 6x + 3 - 6x^2 + 3x + 4 - 6x + 4x - 6x^2$$

$$= -2 + 3 + 4 + 7x - 6x + 3x - 6x + 4x - 6x^2 - 6x^2 - 6x^2$$

$$\frac{dy}{dx} = 5 + 2x - 18x^2$$

9. Differentiate with respect to x: $(x^2 - 3x + 5)(2x - 7)$

Solution

y = $(x^2 - 3x + 5)(2x - 7)$

Let us differentiate this product by applying product rule but without the use u and v. This is done as follows:

$$\frac{dy}{dx} = (x^2 - 3x + 5)\frac{d(2x-7)}{dx} + (2x - 7)\frac{d(x^2 - 3x + 5)}{dx}$$

$$= (x^2 - 3x + 5)2 + (2x - 7)(2x - 3)$$

$$= 2x^2 - 6x + 10 + 4x^2 - 6x - 14x + 21$$

$$\frac{dy}{dx} = 6x^2 - 26x + 31$$

10. If $y = (5x^2 - 3)(2 + \frac{3}{x})$, find $\frac{dy}{dx}$

Solution

$$y = (5x^2 - 3)(2 + \frac{3}{x})$$

$$u = (5x^2 - 3)$$

$$v = (2 + \frac{3}{x})$$

$$\frac{du}{dx} = 10x$$

$$\frac{dv}{dx} = \frac{d(3x^{-1})}{dx} \qquad \text{(Note that } \frac{3}{x} = 3x^{-1})$$

$$= -3x^{-2}$$

$$\frac{dv}{dx} = \frac{-3}{x^2}$$

Hence, $\frac{dy}{dx} = u\frac{dv}{dx} + v\frac{du}{dx}$

$$= (5x^2 - 3)\left(\frac{-3}{x^2}\right) + (2 + \frac{3}{x})10x$$

$$= -15 + \frac{9}{x^2} + 20x + 30$$

$$\frac{dy}{dx} = 15 + 20x + \frac{9}{x^2}$$

Exercise 19

1. If $f(x) = (2x - 1)(3x + 1)$, find $f'(x)$
2. Find the derivative of $y = (3x^2 - 5)(x^2 + 10)$
3. If $y = 5x(3 + x)^{\frac{1}{2}}$ find $\frac{dy}{dx}$
4. Find the derivative of $(x + 5)(x^2 - 7)^3$
5. Differentiate: $y = 2x(3x + 2)(2x^2 - 5)$
6. Find the derivative of $(3x^2 - 1)^3 \sqrt{2x}$

7. Find the derivative of $(9 - x)(x + 3)^{\frac{3}{4}}$

8. If $y = (2 - x)(5 - 3x^2)(x + 3)$, find $\dfrac{dy}{dx}$ by using product rule.

9. Differentiate with respect to x: $(3x^4 - x^2 + 2x)(x - 1)$

10. If $y = (3x^2 - x)(1 + \dfrac{1}{2x})$, find $\dfrac{dy}{dx}$

11. Find the derivative of $x^2(\sqrt{x^5})$

12. Find the derivative of $x^4(2x - 11)^{\frac{2}{3}}$

13. If $y = (7 + x)(1 - x^2)(5x^3 + 1)$, find $\dfrac{dy}{dx}$.

14. Differentiate with respect to x: $(3x^4 - x^3 + x^2 + 2x - 3)(5x + 4)$

15. If $y = (x^3 - 3x + 5)\left(\dfrac{1}{x^5}\right)$, find $\dfrac{dy}{dx}$

CHAPTER 20
QUOTIENT RULE OF DERIVATIVE

If $y = \dfrac{u}{v}$ where u and v are functions of x, then:

$$\frac{dy}{dx} = \frac{v\frac{du}{dx} - u\frac{dv}{dx}}{v^2}$$

This is called the quotient rule of differentiation.

Examples

1. If $y = \dfrac{3x^2 - 8x + 5}{5x - 2}$ find $\dfrac{dy}{dx}$

Solution

$$y = \frac{3x^2 - 8x + 5}{5x - 2}$$

This is of the form $y = \dfrac{u}{v}$. Therefore, we are going to apply product rule.

Let $u = 3x^2 - 8x + 5$

and $v = 5x - 2$

Hence, $\dfrac{du}{dx} = 6x - 8$

$\dfrac{dv}{dx} = 5$

Therefore, $\dfrac{dy}{dx} = \dfrac{v\frac{du}{dx} - u\frac{dv}{dx}}{v^2}$

$$= \frac{(5x - 2)(6x - 8) - (3x^2 - 8x + 5)(5)}{(5x - 2)^2}$$

$$= \frac{30x^2 - 40x - 12x + 16 - 15x^2 - 40x - 25}{(5x - 2)^2}$$

$$\frac{dy}{dx} = \frac{15x^2 - 12x - 9}{(5x - 2)^2}$$

2. Differentiate with respect to x, the function: $\dfrac{3x^2 - 2x}{x + 5}$

Solution

$$y = \frac{3x^2 - 2x}{x + 5}$$

Let $u = 3x^2 - 2x$

and $v = x + 5$

Hence, $\dfrac{du}{dx} = 6x - 2$

$\dfrac{dv}{dx} = 1$

Therefore, $\dfrac{dy}{dx} = \dfrac{v\frac{du}{dx} - u\frac{dv}{dx}}{v^2}$

$= \dfrac{(x+5)(6x-2) - (3x^2-2x)(1)}{(x+5)^2}$

$= \dfrac{6x^2 - 2x + 30x - 10 - 3x^2 + 2x}{(x+5)^2}$

$\dfrac{dy}{dx} = \dfrac{3x^2 + 30x - 10}{(x+5)^2}$

3. Find the differential coefficient of $y = \dfrac{-3}{x^2 + 5}$

Solution

$y = \dfrac{-3}{x^2 + 5}$

$u = -3$

and $v = x^2 + 5$

Hence, $\dfrac{du}{dx} = 0$ (The derivative of a constant is zero)

$\dfrac{dv}{dx} = 2x$

Therefore, $\dfrac{dy}{dx} = \dfrac{v\frac{du}{dx} - u\frac{dv}{dx}}{v^2}$

$= \dfrac{(x^2 + 5) \times 0 - (-3)2x}{(x^2 + 5)^2}$

$= \dfrac{0 - 6x}{(x^2 + 5)^2}$

$\dfrac{dy}{dx} = \dfrac{-6x}{(x^2 + 5)^2}$

4. Differentiate: $y = \dfrac{\sqrt{3 - x}}{\sqrt{3 + x}}$

Solution

$y = \dfrac{\sqrt{3 - x}}{\sqrt{3 + x}}$

$u = \sqrt{3 - x}$

$= (3 - x)^{\frac{1}{2}}$

and $v = \sqrt{3 + x}$

$= (3 + x)^{\frac{1}{2}}$

Hence, $\dfrac{du}{dx} = \dfrac{1}{2}(3 - x)^{\frac{1}{2} - 1} \times -1$ (Note that the derivative of $3 - x$ is -1)

$= -\dfrac{1}{2}(3 - x)^{-\frac{1}{2}}$

229

$$\frac{du}{dx} = \frac{-(3-x)^{-\frac{1}{2}}}{2}$$

$$\frac{dv}{dx} = \frac{1}{2}(3+x)^{\frac{1}{2}-1} \times 1 \quad \text{(Note that the derivative of } 3-x \text{ is } -1)$$

$$= \frac{1}{2}(3+x)^{-\frac{1}{2}}$$

$$\frac{dv}{dx} = \frac{(3+x)^{-\frac{1}{2}}}{2}$$

Therefore, $\dfrac{dy}{dx} = \dfrac{v\frac{du}{dx} - u\frac{dv}{dx}}{v^2}$

$$= \frac{(3+x)^{\frac{1}{2}}\left(\dfrac{-(3-x)^{-\frac{1}{2}}}{2}\right) - (3-x)^{\frac{1}{2}}\left(\dfrac{(3+x)^{-\frac{1}{2}}}{2}\right)}{\left((3+x)^{\frac{1}{2}}\right)^2}$$

Taking out the common terms which are terms with positive exponent in order to factorize the expression gives:

$$\frac{dy}{dx} = \frac{\left((3+x)^{\frac{1}{2}}\right)\left((3-x)^{\frac{1}{2}}\right)\left[\dfrac{-(3-x)^{-1}}{2} - \dfrac{(3+x)^{-1}}{2}\right]}{3+x}$$

Note that in order to obtain the terms in the square bracket, we subtracted the exponents of the factors from the exponents of the original term. For example, $(3+x)^{\frac{1}{2}}\left(\dfrac{-(3-x)^{-\frac{1}{2}}}{2}\right)$ divided by $\left((3+x)^{\frac{1}{2}}\right)\left((3-x)^{\frac{1}{2}}\right)$ gave $\dfrac{-(3-x)^{-1}}{2}$ since the equal terms canceled out and $-\dfrac{1}{2} - \dfrac{1}{2} = -1$, which gave the exponent of -1. Similarly, $(3-x)^{\frac{1}{2}}\left(\dfrac{(3+x)^{-\frac{1}{2}}}{2}\right)$ divided by $\left((3+x)^{\frac{1}{2}}\right)\left((3-x)^{\frac{1}{2}}\right)$ gave $\dfrac{(3+x)^{-1}}{2}$ since the equal terms canceled out and $-\dfrac{1}{2} - \dfrac{1}{2} = -1$, which gave the exponent of -1 as the terms in the square bracket.

Let us now continue with the solution by simplifying further as follows:

$$\frac{dy}{dx} = \frac{\left((3+x)^{\frac{1}{2}}\right)\left((3-x)^{\frac{1}{2}}\right)\left[\dfrac{-(3-x)^{-1}}{2} - \dfrac{(3+x)^{-1}}{2}\right]}{3+x}$$

$$= \frac{\left((3+x)^{\frac{1}{2}}\right)\left((3-x)^{\frac{1}{2}}\right)\left[\dfrac{-1}{2(3-x)} - \dfrac{1}{2(3+x)}\right]}{3+x}$$

$$= \frac{\left((3+x)^{\frac{1}{2}}\right)\left((3-x)^{\frac{1}{2}}\right)\left[\dfrac{-(3+x) - (3-x)}{2(3-x)(3+x)}\right]}{3+x}$$

$$= \frac{\left((3+x)^{\frac{1}{2}}\right)\left((3-x)^{\frac{1}{2}}\right)\left[\frac{-3-x-3+x}{2(3-x)(3+x)}\right]}{3+x}$$

$$= \frac{\left((3+x)^{\frac{1}{2}}\right)\left((3-x)^{\frac{1}{2}}\right)\left[\frac{-6}{2(3-x)(3+x)}\right]}{3+x}$$

$$= \frac{\left((3+x)^{\frac{1}{2}}\right)\left((3-x)^{\frac{1}{2}}\right)\left[\frac{-3}{(3-x)(3+x)}\right]}{3+x}$$

$$= \frac{-3\left((3+x)^{\frac{1}{2}}\right)\left((3-x)^{\frac{1}{2}}\right)}{(3-x)(3+x)(3+x)}$$

$$= \frac{-3\left((3+x)^{\frac{1}{2}}\right)\left((3-x)^{\frac{1}{2}}\right)}{(3+x)^2(3-x)}$$

$$= -3(3+x)^{\frac{1}{2}-2}(3-x)^{\frac{1}{2}-1} \quad \text{(Subtraction of exponents due to the division above)}$$

$$= -3(3+x)^{-\frac{3}{2}}(3-x)^{-\frac{1}{2}}$$

$$= \frac{-3}{(3+x)^{\frac{3}{2}}(3-x)^{\frac{1}{2}}}$$

$$= \frac{-3}{\sqrt{(3+x)^3(3-x)}}$$

5. Differentiate with respect to x: $y = \dfrac{(2x^2-3)^3}{x}$

Solution

$$y = \frac{(2x^2-3)^3}{x}$$
$$u = (2x^2-3)^3$$

and $v = x$

Hence, $\dfrac{du}{dx} = 3(2x^2-3)^{3-1} \times 4x$

$$= 12x(2x^2-3)^2$$

$$\frac{dv}{dx} = 1$$

Therefore, $\dfrac{dy}{dx} = \dfrac{v\frac{du}{dx} - u\frac{dv}{dx}}{v^2}$

$$= \frac{x[12x(2x^2-3)^2] - (2x^2-3)^3 \times 1}{x^2}$$

$$= \frac{12x^2(2x^2-3)^2 - (2x^2-3)^3}{x^2}$$

Factorize the expression by taking out $(2x^2-3)^2$ which has the lower exponent. This gives:

$$\frac{dy}{dx} = \frac{(2x^2 - 3)^2[12x^2 - (2x^2 - 3)]}{x^2}$$

$$= \frac{(2x^2 - 3)^2(12x^2 - 2x^2 + 3)}{x^2}$$

$$\frac{dy}{dx} = \frac{(2x^2 - 3)^2(10x^2 + 3)}{x^2}$$

6. Find the derivative of $\dfrac{\sqrt{(1+2x^2)^3}}{x}$

<u>Solution</u>

$$y = \frac{\sqrt{(1+2x^2)^3}}{x}$$

$$u = \sqrt{(1 + 2x^2)^3}$$

$$= [(1 + 2x^2)^3]^{\frac{1}{2}} \qquad \text{(When the square root sign is removed, we use an exponent of } \tfrac{1}{2}\text{)}$$

$$u = (1 + 2x^2)^{\frac{3}{2}} \qquad \text{(After multiplying the exponents)}$$

$$v = x$$

Hence, $\dfrac{du}{dx} = \dfrac{3}{2}(1 + 2x^2)^{\frac{3}{2} - 1} \text{ x } 4x$ (Note that the derivative of $3 - x$ is -1)

$$= 6x(1 + 2x^2)^{\frac{1}{2}}$$

$$\frac{dv}{dx} = 1$$

Therefore, $\dfrac{dy}{dx} = \dfrac{v\frac{du}{dx} - u\frac{dv}{dx}}{v^2}$

$$= \frac{x\left(6x(1+2x^2)^{\frac{1}{2}}\right) - (1+2x^2)^{\frac{3}{2}} \text{ x } 1}{x^2}$$

Take out $(1 + 2x^2)^{\frac{1}{2}}$ which has the lower exponent and factorize the expression. This gives:

$$\frac{dy}{dx} = \frac{(1+2x^2)^{\frac{1}{2}}[6x^2 - (1+2x^2)]}{x^2} \qquad \left(\text{Note that } \frac{(1+2x^2)^{\frac{3}{2}}}{(1+2x^2)^{\frac{1}{2}}} \text{ gives } (1 + 2x^2) \text{ by subtracting exponents.}\right.$$

$$= \frac{(1+2x^2)^{\frac{1}{2}}(6x^2 - 1 - 2x^2)}{x^2}$$

$$= \frac{(1+2x^2)^{\frac{1}{2}}(4x^2 - 1)}{x^2}$$

$$\frac{dy}{dx} = \frac{(\sqrt{1+2x^2})(4x^2 - 1)}{x^2}$$

232

7. If $y = \dfrac{(4x^3 - 3x^2 + x + 1)^{\frac{1}{2}}}{(x+1)^2}$

Solution

$$y = \dfrac{(4x^3 - 3x^2 + x + 1)^{\frac{1}{2}}}{(x+1)^2}$$

$u = (4x^3 - 3x^2 + x + 1)^{\frac{1}{2}}$ (After multiplying the exponents)

$v = (x + 1)^2$

Hence, $\dfrac{du}{dx} = \dfrac{1}{2}(4x^3 - 3x^2 + x + 1)^{\frac{1}{2} - 1} \times (12x^2 - 6x + 1)$

$$= \dfrac{(12x^2 - 6x + 1)(4x^3 - 3x^2 + x + 1)^{-\frac{1}{2}}}{2}$$

$\dfrac{dv}{dx} = 2(x + 1)^{2-1}$

$\quad\quad = 2(x + 1)$

Therefore, $\dfrac{dy}{dx} = \dfrac{v\frac{du}{dx} - u\frac{dv}{dx}}{v^2}$

$$= \dfrac{\dfrac{(x+1)^2(12x^2 - 6x + 1)(4x^3 - 3x^2 + x + 1)^{-\frac{1}{2}} - (4x^3 - 3x^2 + x + 1)^{\frac{1}{2}}[2(x+1)]}{2}}{[(x+1)^2]^2}$$

Take out the common terms with lower exponents [i.e. $(x + 1)$ and $(4x^3 - 3x^2 + x + 1)^{-\frac{1}{2}}$] and factorize the expression. This gives:

$$\dfrac{dy}{dx} = \dfrac{\dfrac{(x+1)(4x^3 - 3x^2 + x + 1)^{-\frac{1}{2}}[(x+1)(12x^2 - 6x + 1) - (4x^3 - 3x^2 + x + 1)2]}{2}}{[(x+1)^2]^2}$$

Remember to subtract the exponents of the factors from the exponents of the original expression when simplifying. Simplifying further, the expression above gives:

$$\dfrac{dy}{dx} = \dfrac{(x+1)(4x^3 - 3x^2 + x + 1)^{-\frac{1}{2}}[12x^3 - 6x^2 + x + 12x^2 - 6x + 1 - (8x^3 - 6x^2 + 2x + 2)]}{2(x+1)^4}$$

$$= \dfrac{(x+1)(4x^3 - 3x^2 + x + 1)^{-\frac{1}{2}}(12x^3 - 6x^2 + x + 12x^2 - 6x + 1 - 8x^3 + 6x^2 - 2x - 2)}{2(x+1)^4}$$

$$= \dfrac{(x+1)(4x^3 - 3x^2 + x + 1)^{-\frac{1}{2}}(4x^3 + 12x^2 - 7x - 1)}{2(x+1)^4}$$

$$= \dfrac{(4x^3 - 3x^2 + x + 1)^{-\frac{1}{2}}(4x^3 + 12x^2 - 7x - 1)}{2(x+1)^3}$$

Note that $(x + 1)$ cancels out from the numerator and denominator.

Therefore, $\dfrac{dy}{dx} = \dfrac{4x^3 + 12x^2 - 7x - 1}{2(4x^3 - 3x^2 + x + 1)^{\frac{1}{2}}(x+1)^3}$

8. Determine $\dfrac{d}{dx}\left(\dfrac{3 + 2x - x^2}{\sqrt{1+x}}\right)$

Solution

Let $y = \dfrac{3 + 2x - x^2}{\sqrt{1+x}}$

Hence, $u = 3 + 2x - x^2$

$\dfrac{du}{dx} = 2 - 2x$

$v = \sqrt{1 + x}$

$= (1 + x)^{\frac{1}{2}}$

$\dfrac{du}{dx} = \dfrac{1}{2}(1 + x)^{\frac{1}{2} - 1} \times 1$

$= \dfrac{1}{2}(1 + x)^{-\frac{1}{2}}$

Therefore, $\dfrac{dy}{dx} = \dfrac{v\dfrac{du}{dx} - u\dfrac{dv}{dx}}{v^2}$

$= \dfrac{(1 + x)^{\frac{1}{2}}(2 - 2x) - (3 + 2x - x^2)\frac{1}{2}(1 + x)^{-\frac{1}{2}}}{[(1 + x)^{\frac{1}{2}}]^2}$

Take out $(1 + x)^{-\frac{1}{2}}$ as the common factor since it has the lower exponent and factorize the expression. This gives:

$\dfrac{dy}{dx} = \dfrac{(1 + x)^{-\frac{1}{2}}\left[(1+x)(2 - 2x) - (3 + 2x - x^2)\frac{1}{2}\right]}{1+x}$

$= \dfrac{(1 + x)^{-\frac{1}{2}}\left[\dfrac{2(1+x)(2 - 2x) - (3 + 2x - x^2)}{2}\right]}{1+x}$

$= \dfrac{(1 + x)^{-\frac{1}{2}}[(2 + 2x)(2 - 2x) - 3 - 2x + x^2]}{2(1+x)}$

$= \dfrac{(1 + x)^{-\frac{1}{2}}(4 - 4x + 4x - 4x^2 - 3 - 2x + x^2)}{2(1+x)}$

$$= \frac{1 - 2x - 3x^2}{2(1+x)^{\frac{1}{2}}(1+x)} \quad \text{[When } (1+x)^{-\frac{1}{2}} \text{ is taken to the denominator it becomes } (1+x)^{\frac{1}{2}}]$$

$$\frac{dy}{dx} = \frac{1 - 2x - 3x^2}{2(1+x)^{\frac{3}{2}}} \quad \text{(The exponents of same terms have been added together, i.e. } \frac{1}{2} + 1 = \frac{3}{2})$$

Exercise 20

1. If $y = \dfrac{x^2 - 5x + 1}{x - 1}$ find $\dfrac{dy}{dx}$

2. Differentiate with respect to x, the function: $\dfrac{4x^2 - x}{2x + 3}$

3. Find the differential coefficient of $y = \dfrac{-7}{3x^2 + 1}$

4. Differentiate: $y = \dfrac{\sqrt{1+x}}{\sqrt{1-x}}$

5. Differentiate with respect to x: $y = \dfrac{(x^3 - 2)^2}{x^2}$

6. Find the derivative of $\dfrac{\sqrt{2 + 3x^2}}{x^3}$

7. If $y = \dfrac{(x^2 - x - 4)^{\frac{1}{3}}}{(2x + 1)^2}$

8. Determine $\dfrac{d}{dx}\left(\dfrac{x - 2x^3}{\sqrt{2-x}}\right)$

9. Find the differential coefficient of $y = -\dfrac{1}{1 - 3x^2}$

10. Differentiate: $y = \dfrac{2 + x}{2 - x}$

CHAPTER 21
DERIVATIVE OF PARAMETRIC EQUATIONS

If y = f(t) and x = g(t) are two different functions of a common variable, t, then the two equations are called parametric equations. The variable t, is the parameter.

The derivative of a parametric equation such as the one stated above is obtained as follows:

$$\frac{dy}{dx} = \frac{\frac{dy}{dt}}{\frac{dx}{dt}}$$

Examples

1. If y = 5 + t² and x = 3 − 2t², find $\frac{dy}{dx}$

Solution

$$y = 5 + t^2$$

Hence, $\frac{dy}{dt} = 2t$

$$x = 3 - 2t^2$$

Hence, $\frac{dx}{dt} = -4t$

Therefore, $\dfrac{dy}{dx} = \dfrac{\frac{dy}{dt}}{\frac{dx}{dt}}$

$$= \frac{2t}{-4t}$$

$$= -\frac{1}{2} \qquad \text{(t cancels out)}$$

2. Find $\frac{dy}{dx}$ of the functions below which are expressed in the parametric form.

$$x = \frac{5}{t^2} \quad \text{and} \quad y = 2t^5 - 3$$

Solution

$$x = \frac{5}{t^2}$$

$$x = 5t^{-2}$$

$$\frac{dx}{dt} = -10t^{-3}$$

$$y = 2t^5 - 3$$

$$\frac{dy}{dt} = 10t^4$$

Therefore, $\dfrac{dy}{dx} = \dfrac{\frac{dy}{dt}}{\frac{dx}{dt}}$

$$= \frac{10t^4}{-10t^{-3}}$$
$$= -t^{4-(-3)} \quad \text{(10 cancels out)}$$
$$= -t^{4+3}$$
$$= -t^7$$

3. If $V = \frac{4}{3}\pi r^3$ and $A = \pi r^2$, find $\frac{dA}{dV}$

Solution

$$V = \frac{4}{3}\pi r^3$$

$$\frac{dV}{dr} = 3(\frac{4}{3})\pi r^2$$

$$= 4\pi r^2$$

$$A = \pi r^2$$

$$\frac{dA}{dr} = 2\pi r$$

$$\frac{dA}{dV} = \frac{\frac{dA}{dr}}{\frac{dV}{dr}}$$

$$= \frac{2\pi r}{4\pi r^2}$$

$$\frac{dA}{dV} = \frac{1}{2r}$$

4. Determine the derivative of the curve defined by the equations: $x = t^2 - 4t$ and $y = 2t^3 - 7t$.

Solution

$$y = 2t^3 - 7t$$

Hence, $\frac{dy}{dt} = 6t^2 - 7$

$$x = t^2 - 4t$$

Hence, $\frac{dx}{dt} = 2t - 4$

Therefore, $\frac{dy}{dx} = \frac{\frac{dy}{dt}}{\frac{dx}{dt}}$

$$= \frac{6t^2 - 7}{2t - 4}$$

5. Given that $v = u + at$ and $s = ut + \frac{1}{2}at^2$. Find $\frac{dv}{ds}$ if u and a are constants.

Solution

$$v = u + at$$

$$\frac{dv}{dt} = a$$

$$s = ut + \frac{1}{2}at^2$$

$$\frac{ds}{dt} = u + (2 \times \frac{1}{2} \times at)$$

$$= u + at$$

Hence, $\dfrac{dv}{ds} = \dfrac{\frac{dv}{dt}}{\frac{ds}{dt}}$

$$= \frac{a}{u + at}$$

6. If $y = \dfrac{t}{t-2}$ and $x = \dfrac{1}{t+1}$, find $\dfrac{dy}{dx}$.

Solution

$$y = \frac{t}{t-2}$$

$$\frac{dy}{dt} = \frac{(t-2)(1) - t(1)}{(t-2)^2} \qquad \text{(Use of quotient rule where u = t and v = t − 2)}$$

$$= \frac{t - 2 - t}{(t-2)^2}$$

$$\frac{dy}{dt} = \frac{-2}{(t-2)^2}$$

$$x = \frac{1}{t+1}$$

$$\frac{dx}{dt} = \frac{(t+1)(0) - 1(1)}{(t+1)^2} \qquad \text{(Use of quotient rule where u = 1 and v = t + 1)}$$

$$= \frac{-1}{(t+1)^2}$$

Hence, $\dfrac{dy}{dx} = \dfrac{\frac{dy}{dt}}{\frac{dx}{dt}}$

$$= \frac{\frac{-2}{(t-2)^2}}{\frac{-1}{(t+1)^2}}$$

$$= \frac{-2}{(t-2)^2} \times \frac{(t+1)^2}{-1}$$

$$\frac{dy}{dx} = \frac{2(t+1)^2}{(t-2)^2}$$

7. The parametric equations of the motion of a stone are: $y = 12 + 3t - 2t^2$ and $x = 5t$. Find $\dfrac{dy}{dx}$.

Solution

$$y = 12 + 3t - 2t^2$$

$$\frac{dy}{dt} = 3 - 4t$$

$$x = 5t$$

$$\frac{dx}{dt} = 5$$

$$\frac{dy}{dx} = \frac{\frac{dy}{dt}}{\frac{dx}{dt}}$$

$$= \frac{3 - 4t}{5}$$

8. If the parametric equations of a parabola are $y = \frac{2mt^2}{1 + t^2}$ and $x = \frac{2m}{1 + t^2}$ where m is a constant, find $\frac{dy}{dx}$.

<u>Solution</u>

$$y = \frac{2mt^2}{1 + t^2}$$

$$\frac{dy}{dt} = \frac{(1 + t^2)(4mt) - 2mt^2(2t)}{(1 + t^2)^2} \qquad \text{(Use of quotient rule where u = } 2mt^2 \text{ and v = 1 + } t^2)$$

$$= \frac{4mt + 4mt^3 - 4mt^3}{(1 + t^2)^2}$$

$$\frac{dy}{dt} = \frac{4mt}{(1 + t^2)^2}$$

$$x = \frac{2m}{1 + t^2}$$

$$\frac{dx}{dt} = \frac{(1 + t^2)(0) - 2m(2t)}{(1 + t^2)^2} \qquad \text{(Use of quotient rule where u = 1 and v = t + 1)}$$

$$= \frac{-4mt}{(1 + t^2)^2}$$

Hence, $\frac{dy}{dx} = \frac{\frac{dy}{dt}}{\frac{dx}{dt}}$

$$= \frac{\frac{4mt}{(1 + t^2)^2}}{\frac{-4mt}{(1 + t^2)^2}}$$

$$= \frac{4mt}{(1 + t^2)^2} \quad \text{x} \quad \frac{(1 + t^2)^2}{-4mt}$$

$$\frac{dy}{dx} = -1 \qquad \text{(Same terms cancels out)}$$

Exercise 21

1. If $y = t^3 - 5$ and $x = t^2 + 1$ find $\dfrac{dy}{dx}$

2. Find $\dfrac{dy}{dx}$ of the functions below which are expressed in the parametric form:

$x = \dfrac{1}{4 - t^3}$ and $y = 3 + t^2$

3. If $V = \dfrac{1}{3}\pi r^2$ and $A = \pi r l + \pi r^2$, find $\dfrac{dV}{dA}$

4. Find the derivative of the curve defined by the equations: $x = 5t^2 - t + 3$ and $y = t^2 - t - 1$.

5. Given that $F = \dfrac{m(v - u)}{t}$ and $s = \dfrac{(u+v)t}{2}$. Find $\dfrac{dF}{ds}$ if u, v and m are constants.

6. If $y = \dfrac{5}{t^2 - 1}$ and $x = \dfrac{3}{t^4 + 1}$, find $\dfrac{dy}{dx}$

7. The parametric equations of the motion of a stone are: $y = t - 3t^4$ and $x = t^2 + 3$. Find $\dfrac{dy}{dx}$.

8. If the parametric equations of a parabola are $y = \dfrac{at^3}{2}$ and $x = \dfrac{at}{5}$ where a is a constant, find $\dfrac{dy}{dx}$.

9. If $V = \dfrac{1}{3}\pi r^3$ and $A = 4\pi r^2$, find $\dfrac{dA}{dV}$

10. Given that $G = 2m + s^2$ and $H = m^2 - \dfrac{3}{s}$. Find $\dfrac{dG}{dH}$ if m is a constant.

CHAPTER 22
DERIVATIVE OF IMPLICIT FUNCTIONS

In the function y = f(x), y is said to be expressed explicitly in terms of x. However, in expressions such as $2xy - x^2y = 5$, the relationship between y and x is said to be implicit.

In order to differentiate implicit functions, y is differentiated just like x but with the addition of $\frac{dy}{dx}$ along with the value obtained.

Examples

1. Differentiate implicitly, the expression: $2x^2 + y^2 = 9$.

<u>Solution</u>

$$2x^2 + y^2 = 9$$

Follow the rule of differentiation and add $\frac{dy}{dx}$ to the value obtained whenever you differentiate y. Hence we differentiate each term in the expression above as follows:

$$\frac{d(2x^2)}{dx} + \frac{d(2y^2)}{dx} = \frac{d(9)}{dx}$$

$$4x + 2y\frac{dy}{dx} = 0 \quad \text{(The derivative of } y^2 \text{ is 2y and the addition of } \frac{dy}{dx} \text{ to 2y gives } 2y\frac{dy}{dx}\text{)}$$

We now make $\frac{dy}{dx}$ the subject of the formula as follows:

$$2y\frac{dy}{dx} = -4x$$

$$\frac{dy}{dx} = \frac{-4x}{2y} \quad \text{(When both sides are divided by 2y)}$$

$$\frac{dy}{dx} = \frac{-2x}{y}$$

2. If $x^3 + y^3 = 18xy$, find $\frac{dy}{dx}$

<u>Solution</u>

$$x^3 + y^3 = 18xy$$

$$3x^2 + 3y^2\frac{dy}{dx} = (18x \times 1\frac{dy}{dx}) + (y \times 18)$$

Note that $18xy$ is differentiated by using product rule where $18x$ is taken as u while y is take as v. Also, the derivative of y is what gave us $1\frac{dy}{dx}$. The above differentiation now simplifies to:

$$3x^2 + 3y^2\frac{dy}{dx} = 18x\frac{dy}{dx} + 18y$$

Collect terms in $\frac{dy}{dx}$ on one side in order to make $\frac{dy}{dx}$ the subject of the formula as follows:

$$3y^2\frac{dy}{dx} - 18x\frac{dy}{dx} = 18y - 3x^2$$

Factorizing the left hand side gives:

$$\frac{dy}{dx}(3y^2 - 18x) = 18y - 3x^2$$

Divide both sides by $3y^2 - 18x$. This gives:

$$\frac{dy}{dx} = \frac{18y - 3x^2}{3y^2 - 18x}$$

$$= \frac{3(6y - x^2)}{3(y^2 - 6x)}$$

$$\frac{dy}{dx} = \frac{6y - x^2}{y^2 - 6x} \qquad \text{(3 cancels out)}$$

3. Find $\frac{dy}{dx}$ given that $x^2y^2 - 3xy + 4xy^3 = 4$

Solution

$$x^2y^2 - 3xy + 4xy^3 = 4$$

Apply product rule to x^2y^2, $3xy$ and $4xy^3$ and differentiate appropriately as follows:

$$(x^2 \times 2y\frac{dy}{dx}) + (y^2 \times 2x) - [(3x \times 1\frac{dy}{dx}) + (y \times 3)] + (4x \times 3y^2\frac{dy}{dx}) + (y^3 \times 4) = 0$$

$$2x^2y\frac{dy}{dx} + 2xy^2 - (3x\frac{dy}{dx} + 3y) + 12xy^2\frac{dy}{dx} + 4y^3 = 0$$

$$2x^2y\frac{dy}{dx} + 2xy^2 - 3x\frac{dy}{dx} - 3y + 12xy^2\frac{dy}{dx} + 4y^3 = 0$$

$$2x^2y\frac{dy}{dx} - 3x\frac{dy}{dx} + 12xy^2\frac{dy}{dx} + 2xy^2 - 3y + 4y^3 = 0$$

$$2x^2y\frac{dy}{dx} - 3x\frac{dy}{dx} + 12xy^2\frac{dy}{dx} = 3y - 2xy^2 - 4y^3$$

Factorizing the left hand side gives:

$$\frac{dy}{dx}(2x^2y - 3x + 12xy^2) = 3y - 2xy^2 - 4y^3$$

$$\frac{dy}{dx} = \frac{3y - 2xy^2 - 4y^3}{2x^2y - 3x + 12xy^2}$$

4. Differentiate $x^4 + 6x^2y^2 - 5 = 0$ implicitly with respect to x.

Solution

$$x^4 + 6x^2y^2 - 5 = 0$$

We differentiate accordingly and apply product rule to $6x^2y^2$ as follows:

$$4x^3 + (6x^2 \times 2y\frac{dy}{dx}) + (y^2 \times 12x) - 0 = 0$$

$$4x^3 + 12x^2y\frac{dy}{dx} + 12xy^2 = 0$$

$$12x^2y\frac{dy}{dx} = -4x^3 - 12xy^2$$

$$\frac{dy}{dx} = \frac{-4x^3 - 12xy^2}{12x^2y}$$

$$= \frac{-4x(x^2 + 3y^2)}{12x^2y}$$

$$\frac{dy}{dx} = \frac{-(x^2 + 3y^2)}{3xy} \qquad \text{(4 and } x \text{ cancels out)}$$

5. Find $\dfrac{dy}{dx}$ if $\dfrac{x^2}{16} + \dfrac{y^2}{25} = 1$

Solution

$$\frac{x^2}{16} + \frac{y^2}{25} = 1$$

$$\frac{2x}{16} + \frac{2y}{25}\frac{dy}{dx} = 0$$

$$\frac{2y}{25}\frac{dy}{dx} = -\frac{2x}{16}$$

$$\frac{2y}{25}\frac{dy}{dx} = -\frac{x}{8}$$

$$\frac{dy}{dx} = \frac{-\frac{x}{8}}{\frac{2y}{25}}$$

$$= -\frac{x}{8} \times \frac{25}{2y}$$

$$\frac{dy}{dx} = -\frac{25x}{16y}$$

Exercise 22

1. Differentiate implicitly, the expression: $5x^3 + 3y = y^2$.

2. If $2x^2 + 3y^3 = 10$, find $\dfrac{dy}{dx}$

3. Find $\dfrac{dy}{dx}$ given that $xy - 3xy^2 + 4x^3 = 4$

4. Differentiate $2x^2 + xy^2 - 5 = x^3$ implicitly with respect to x.

5. Find $\dfrac{dy}{dx}$ if $\dfrac{x^3}{3} - \dfrac{y^2}{2} = 0$

6. Differentiate implicitly, the expression: $2x^2y + 5y = 1$.

7. If $2x^2y + 4y^3 = 2y$, find $\dfrac{dy}{dx}$

8. Find $\dfrac{dy}{dx}$ given that $5y^2 - 4xy = y$

243

9. Differentiate $x + y^2 - x^2y^2 = 7$ implicitly with respect to x.

10. Find $\dfrac{dy}{dx}$ if $x^5 - \dfrac{3}{y} = 2$

CHAPTER 23
DERIVATIVE OF TRIGONOMETRIC FUNCTIONS

The derivatives of trigonometric functions are as given below:

If y = sin$x$, then $\dfrac{dy}{dx}$ = cos$x$

If y = cos$x$, then $\dfrac{dy}{dx}$ = $-$ sin$x$

If y = tan$x$, then $\dfrac{dy}{dx}$ = sec^2x

If y = cot$x$, then $\dfrac{dy}{dx}$ = $-$ cosec^2x

If y = sec$x$, then $\dfrac{dy}{dx}$ = sec$x$tan$x$

If y = cosec$x$, then $\dfrac{dy}{dx}$ = $-$ cot$x$cosec$x$

Example

1. Find the derivative of cos5x

<u>Solution</u>

Let y = cos5x

We have to use chain rule from composite function due to 5x which is a function of x.

Hence, let u = 5x

Therefore, y = cosu   (By replacing 5x with u)

$\dfrac{du}{dx} = 5$

$\dfrac{dy}{du} = -$ sinu   (Recall that the derivative of cos$x$ is $-$sin$x$)

Therefore, $\dfrac{dy}{dx} = \dfrac{dy}{du}$ x $\dfrac{du}{dx}$ (Chain rule)

      $= -$ sinu x 5

      $= -$ 5sinu

$\dfrac{dy}{dx} = -$ 5sin5x (Since u = 5x)

2. If y = sin$\dfrac{1}{2}$$x$, find $\dfrac{dy}{dx}$

<u>Solution</u>

   y = sin$\dfrac{1}{2}$$x$

Let u = $\dfrac{1}{2}$$x$

Therefore, y = sinu   (By replacing $\dfrac{1}{2}$$x$ with u)

$$\frac{du}{dx} = \frac{1}{2}$$

$$\frac{dy}{du} = \cos u \qquad \text{(Recall that the derivative of } \sin x \text{ is } \cos x\text{)}$$

Therefore, $\dfrac{dy}{dx} = \dfrac{dy}{du} \times \dfrac{du}{dx} \qquad$ (Chain rule)

$$= \cos u \times \frac{1}{2}$$

$$= \frac{1}{2}\cos u$$

$$\frac{dy}{dx} = \frac{1}{2}\cos\frac{1}{2}x \qquad \text{(Since } u = \frac{1}{2}x\text{)}$$

3. Find the derivative of $5\cos 3x$

<u>Solution</u>

$\qquad y = 5\cos 3x$

Let $u = 3x$

Therefore, $y = 5\cos u$

$$\frac{du}{dx} = 3$$

$$\frac{dy}{du} = 5(-\sin u) \qquad \text{(Note that the constant term i.e. 5 should be used to multiply the derivative)}$$

$$= -5\sin u$$

Therefore, $\dfrac{dy}{dx} = \dfrac{dy}{du} \times \dfrac{du}{dx}$

$$= -5\sin u \times 3$$

$$= -15\sin u$$

$$\frac{dy}{dx} = -15\sin 3x \qquad \text{(Since } u = 3x\text{)}$$

4. Find the derivative of $\sin^2 x$

<u>Solution</u>

$\qquad y = \sin^2 x$

Note that $\sin^2 x = \sin x \times \sin x$

Hence, let $u = \sin x$

Therefore, $y = u^2 \quad$ (i.e. $u \times u$ from $\sin x \times \sin x$)

$$\frac{du}{dx} = \cos x$$

$$\frac{dy}{du} = 2u$$

Therefore, $\dfrac{dy}{dx} = \dfrac{dy}{du} \times \dfrac{du}{dx}$

$$= 2u \times \cos x$$

$$= 2u\cos x$$

$$\frac{dy}{dx} = 2\sin x\cos x \qquad \text{(Since } u = \sin x\text{)}$$

5. Differentiate with respect to x: $y = \tan 2x$

<u>Solution</u>

$$y = \tan 2x$$

$$\frac{dy}{dx} = \sec^2 2x \ \times \ \frac{d(2x)}{dx} \qquad \text{(Note that the derivative of } \tan x \text{ is } \sec^2 x\text{)}$$

$$= \sec^2 2x \ \times \ 2$$

$$= 2\sec^2 2x$$

6. If $y = \cos^3 6x^2$, differentiate y with respect to x.

<u>Solution</u>

$$y = \cos^3 6x^2$$

Let us solve this problem without the use of v as follows:

Let $u = \cos 6x^2$

Hence, $y = u^3$

$$\frac{du}{dx} = 12x(-\sin 6x^2) \qquad \text{(Note that } 12x \text{ is from the derivative of } 6x^2\text{)}$$

$$= -12x\sin 6x^2$$

$$\frac{dy}{du} = 3u^2$$

Therefore, $\dfrac{dy}{dx} = \dfrac{dy}{du} \ \times \ \dfrac{du}{dx}$

$$= 3u^2 \ \times \ (-12x\sin 6x^2)$$

$$= -36xu^2\sin 6x^2$$

$$= -36x(\cos 6x^2)^2\sin 6x^2 \qquad \text{(u has been replaced with } \cos 6x^2\text{)}$$

$$\frac{dy}{dx} = -36x\cos^2 6x^2\sin 6x^2$$

7. Find the derivative of $y = \sec 3x$

<u>Solution</u>

$$y = \sec 3x$$

$$\frac{dy}{dx} = \sec 3x\tan 3x \ \times \ \frac{d(3x)}{dx} \qquad \text{(Note that the derivative of } \sec x \text{ is } \sec x\tan x\text{)}$$

$$= \sec 3x\tan 3x \ \times \ 3$$

$$\frac{dy}{dx} = 3\sec 3x\tan 3x$$

8. Find the derivative of $\operatorname{cosec} 4x^3$

<u>Solution</u>

$$y = \text{cosec}\,4x^3$$

$$\frac{dy}{dx} = -\text{cosec}\,4x^3 \cot 4x^3 \times \frac{d(4x^3)}{dx}$$ (Note that the derivative of cosec is −cosec cot)

$$= -\text{cosec}\,4x^3 \cot 4x^3 \times 12x^2$$

$$= -12x^2 \text{cosec}\,4x^3 \cot 4x^3$$

9. Find the derivative of $\cot^2 2x^4$

<u>Solution</u>

$$y = \cot^2 2x^4$$

This can also be written as: $y = \cot 2x^4 \times \cot 2x^4$

Let $u = 2x^4$ (Take the function of x)

Also, let $v = \cot u$ (Take a function of u without taking the exponent)

Hence, $y = v^2$ (Since $v^2 = (\cot u)^2 = (\cot 2x^4)^2 = \cot^2 2x^4$. Hence, $y = v^2$)

$$\frac{du}{dx} = 8x^3$$

$$\frac{dv}{du} = -\text{cosec}^2 u$$

$$\frac{dy}{dv} = 2v$$

Therefore, $\dfrac{dy}{dx} = \dfrac{dy}{dv} \times \dfrac{dv}{du} \times \dfrac{du}{dx}$

$$= 2v \times -\text{cosec}^2 u \times 8x^3$$

$$= -16x^3 \, v \, \text{cosec}^2 u$$

$$= -16x^3 \cot u \, \text{cosec}^2 u$$ (Since v = cot u)

Substituting in the original value of u gives:

$$\frac{dy}{dx} = -16x^3 \cot 2x^4 \text{cosec}^2 2x^4$$

10. Find the derivative of $5x\sin 2x$

<u>Solution</u>

$$y = 5x\sin 2x$$

We are going to apply the product rule of differentiation.

Let $u = 5x$

and $v = \sin 2x$

$$\frac{du}{dx} = 5$$

$$\frac{dv}{dx} = 2\cos 2x$$ (Note that the differentiation of $2x$ gives 2)

Hence, $\dfrac{dy}{dx} = u\dfrac{dv}{dx} + v\dfrac{du}{dx}$

$$= (5x \times 2\cos 2x) + (\sin 2x \times 5)$$

$$= 10x\cos2x + 5\sin2x$$

$$\frac{dy}{dx} = 5(2x\cos2x + \sin2x)$$

11. Find $\frac{dy}{dx}$ if $y = \frac{1}{x}\sec x$

Solution

$$y = \frac{1}{x}\sec x$$

We apply product rule as follows:

$$u = \frac{1}{x}$$

$$= x^{-1}$$

$$v = \sec3x$$

$$\frac{du}{dx} = -1x^{-2}$$

$$= \frac{-1}{x^2}$$

$$\frac{dv}{dx} = \sec3x\tan3x \times 3$$

$$= 3\sec3x\tan3x$$

Note that the derivative of $\sec x$ is $\sec x\tan x$ and 3 is from the derivative of $3x$

Hence, $\frac{dy}{dx} = u\frac{dv}{dx} + v\frac{du}{dx}$

$$= \frac{1}{x}(3\sec3x\tan3x) + \sec3x\left(\frac{-1}{x^2}\right)$$

$$\frac{dy}{dx} = \frac{3}{x}\sec3x\tan3x - \frac{1}{x^2}\sec3x$$

12. Find the derivative of $3\csc x^6$

Solution

$$y = 3\csc x^6$$

Let $u = x^6$

Hence, $y = 3\csc u$

$$\frac{du}{dx} = 6x^5$$

$$\frac{dy}{du} = -3\cot u\csc u \quad \text{(The constant term i.e. 5 should be used to multiply the derivative)}$$

Therefore, $\frac{dy}{dx} = \frac{dy}{du} \times \frac{du}{dx}$

$$= -3\cot u\csc u \times 6x^5$$

$$= -18x^5\cot u\csc u$$

$$\frac{dy}{dx} = -18x^5\cot x^6\csc x^6 \quad \text{(Since } u = x^6\text{)}$$

13. If $y = \dfrac{1 + \cos 2x}{\sin 2x}$ find $\dfrac{dy}{dx}$.

Solution

$$y = \frac{1 + \cos 2x}{\sin 2x}$$

We have to apply quotient rule on this as follows:

$u = 1 + \cos 2x$

$v = \sin 2x$

$\dfrac{du}{dx} = -2\sin 2x$

$\dfrac{dv}{dx} = 2\cos 2x$

Hence, $\dfrac{dy}{dx} = \dfrac{v\frac{du}{dx} - u\frac{dv}{dx}}{v^2}$ (Quotient rule)

$$= \frac{\sin 2x(-2\sin 2x) - (1 + \cos 2x)(2\cos 2x)}{(\sin 2x)^2}$$

$$= \frac{-2\sin^2 2x - (2\cos 2x + 2\cos^2 2x)}{\sin^2 2x}$$ (Note that $(\sin 2x)^2 = \sin^2 2x$)

$$= \frac{-2\sin^2 2x - 2\cos 2x - 2\cos^2 2x}{\sin^2 2x}$$

$$= \frac{-2\sin^2 2x - 2\cos^2 2x - 2\cos 2x}{\sin^2 2x}$$

$$= \frac{-2(\sin^2 2x + \cos^2 2x) - 2\cos 2x}{\sin^2 2x}$$

$$= \frac{-2(1) - 2\cos 2x}{\sin^2 2x}$$ (Note that $\sin^2 x + \cos^2 x = 1$, hence $\sin^2 2x + \cos^2 2x = 1$)

$$\frac{dy}{dx} = \frac{-2(1 + \cos 2x)}{\sin^2 2x}$$

14. Differentiate $y = \dfrac{\sin^2 x}{x}$

Solution

$$y = \frac{\sin^2 x}{x}$$

This is also quotient rule.

Therefore, $u = \sin^2 x$

$v = x$

Let us follow a direct and systematic way of differentiating trigonometric functions.

Hence, in order to differentiate $\sin 2x$:

First, differentiate the exponent of $\sin^2$ without changing the trigonometric term. This gives:

$2\sin^{2-1}$

$= 2\sin$

250

Then add the term in x. This gives:

    $2\sin x$

The next step is to differentiate sin which gives cos. Also add the term in x to obtain $\cos x$.

Finally, differentiate the term in x. Hence, differentiating x gives 1.

Now multiply the three terms obtained in the three steps above. This gives:

    $2\sin x \;\times\; \cos x \;\times\; 1$

    $= 2\sin x \cos x$

Hence, $\dfrac{du}{dx} = 2\sin x \cos x$ (As obtained above)

Since, $v = x$

Then, $\dfrac{dv}{dx} = 1$

Hence, $\dfrac{dy}{dx} = \dfrac{v\dfrac{du}{dx} - u\dfrac{dv}{dx}}{v^2}$

$= \dfrac{x(2\sin x\cos x) - \sin^2 x(1)}{x^2}$

$= \dfrac{2x\sin x\cos x - \sin^2 x}{x^2}$

$= \dfrac{\sin x(2x\cos x - \sin x)}{x^2}$

15. Differentiate with respect to x: $\dfrac{\sec 2x}{x^3 + 1}$

Solution

$y = \dfrac{\sec 2x}{x^3 + 1}$

$u = \sec 2x$

$v = x^3 + 1$

Hence, $\dfrac{du}{dx} = 2\sec 2x\tan 2x$

Then, $\dfrac{dv}{dx} = 3x^2$

Hence, $\dfrac{dy}{dx} = \dfrac{v\dfrac{du}{dx} - u\dfrac{dv}{dx}}{v^2}$

$= \dfrac{(x^3 + 1)(2\sec 2x\tan 2x) - \sec 2x(3x^2)}{(x^3 + 1)^2}$

$= \dfrac{(x^3 + 1)(2\sec 2x\tan 2x) - 3x^2\sec 2x}{(x^3 + 1)^2}$

16. Find the derivative of $\dfrac{\cos \sqrt{x}}{1 + x}$

<u>Solution</u>

$$y = \frac{\cos\sqrt{x}}{1+x}$$

$$u = \cos\sqrt{x}$$

$$= \cos x^{\frac{1}{2}}$$

$$v = 1 + x$$

Hence, $\dfrac{du}{dx} = \dfrac{d(x^{\frac{1}{2}})}{dx} \times \dfrac{d(\cos)}{dx}$

Note that $\dfrac{d(\cos)}{dx}$ means the derivative of cos which gives $-\sin$, and then $x^{\frac{1}{2}}$ is added to it to give $-\sin x^{\frac{1}{2}}$. Hence we continue as follows:

$$\frac{du}{dx} = \frac{1}{2} x^{-\frac{1}{2}} \times -\sin x^{\frac{1}{2}}$$

$$= \frac{1}{2x^{\frac{1}{2}}} \times -\sin x^{\frac{1}{2}}$$

$$\frac{du}{dx} = \frac{-\sin x^{\frac{1}{2}}}{2x^{\frac{1}{2}}}$$

Also, $\dfrac{dv}{dx} = 1$

Hence, $\dfrac{dy}{dx} = \dfrac{v\dfrac{du}{dx} - u\dfrac{dv}{dx}}{v^2}$

$$= \frac{(1+x)\left(\dfrac{-\sin x^{\frac{1}{2}}}{2x^{\frac{1}{2}}}\right) - \cos x^{\frac{1}{2}}(1)}{(1+x)^2}$$

$$= \frac{-\sin\sqrt{x}}{2\sqrt{x}(1+x)} - \frac{\cos\sqrt{x}}{(1+x)^2} \qquad \text{(When the fractions are separated)}$$

$$\frac{dy}{dx} = \frac{-(1+x)\sin\sqrt{x} - 2\sqrt{x}\cos\sqrt{x}}{2\sqrt{x}(1+x)^2} \qquad \text{(When the fractions are combined)}$$

17. Find $\dfrac{dy}{dx}$ if $y = \dfrac{1-x^2}{1+\cos x}$

<u>Solution</u>

$$y = \frac{1-x^2}{1+\cos x}$$

$$u = 1 - x^2$$

$$v = 1 + \cos x$$

$$\frac{du}{dx} = -2x$$

$$\frac{dv}{dx} = -\sin x$$

Hence, $\dfrac{dy}{dx} = \dfrac{v\frac{du}{dx} - u\frac{dv}{dx}}{v^2}$

$= \dfrac{(1+\cos x)(-2x) - (1-x^2)(-\sin x)}{(1+\cos x)^2}$

$\dfrac{dy}{dx} = \dfrac{-2x(1+\cos x) + (1-x^2)\sin x}{(1+\cos x)^2}$

Note that the negative sign from $-\sin x$ changed the negative sign at the middle to a positive sign since negative sign multiplied by negative sign gives a positive sign.

18. Find the derivative of $8x\sin x^2$

Solution

$y = 8x\sin x^2$

We apply product rule as follows:

$u = 8x$

$\dfrac{du}{dx} = 8$

$v = \sin x^2$

$\dfrac{dv}{dx} = \dfrac{d(\sin)}{dx} \times \dfrac{d(x^2)}{dx}$

$= \cos x^2 \times 2x$

$= 2x\cos x^2$

Note that $\dfrac{d(\sin)}{dx}$ gives cos which result to $\cos x^2$ when x^2 from the question is added

Hence, $\dfrac{dy}{dx} = u\dfrac{dv}{dx} + v\dfrac{du}{dx}$

$= 8x(2x\cos x^2) + \sin x^2(8)$

$= 16x^2\cos x^2 + 8\sin x^2$

$\dfrac{dy}{dx} = 8(2x^2\cos x^2 + \sin x^2)$

19. Differentiate with respect to x: $\sec 2x\sin^3 2x$

Solution

$y = \sec 2x\sin^3 2x$

We apply product rule as follows:

$u = \sec 2x$

$\dfrac{du}{dx} = \dfrac{d(\sec)}{dx} \times \dfrac{d(2x)}{dx}$

$= \sec 2x\tan 2x \times 2$

$\dfrac{du}{dx} = 2\sec 2x\tan 2x$

Note that $\dfrac{d(\sec)}{dx}$ gives sectan, but remember to add $2x$ after sec and tan respectively to obtain

sec2xtan2x

$\quad$ v = $\sin^3 2x$

In order to directly differentiate a trigonometric term with exponent like this (i.e. $\sin^3$) we have three solutions to multiply as follows:

First solution: consider only the exponent and differentiate $\sin^3$. This gives:

$\quad \dfrac{d(\sin^3)}{dx} = 3\sin^2$

We now add $2x$ from the question to obtain $3\sin^2 2x$

Second solution: $\dfrac{d(\sin)}{dx} = \cos$ $\;$ which gives $\cos 2x$

Third solution: $\dfrac{d(2x)}{dx} = 2$

Multiply theses three solutions to give the derivative of v = $\sin^3 2x$ as follows:

$\quad \dfrac{dv}{dx} = 3\sin^2 2x \; \text{x} \; \cos 2x \; \text{x} \; 2$

$\qquad = 6\sin^2 2x\cos 2x$

Now, $\dfrac{dy}{dx} = u\dfrac{dv}{dx} + v\dfrac{du}{dx}$

$\quad = \sec 2x(6\sin^2 2x\cos 2x) + \sin^3 2x(2\sec 2x\tan 2x)$

$\quad = \dfrac{1}{\cos 2x}(6\sin^2 2x\cos 2x) \; + \; \sin^3 2x(2\dfrac{1}{\cos 2x}\dfrac{\sin 2x}{\cos 2x})$ $\quad$ (Note that $\sec 2x = \dfrac{1}{\cos 2x}$ and $\tan 2x = \dfrac{\sin 2x}{\cos 2x}$)

$\quad = 6\sin^2 2x + 2\sin^2 2x(\dfrac{\sin 2x}{\cos 2x}\dfrac{\sin 2x}{\cos 2x})$

Note that one $\sin 2x$ has been taken out of $\sin^3 2x$ and placed inside the bracket.

$\quad = 6\sin^2 2x + 2\sin^2 2x(\tan 2x\tan x)$ $\quad$ (Since $\dfrac{\sin 2x}{\cos 2x} = \tan 2x$)

$\quad = 6\sin^2 2x + 2\sin^2 2x(\tan^2 2x)$

$\dfrac{dy}{dx} = 2\sin^2 2x(3 + \tan^2 2x)$ $\quad$ (After factorization)

20. Differentiate with respect to x: $\dfrac{\sec x - \tan x}{\sec x + \tan x}$

Solution

$\quad y = \dfrac{\sec x - \tan x}{\sec x + \tan x}$

$\quad u = \sec x - \tan x$

$\quad \dfrac{du}{dx} = \sec x\tan x - \sec^2 x$

$\quad v = \sec x + \tan x$

$\quad \dfrac{dv}{dx} = \sec x\tan x + \sec^2 x$

$$\frac{dy}{dx} = \frac{v\frac{du}{dx} - u\frac{dv}{dx}}{v^2}$$

$$= \frac{(\sec x + \tan x)(\sec x\tan x - \sec 2x) - (\sec x - \tan x)(\sec x\tan x + \sec 2x)}{(\sec x + \tan x)^2}$$

Expanding bracket in the numerator gives:

$$\frac{dy}{dx} = \frac{\sec^2 x\tan x - \sec^3 x + \sec x\tan^2 x - \sec^2 x\tan x - [\sec^2 x\tan x + \sec^3 x - \sec x\tan^2 x - \sec^2 x\tan x]}{(\sec x + \tan x)^2}$$

$$= \frac{\sec^2 x\tan x - \sec^3 x + \sec x\tan^2 x - \sec^2 x\tan x - \sec^2 x\tan x - \sec^3 x + \sec x\tan^2 x + \sec^2 x\tan x]}{(\sec x + \tan x)^2}$$

$$= \frac{2\sec x\tan^2 x - 2\sec^3 x}{(\sec x + \tan x)^2} \quad \text{(Note that all the sec}^2x\text{tan}x \text{ have cancelled out)}$$

$$= \frac{2\sec x(\tan^2 x - \sec^2 x)}{(\sec x + \tan x)^2}$$

$$= \frac{2\sec x(-1)}{(\sec x + \tan x)^2} \quad \text{(Note that tan}^2x - \text{sec}^2x = -1)$$

$$= \frac{-2\sec x}{(\sec x + \tan x)^2}$$

$$\frac{dy}{dx} = -\frac{2\sec x}{(\sec x + \tan x)^2}$$

21. Find the derivative of $\sqrt{\dfrac{\cos 2x}{1 + \sin 2x}}$

Solution

$$y = \sqrt{\frac{\cos 2x}{1 + \sin 2x}}$$

This can also be written as:

$$y = \frac{(\cos 2x)^{\frac{1}{2}}}{(1 + \sin 2x)^{\frac{1}{2}}}$$

$$u = (\cos 2x)^{\frac{1}{2}}$$

Using the chain rule, we obtain $\dfrac{du}{dx}$ as follows:

$$\frac{du}{dx} = \frac{1}{2}(\cos 2x)^{\frac{1}{2}-1} \times \frac{d(\cos 2x)}{dx}$$

$$= \frac{1}{2}(\cos 2x)^{-\frac{1}{2}} \times -2\sin 2x$$

$$= \frac{-2\sin 2x}{2(\cos 2x)^{\frac{1}{2}}}$$

$$= \frac{-\sin 2x}{(\cos 2x)^{\frac{1}{2}}}$$

$$v = (1 + \sin 2x)^{\frac{1}{2}}$$

$$\frac{dv}{dx} = \frac{1}{2}(1 + \sin 2x)^{\frac{1}{2}-1} \text{ x } \frac{d(1+\sin 2x)}{dx}$$

$$= \frac{1}{2}(1 + \sin 2x)^{-\frac{1}{2}} \text{ x } 2\cos 2x$$

$$= \frac{2\cos 2x}{2(1 + \sin 2x)^{\frac{1}{2}}}$$

$$= \frac{\cos 2x}{(1 + \sin 2x)^{\frac{1}{2}}}$$

$$\frac{dy}{dx} = \frac{v\frac{du}{dx} - u\frac{dv}{dx}}{v^2}$$

$$= \frac{(1 + \sin 2x)^{\frac{1}{2}}\left(\dfrac{-\sin 2x}{(\cos 2x)^{\frac{1}{2}}}\right) - (\cos 2x)^{\frac{1}{2}}\left(\dfrac{\cos 2x}{(1 + \sin 2x)^{\frac{1}{2}}}\right)}{[(1 + \sin 2x)^{\frac{1}{2}}]^2}$$

$$= \frac{\left(\dfrac{-\sin 2x(1 + \sin 2x)^{\frac{1}{2}}}{(\cos 2x)^{\frac{1}{2}}}\right) - \left(\dfrac{\cos 2x(\cos 2x)^{\frac{1}{2}}}{(1 + \sin 2x)^{\frac{1}{2}}}\right)}{1 + \sin 2x}$$

$$= \frac{\left(\dfrac{-\sin 2x(1 + \sin 2x) - \cos 2x(\cos 2x)}{(\cos 2x)^{\frac{1}{2}}(1 + \sin 2x)^{\frac{1}{2}}}\right)}{1 + \sin 2x}$$

Note that $(\cos 2x)^{\frac{1}{2}}$ x $(\cos 2x)^{\frac{1}{2}}$ = $\cos 2x$ (Since the exponents are added). Similarly:

$(1 + \sin 2x)^{\frac{1}{2}}$ x $(1 + \sin 2x)^{\frac{1}{2}}$ = $1 + \sin 2x$. Simplifying the above expression further, gives:

$$\frac{dy}{dx} = \frac{-\sin 2x(1 + \sin 2x) - \cos^2 2x}{(\cos 2x)^{\frac{1}{2}}(1 + \sin 2x)^{\frac{1}{2}}(1 + \sin 2x)}$$

$$= \frac{-\sin 2x - \sin^2 2x - \cos^2 2x}{(\cos 2x)^{\frac{1}{2}}(1 + \sin 2x)^{\frac{1}{2}}(1 + \sin 2x)}$$

$$= \frac{-\sin 2x - (\sin^2 2x + \cos^2 2x)}{(\cos 2x)^{\frac{1}{2}}(1 + \sin 2x)^{\frac{1}{2}}(1 + \sin 2x)}$$

$$= \frac{-\sin 2x - 1}{(\cos 2x)^{\frac{1}{2}}(1 + \sin 2x)^{\frac{1}{2}}(1 + \sin 2x)} \qquad \text{(Note that } \sin^2 2x + \cos^2 2x = 1\text{)}$$

$$= \frac{-(1 + \sin 2x)}{(\cos 2x)^{\frac{1}{2}}(1 + \sin 2x)^{\frac{1}{2}}(1 + \sin 2x)}$$

$$= \frac{-1}{(\cos 2x)^{\frac{1}{2}}(1 + \sin 2x)^{\frac{1}{2}}} \qquad \text{(Note that } 1 + \sin 2x \text{ cancels out)}$$

22. Differentiate with respect to x: $\sin x - 2x\cos x$

Solution

$$y = \sin x - 2x\cos x$$

Treat $2x\cos x$ using product rule.

$$\frac{dy}{dx} = \frac{d(\sin x)}{dx} - \left(2x\frac{d(\cos x)}{dx} + \cos x\frac{d(2x)}{dx}\right)$$
$$= \cos x - [2x(-\sin x) + \cos x(2)]$$
$$= \cos x + 2x\sin x - 2\cos x$$
$$= 2x\sin x - \cos x$$

23. Find the derivative of $\cos^5 3x^4$

Solution

$$y = \cos^5 3x^4$$

A direct way of differentiating this problem is applied as follows:

$$\frac{dy}{dx} = \frac{d(\cos^5)}{dx} \times \frac{d(\cos)}{dx} \times \frac{d(3x^4)}{dx}$$
$$= 5\cos^4 3x^4 \times (-\sin 3x^4) \times 12x^3$$

Note that the derivative of $\cos^5$ gives $5\cos^4$ (do not change the trigonometric term, i.e. cos) and then the addition of $3x^4$ from the question gives $5\cos^4 3x^4$.

Similarly the derivative of cos gives $-\sin$ and the addition of $3x^4$ from the question gives $-\sin 3x^4$.

Hence, multiplying the terms above gives:

$$\frac{dy}{dx} = -60x^3\cos^4 3x^4\sin 3x^4$$

24. Differentiate with respect to x: $\sin^8 15x^6$

Solution

$$y = \sin^8 15x^6$$

We can also differentiate this problem directly as follows:

$$\frac{dy}{dx} = \frac{d(\sin^8)}{dx} \times \frac{d(\sin)}{dx} \times \frac{d(15x^6)}{dx}$$
$$= 8\sin^7 15x^6 \times \cos 15x^6 \times 90x^5$$

Note that the derivative of $\sin^8$ gives $8\sin^7$ (in this case do not change the trigonometric term, i.e. sin) and then the addition of $15x^6$ from the question gives $8\sin^7 15x^6$.

Similarly the derivative of sin gives cos and the addition of $15x^6$ from the question gives $\cos 15x^6$. Hence, we continue as follows:

$$\frac{dy}{dx} = 8\sin^7 15x^6 \times \cos 15x^6 \times 90x^5$$
$$= 720x^5\sin^7 15x^6\cos 15x^6$$

Exercise 23

1. Find the derivative of $\tan 2x$

2. If $y = \cos \dfrac{1}{5} x$, find $\dfrac{dy}{dx}$

3. Find the derivative of $10\cos 5x$

4. Find the derivative of $\sin^3 x$

5. Differentiate with respect to x: $y = \tan^2 x$

6. If $y = \sin^4 3x^5$, differentiate y with respect to x.

7. Find the derivative of $y = \csc 6x^2$

8. Find the derivative of $\sec 2x^5$

9. Find the derivative of $\tan 3x^2$

10. Find the derivative of $x^2 \cos 3x$

11. Find $\dfrac{dy}{dx}$ if $y = \dfrac{3}{x^3} \sin 2x$

12. Find the derivative of $12\sec x^4$

13. If $y = \dfrac{2 - \sin 5x}{\tan x}$ find $\dfrac{dy}{dx}$.

14. Differentiate $y = \dfrac{\sin^3 x}{2x}$

15. Differentiate with respect to x: $\dfrac{\cot 2x}{x + 3}$

16. Find the derivative of $\dfrac{\sin \sqrt[3]{3}}{5x}$

17. Find $\dfrac{dy}{dx}$ if $y = \dfrac{x^2 - 3}{\sec 2x}$

18. Find the derivative of $3x^2 \cos 3x^2$

19. Differentiate with respect to x: $\cot x \cos^2 x$

20. Differentiate with respect to x: $\dfrac{\sin x - \cos x}{\sin x + \cos x}$

21. Find the derivative of $\dfrac{\sec 4x}{\cos 4x - 2}$

22. Differentiate with respect to x: $\cos 3x - x^2 \sec x$

23. Find the derivative of $\sin^9 x^3$

24. Differentiate with respect to x: $\sin 6x^{\frac{1}{2}}$

25. Find $\dfrac{dy}{dx}$ if $y = \dfrac{1}{x^2} \sin x^3$

26. Find the derivative of $\cos^3 2x^5$

27. If $y = \dfrac{\cos 10x}{\sin 2x}$ find $\dfrac{dy}{dx}$.

28. Differentiate $y = \dfrac{3\sin x^4}{2x}$

29. Differentiate with respect to x: $\dfrac{\tan 5x}{2x - 1}$

30. Find the derivative of $\dfrac{\tan^2 x}{2x}$

CHAPTER 24
DERIVATIVE OF INVERSE FUNCTIONS

If the derivative of a function is given by $\dfrac{dy}{dx}$, then the derivative of the inverse function is given

by: $\dfrac{1}{\frac{dy}{dx}} = \dfrac{dx}{dy}$

Or, $\dfrac{dy}{dx} = \dfrac{1}{\frac{dx}{dy}}$

Examples

1. Find $\dfrac{dx}{dy}$ if y = $\sqrt[3]{x}$

<u>Solution</u>

Method 1

$$y = \sqrt[3]{x}$$

Or $y = x^{\frac{1}{3}}$

Let us make x the subject of the formula. The inverse of $\dfrac{1}{3}$ is 3. Hence raise both sides to the

exponent 3 as follows:

$$y^3 = \left(x^{\frac{1}{3}}\right)^3$$

$$y^3 = x^1 \qquad \text{(Note that 1 was obtained from } \tfrac{1}{3} \text{ x 3)}$$

Hence, $x = y^3$

Therefore, $\dfrac{dx}{dy} = 3y^2$

Method 2

$$y = \sqrt[3]{x}$$

Or $y = x^{\frac{1}{3}}$

$$\frac{dy}{dx} = \frac{1}{3}x^{\frac{1}{3}-1}$$

$$= \frac{1}{3}x^{-\frac{2}{3}}$$

$$\frac{dy}{dx} = \frac{1}{3x^{\frac{2}{3}}}$$

Hence, $\dfrac{dx}{dy} = \dfrac{1}{\frac{dy}{dx}}$

$$= \frac{3x^{\frac{2}{3}}}{1} \qquad \text{(This means the inverse of } \dfrac{1}{3x^{\frac{2}{3}}} \text{)}$$

$$= 3x^{\frac{2}{3}}$$

$$= 3(\sqrt[3]{x})^2 \qquad \text{[Recall from indices that } x^{\frac{a}{b}} = (\sqrt[b]{x})^a]$$

$$\frac{dx}{dy} = 3y^2 \qquad \text{(Since } y = \sqrt[3]{x})$$

2. If $y = \sqrt[5]{2x - 3}$ find $\dfrac{dx}{dy}$

<u>Solution</u>

$$y = \sqrt[5]{2x - 3}$$

Or $y = (2x - 3)^{\frac{1}{5}}$

Let us make x the subject of the formula. The inverse of $\dfrac{1}{5}$ is 5. Hence raise both sides to the exponent 5 as follows:

$$y^5 = [(2x - 3)^{\frac{1}{5}}]^5$$

$$y^5 = 2x - 3 \qquad \text{(Note that } \frac{1}{5} \times 5 = 1, \text{ which cancels the fractional exponent)}$$

Hence, $2x = y^5 + 3$

$$x = \frac{y^5 + 3}{2}$$

$$x = \frac{y^5}{2} + \frac{3}{2} \qquad \text{(When we separate into fractions by dividing each part by the denominator)}$$

Therefore, $\dfrac{dx}{dy} = \dfrac{5y^4}{2}$

3. If $y = x^3 - 5$. Find $\dfrac{dx}{dy}$

<u>Solution</u>

$$y = x^3 - 5$$

$$y + 5 = x^3$$

$$x^3 = y + 5$$

$$x = (y + 5)^{\frac{1}{3}} \qquad \text{(By raising both sides to an exponent of the inverse of 3 which is } \frac{1}{3})$$

$$\frac{dx}{dy} = \frac{1}{3}(y + 5)^{\frac{1}{3} - 1} \times 1$$

Note that the 1 was obtained from the derivative of y + 5 since chain rule was used.

$$\frac{dx}{dy} = \frac{1}{3}(y + 5)^{-\frac{2}{3}}$$

$$\frac{dx}{dy} = \frac{1}{3(y+5)^{\frac{2}{3}}}$$

Recall from indices that $x^{\frac{a}{b}} = (\sqrt[b]{x})^a$. Applying this rule gives:

$$\frac{dx}{dy} = \frac{1}{3[\sqrt[3]{(y+5)^2}]}$$

4. If $y = \frac{1}{2}x^4 + 3$, find $\frac{dx}{dy}$

Solution

$$y = \frac{1}{2}x^4 + 3$$

$$y - 3 = \frac{1}{2}x^4$$

$$2(y - 3) = x^4$$

$$x^4 = 2y - 6$$

$x = (2y - 6)^{\frac{1}{4}}$ (This is obtained by raising both sided to an exponent of the inverse of 4, i.e. $\frac{1}{4}$)

Hence we use chain rule to determine $\frac{dx}{dy}$ as follows:

$\frac{dx}{dy} = \frac{1}{4}(2y - 6)^{\frac{1}{4} - 1} \times 2$ (Note that 2 is from the derivative of $2y - 6$)

$$\frac{dx}{dy} = \frac{1}{2}(2y - 6)^{-\frac{3}{4}}$$

$$\frac{dx}{dy} = \frac{1}{2(2y - 6)^{\frac{3}{4}}}$$

Hence, $\frac{dx}{dy} = \frac{1}{2[\sqrt[4]{(2y-6)^3}]}$ [This is obtained from the law of indices given by: $x^{\frac{a}{b}} = (\sqrt[b]{x})^a$]

5. If $y = \frac{x + 2}{x}$, find $\frac{dx}{dy}$.

Solution

$$y = \frac{x + 2}{x}$$

$xy = x + 2$ (When we cross multiply)

$xy - x = 2$

Factorizing the left hand side gives:

$x(y - 1) = 2$

$$x = \frac{2}{y - 1}$$

$x = 2(y - 1)^{-1}$ (Take note of the use of negative exponent when the denominator goes up)

Hence we use chain rule to determine $\frac{dx}{dy}$ as follows:

$\frac{dx}{dy} = -1 \times 2(y - 1)^{-1-1} \times 1$ (Note that 1 is from the derivative of $y - 1$)

$$\frac{dx}{dy} = -2(y-1)^{-2}$$

$$\frac{dx}{dy} = \frac{-2}{(y-1)^2}$$

6. Find $\frac{dx}{dy}$ if $y = \frac{1}{x+2}$

<u>Solution</u>

$$y = \frac{1}{x+2}$$

$$x + 2 = \frac{1}{y}$$

$$x = \frac{1}{y} - 2$$

$$x = y^{-1} - 2$$

$$\frac{dx}{dy} = -1 \times y^{-1-1}$$

$$= -y^{-2}$$

$$\frac{dx}{dy} = \frac{-1}{y^2}$$

7. Find $\frac{dx}{dy}$ if $y = x^{\frac{2}{3}}$

<u>Solution</u>

$$y = x^{\frac{2}{3}}$$

Raise both sides to the exponent $\frac{3}{2}$ i.e. the inverse of $\frac{2}{3}$. This gives:

$$y^{\frac{3}{2}} = (x^{\frac{2}{3}})^{\frac{3}{2}}$$

$$y^{\frac{3}{2}} = x \quad \text{(Note that } \frac{2}{3} \times \frac{3}{2} = 1, \text{ and } x^1 = x\text{)}$$

$$x = y^{\frac{3}{2}}$$

$$\frac{dx}{dy} = \frac{3}{2} \times y^{\frac{3}{2}-1}$$

$$= \frac{3}{2} y^{\frac{1}{2}}$$

$$\frac{dx}{dy} = \frac{3\sqrt{y}}{2}$$

8. If $y = (5x + 7)^{\frac{3}{10}}$ find $\frac{dx}{dy}$.

<u>Solution</u>

$$y = (5x + 7)^{\frac{3}{10}}$$

Raise both sides to the exponent $\dfrac{10}{3}$ i.e. the inverse of $\dfrac{3}{10}$. This gives:

$$y^{\frac{10}{3}} = [(5x + 7)^{\frac{3}{10}}]^{\frac{10}{3}}$$

$$y^{\frac{10}{3}} = 5x + 7 \quad \text{(Note that } \frac{3}{10} \times \frac{10}{3} = 1)$$

$$y^{\frac{10}{3}} - 7 = 5x$$

$$x = \frac{y^{\frac{10}{3}} - 7}{5}$$

$$= \frac{y^{\frac{10}{3}}}{5} - \frac{7}{5}$$

$$x = \frac{1}{5}y^{\frac{10}{3}} - \frac{7}{5}$$

$$\frac{dx}{dy} = \frac{10}{3} \times \frac{1}{5}y^{\frac{10}{3} - 1}$$

$$= \frac{2}{3}y^{\frac{7}{3}}$$

$$\frac{dx}{dy} = \frac{2\sqrt[3]{y^7}}{3}$$

Exercise 24

1. Find $\dfrac{dx}{dy}$ if $y = 7x^3$

2. If $y = \sqrt[3]{1 + x^2}$ find $\dfrac{dx}{dy}$

3. If $y = 2x^5 - 3$. Find $\dfrac{dx}{dy}$

4. If $y = \dfrac{2}{3}x^3 - 9$, find $\dfrac{dx}{dy}$

5. If $y = \dfrac{2x + 1}{x}$, find $\dfrac{dx}{dy}$.

6. Find $\dfrac{dx}{dy}$ if $y = \dfrac{3}{x^2 - 5}$

7. Find $\dfrac{dx}{dy}$ if $y = 2x^{\frac{1}{4}}$

8. If $y = (x - 3)^{\frac{1}{5}}$ find $\dfrac{dx}{dy}$.

9. If $y = \dfrac{4x + 5}{x}$, find $\dfrac{dx}{dy}$.

10. Find $\dfrac{dx}{dy}$ if $y = \dfrac{1}{x^3 + 8}$

CHAPTER 25
DERIVATIVES OF INVERSE TRIGONOMETRIC FUNCTIONS

Recall that if $\sin x = y$, then $x = \sin^{-1} y$. This is referred to as inverse trigonometric function. The derivatives of inverse trigonometric functions are given below.

If $y = \sin^{-1} x$, then $\dfrac{dy}{dx} = \dfrac{1}{\sqrt{1-x^2}}$

If $y = \cos^{-1} x$, then $\dfrac{dy}{dx} = \dfrac{-1}{\sqrt{1-x^2}}$

If $y = \tan^{-1} x$, then $\dfrac{dy}{dx} = \dfrac{1}{1+x^2}$

If $y = \cot^{-1} x$, then $\dfrac{dy}{dx} = \dfrac{-1}{1+x^2}$

If $y = \sec^{-1} x$, then $\dfrac{dy}{dx} = \dfrac{1}{x\sqrt{x^2-1}}$

If $y = \operatorname{cosec}^{-1} x$, then $\dfrac{dy}{dx} = \dfrac{-1}{x\sqrt{x^2-1}}$

Note that $\sin^{-1} x$ can also be written as $\arcsin x$. Other inverse function can be written in a similar way.

Examples

1. If $y = \cos^{-1} x$ find the $\dfrac{dy}{dx}$.

<u>Solution</u>

$$y = \cos^{-1} x$$

$$\frac{dy}{dx} = \frac{-1}{\sqrt{1-x^2}}$$

2. Find the derivative of $\sin^{-1} 3x$

<u>Solution</u>

$$y = \sin^{-1} 3x$$

Let us use the chain rule to solve this problem

Let $u = 3x$

Hence, $y = \sin^{-1} u$

$$\frac{du}{dx} = 3$$

$$\frac{dy}{du} = \frac{1}{\sqrt{1-u^2}}$$

Therefore, $\dfrac{dy}{dx} = \dfrac{dy}{du} \times \dfrac{du}{dx}$ \qquad (Chain rule)

$$= \frac{1}{\sqrt{1-u^2}} \times 3$$

$$= \frac{3}{\sqrt{1-u^2}}$$

$$= \frac{3}{\sqrt{1-(3x)^2}} \qquad \text{(Since } u = 3x)$$

$$\frac{dy}{dx} = \frac{3}{\sqrt{1-9x^2}}$$

3. Find the derivative of $\cot^{-1}x^2$

Solution

$$y = \cot^{-1}x^2$$

Let $u = x^2$

Hence, $y = \cot^{-1}u$

$$\frac{du}{dx} = 2x$$

$$\frac{dy}{du} = \frac{-1}{1+u^2}$$

Therefore, $\dfrac{dy}{dx} = \dfrac{dy}{du} \times \dfrac{du}{dx}$

$$= \frac{-1}{1+u^2} \times 2x$$

$$= \frac{-2x}{1+u^2}$$

$$= \frac{-2x}{1+(x^2)^2} \qquad \text{(since } u = x^2)$$

$$\frac{dy}{dx} = \frac{-2x}{1+x^4}$$

4. If $y = \sec^{-1}2x^3$, find $y = \dfrac{dy}{dx}$

Solution

$$y = \sec^{-1}2x^3$$

Let $u = 2x^3$

Hence, $y = \sec^{-1}u$

$$\frac{du}{dx} = 6x^2$$

$$\frac{dy}{du} = \frac{1}{u\sqrt{u^2-1}}$$

Therefore, $\dfrac{dy}{dx} = \dfrac{dy}{du} \times \dfrac{du}{dx}$

$$= \frac{1}{u\sqrt{u^2 - 1}} \times 6x^2$$

$$= \frac{6x^2}{u\sqrt{u^2 - 1}}$$

$$= \frac{6x^2}{2x^3\sqrt{(2x^3)^2 - 1}} \qquad \text{(since } u = 2x^3\text{)}$$

$$\frac{dy}{dx} = \frac{3}{x\sqrt{4x^6 - 1}} \qquad \text{(Note that } \frac{6x^2}{2x^3} = \frac{3}{x}\text{)}$$

5. Find the derivative $x^2\tan^{-1}x$

<u>Solution</u>

$$y = x^2\tan^{-1}x$$

we apply the product rule as follows:

$$u = x^2$$

$$v = \tan^{-1}x$$

$$\frac{du}{dx} = 2x$$

$$\frac{dv}{dx} = \frac{1}{1 + x^2}$$

$$\frac{dy}{dx} = u\frac{dv}{dx} + v\frac{du}{dx} \qquad \text{(product rule)}$$

$$= x^2\left(\frac{1}{1 + x^2}\right) + \tan^{-1}x(2x)$$

$$\frac{dy}{dx} = \frac{x^2}{1 + x^2} + 2x\tan^{-1}x$$

6. If $y = 3x - 1 \operatorname{cosec}^{-1}x^3$, find $\dfrac{dy}{dx}$.

<u>Solution</u>

$$y = 3x - 1 \operatorname{cosec}^{-1}x^3$$

$$u = 3x - 1$$

$$\frac{du}{dx} = 3$$

$$v = \operatorname{cosec}^{-1}x^3$$

$$\frac{dv}{dx} = \frac{-1}{x^3\sqrt{(x^3)^2 - 1}} \times 3x^2 \qquad \text{(From use of chain rule. Note that } 3x^2 \text{ is from the derivative of } x^3\text{)}$$

$$= \frac{-3x^2}{x^3\sqrt{x^6 - 1}}$$

$$\frac{dv}{dx} = \frac{-3}{x\sqrt{x^6 - 1}}$$

$$\frac{dy}{dx} = u\frac{dv}{dx} + v\frac{du}{dx} \qquad \text{(product rule)}$$

$$= 3x - 1\left(\frac{-3}{x\sqrt{x^6-1}}\right) + \text{cosec}^{-1}x^3(3)$$

$$\frac{dy}{dx} = \frac{-3(3x-1)}{x\sqrt{x^6-1}} + 3\text{cosec}^{-1}x^3$$

7. Find the derivative of $\cos^{-1}(5x - 3)$

Solution

$y = \cos^{-1}(5x - 3)$

Let $u = 5x - 3$

Hence, $y = \cos^{-1}u$

$$\frac{du}{dx} = 5$$

$$\frac{dy}{du} = \frac{-1}{\sqrt{1-u^2}}$$

Therefore, $\dfrac{dy}{dx} = \dfrac{dy}{du} \times \dfrac{du}{dx}$

$$= \frac{-1}{\sqrt{1-u^2}} \times 5$$

$$= \frac{-5}{\sqrt{1-u^2}}$$

$$= \frac{-5}{\sqrt{1-(5x-3)^2}} \qquad \text{(Since } u = 5x - 3)$$

Expanding the bracket gives:

$$\frac{dy}{dx} = \frac{-5}{\sqrt{1-(25x^2-15x-15x+9)}}$$

$$= \frac{-5}{\sqrt{1-25x^2+15x+15x-9}}$$

$$\frac{dy}{dx} = \frac{-5}{\sqrt{-25x^2+30x-8}}$$

8. If $y = \tan^{-1}\left(\dfrac{1}{m}\right)$, find $\dfrac{dy}{dm}$

Solution

$$y = \tan^{-1}\left(\frac{1}{m}\right)$$

Let $u = \dfrac{1}{m} = m^{-1}$

Hence, $y = \tan^{-1}u$

$$\frac{du}{dm} = -1m^{-2}$$

$$\frac{du}{dm} = \frac{-1}{m^2}$$

$$\frac{dy}{du} = \frac{1}{1+u^2}$$

269

Therefore, $\dfrac{dy}{dm} = \dfrac{dy}{du} \times \dfrac{du}{dm}$

$= \dfrac{1}{1+u^2} \times \dfrac{-1}{m^2}$

$= \dfrac{1}{1+\left(\frac{1}{m}\right)^2} \times \dfrac{-1}{m^2}$

$= \dfrac{-1}{m^2\left[1+\left(\frac{1}{m}\right)^2\right]}$

$= \dfrac{-1}{m^2\left(1+\frac{1}{m^2}\right)}$

$\dfrac{dy}{dm} = \dfrac{-1}{m^2+1}$

9. If $y = \sin 3x$, find $\dfrac{dx}{dy}$.

Solution

$y = \sin 3x$

$\sin^{-1}y = 3x$ \qquad (Recall that if $\sin a = b$, then $a = \sin^{-1}b$)

$x = \dfrac{\sin^{-1}y}{3}$

$x = \dfrac{1}{3}\sin^{-1}y$

Hence, $\dfrac{dx}{dy} = \dfrac{1}{3}\dfrac{1}{\sqrt{1-y^2}}$

$\dfrac{dx}{dy} = \dfrac{1}{3\sqrt{1-y^2}}$

Note that this example asked us to find $\dfrac{dx}{dy}$ and not $\dfrac{dy}{dx}$

10. If $y = (\cos 5x)^2$ find $\dfrac{dx}{dy}$.

Solution

$y = (\cos 5x)^2$

This can also be written as $\cos^2 5x$. Hence:

$y = \cos^2 5x$.

Or, $\cos^2 5x = y$

taking the square root of both sides gives:

$\cos 5x = \sqrt{y}$

$5x = \cos^{-1}\sqrt{y}$

$$x = \frac{\cos^{-1}\sqrt{y}}{5}$$

$$= \frac{1}{5}\cos^{-1}\sqrt{y}$$

$$x = \frac{1}{5}\cos^{-1}y^{\frac{1}{2}}$$

Hence, $\dfrac{dx}{dy} = \dfrac{1}{5} \dfrac{-1}{\sqrt{1-(y^{\frac{1}{2}})^2}} \times \dfrac{1}{2}y^{-\frac{1}{2}}$ (Note that $\dfrac{1}{2}y^{-\frac{1}{2}}$ is from the derivative of $y^{\frac{1}{2}}$)

$$= \frac{-1}{5\sqrt{1-y}} \times \frac{1}{2y^{\frac{1}{2}}}$$

$$= \frac{-1}{10y^{\frac{1}{2}}\sqrt{1-y}}$$

$$= \frac{-1}{10\sqrt{y}\sqrt{1-y}}$$

$$\frac{dx}{dy} = \frac{-1}{10\sqrt{y(1-y)}}$$

Exercise 25

1. If $y = \sin^{-1}2x$ find the $\dfrac{dy}{dx}$.

2. Find the derivative of $\cos^{-1}x$

3. Find the derivative of $\cot^{-1}3x^2$

4. If $y = \cot^{-1}x^4$, find $y = \dfrac{dy}{dx}$

5. Find the derivative $3x\sec^{-1}x$

6. If $y = x^2 + 3\tan^{-1}x$, find $\dfrac{dy}{dx}$.

7. Find the derivative of $\tan^{-1}(x + 5)$

8. If $y = \cos^{-1}\left(\dfrac{1}{x}\right)$, find $\dfrac{dy}{dx}$

9. If $y = \cos5x$, find $\dfrac{dx}{dy}$.

10. If $y = (\sec x)^2$ find $\dfrac{dx}{dy}$.

11. Find the derivative $5x\tan^{-1}3x$

12. If $y = x^2 + 4\sin^{-1}x^5$, find $\dfrac{dy}{dx}$.

13. Find the derivative of $\csc^{-1}x^3$

14. If $y = \tan 2x$, find $\dfrac{\mathrm{d}x}{\mathrm{d}y}$.

15. If $y = \sin^3 5x$, find $\dfrac{\mathrm{d}x}{\mathrm{d}y}$.

CHAPTER 26
DERIVATIVES OF HYPERBOLIC FUNCTIONS

Hyperbolic functions are functions in calculus which are expressed by the combination of the exponential functions e^x and e^{-x}. The derivatives of the six main hyperbolic functions are as given below.

1. If $y = \sinh x$, then $\dfrac{dy}{dx} = \cosh x$

2. If $y = \cosh x$, then $\dfrac{dy}{dx} = \sinh x$

3. If $y = \tanh x$, then $\dfrac{dy}{dx} = \operatorname{sech}^2 x$

4. If $y = \coth x$, then $\dfrac{dy}{dx} = -\operatorname{cosech}^2 x$

5. If $y = \operatorname{sech} x$, then $\dfrac{dy}{dx} = -\operatorname{sech} x \tanh x$

6. If $y = \operatorname{cosech} x$, then $\dfrac{dy}{dx} = -\operatorname{cosech} x \coth x$

Note that $\sinh x = \dfrac{e^x - e^{-x}}{2}$ and $\cosh x = \dfrac{e^x + e^{-x}}{2}$

Examples

1. If $y = \cosh x - 5\sinh x$, find $\dfrac{dy}{dx}$.

Solution

$$y = \cosh x - 5\sinh x$$

$$\frac{dy}{dx} = \sinh x - 5\cosh x$$

2. Find the derivative of $2x^3 \coth x$

Solution

$$y = 2x^3 \coth x$$

Using product rule gives $\dfrac{dy}{dx}$ as follows:

$$\frac{dy}{dx} = 2x^3\left[\frac{d(\coth x)}{dx}\right] + \coth x\left[\frac{d(2x^3)}{dx}\right]$$

$$= 2x^3(-\operatorname{cosech}^2 x) + \coth x(6x^2)$$

$$= -2x^3\operatorname{cosech}^2 x + 6x^2\coth x$$

$$\frac{dy}{dx} = -2x^2(x\operatorname{cosech}^2 x - 3\coth x)$$

Take note of the change in sign of the term in the bracket. This is due to the negative sign outside the bracket. Expanding the bracket gives the original expression that was factorized.

3. If $y = \dfrac{\cosh x}{x^2 + 1}$ find $\dfrac{dy}{dx}$.

Solution

Let us use product rule to obtained $\dfrac{dy}{dx}$ as follows:

$$\frac{dy}{dx} = \frac{(x^2+1)\left[\frac{d(\cosh x)}{dx}\right] - \cosh x\left[\frac{d(x^2+1)}{dx}\right]}{(x^2+1)^2}$$

$$= \frac{(x^2+1)(\sinh x) - \cosh x(2x)}{(x^2+1)^2}$$

$$\frac{dy}{dx} = \frac{\sinh x(x^2+1) - 2x\cosh x}{(x^2+1)^2}$$

4. Find the derivative of $(\sinh 3x)^2$

Solution

$\quad y = (\sinh 3x)^2 \qquad$ (This can also be written as $\sinh^2 3x$)

Let $u = \sinh 3x$

Hence, $y = u^2$

$\quad \dfrac{du}{dx} = 3\cosh 3x \qquad$ (Note that 3 is from the derivative of $3x$)

$\quad \dfrac{dy}{du} = 2u$

$\quad \dfrac{dy}{dx} = \dfrac{dy}{du} \times \dfrac{du}{dx}$

$\qquad = 2u \times 3\cosh 3x$

$\qquad = 6u\cosh 3x$

$\qquad = 6\sinh 3x \cosh 3x \qquad$ (u has been replaced with $\sinh 3x$)

5. Find the derivative of $\sinh^4 2x^3$

Solution

$\quad y = \sinh^4 2x^3$

Let $u = \sinh 2x^3$

Hence, $y = u^4$

$\quad \dfrac{du}{dx} = 6x^2\cosh 2x^3 \qquad$ (Note that $6x^2$ is from the derivative of $2x^3$)

$\quad \dfrac{dy}{du} = 4u^3$

$\quad \dfrac{dy}{dx} = \dfrac{dy}{du} \times \dfrac{du}{dx}$

$\qquad = 4u^3 \times 6x^2\cosh 2x^3$

$\qquad = 24x^2 u^3\cosh 2x^3$

$\quad \dfrac{dy}{dx} = 24x^2\sinh^3 2x^3\cosh 2x^3 \qquad$ (u has been replaced with $\sinh 2x^3$)

6. Find the derivative of $\text{cosech}4x^3$

Solution

$y = \text{cosech}4x^3$

$\dfrac{dy}{dx} = -\text{cosech}4x^3\text{coth}4x^3 \times \dfrac{d(4x^3)}{dx}$ (Note that the derivative of cosech is $-$cosechcoth)

$= -\text{cosech}4x^3\text{coth}4x^3 \times 12x^2$

$= -12x^2\text{cosech}4x^3\text{coth}4x^3$

7. Find the derivative of $\text{coth}^2 2x^4$

Solution

$y = \text{coth}^2 2x^4$

Let $u = \text{coth}2x^4$

Hence, $y = u^2$

$\dfrac{du}{dx} = -\text{cosech}^2 2x^4 \times \dfrac{d(2x^4)}{dx}$

$= -\text{cosech}^2 2x^4 \times 8x^3$

$\dfrac{du}{dx} = -8x^3\text{cosech}^2 2x^4$

$\dfrac{dy}{du} = 2u$

Therefore, $\dfrac{dy}{dx} = \dfrac{dy}{du} \times \dfrac{du}{dx}$

$= 2u \times (-8x^3\text{cosech}^2 2x^4)$

$= -16x^3 u\,\text{cosech}^2 2x^4$

$\dfrac{dy}{dx} = -16x^3\text{coth}2x^4\text{cosech}^2 2x^4$ (Since $u = \text{coth}2x^4$)

8. Find the derivative of $5x\text{sinh}2x$

Solution

$y = 5x\text{sinh}2x$

We apply the product rule of differentiation as follows:

Let $u = 5x$

and $v = \text{sinh}2x$

$\dfrac{du}{dx} = 5$

$\dfrac{dv}{dx} = 2\text{cosh}2x$ (Note that the differentiation of $2x$ gives 2)

Hence, $\dfrac{dy}{dx} = u\dfrac{dv}{dx} + v\dfrac{du}{dx}$

$= (5x \times 2\text{cosh}2x) + (\text{sinh}2x \times 5)$

$= 10x\text{cosh}2x + 5\text{sinh}2x$

$$\frac{dy}{dx} = 5(2x\cosh2x + \sinh2x)$$

9. If $y = \dfrac{1 + \cosh 2x}{\sinh 2x}$ find $\dfrac{dy}{dx}$.

Solution

$$y = \frac{1 + \cosh 2x}{\sinh 2x}$$

We apply quotient rule as follows:

$u = 1 + \cosh2x$

$v = \sinh2x$

$$\frac{du}{dx} = 2\sinh2x$$

$$\frac{dv}{dx} = 2\cosh2x$$

Hence, $\dfrac{dy}{dx} = \dfrac{v\frac{du}{dx} - u\frac{dv}{dx}}{v^2}$ (Quotient rule)

$$= \frac{\sinh 2x(2\sinh 2x) - (1 + \cosh 2x)(2\cosh 2x)}{(\sinh 2x)^2}$$

$$= \frac{2\sin^2 2x - (2\cosh 2x + 2\cosh^2 2x)}{\sinh^2 2x} \qquad \text{(Note that } (\sinh2x)^2 = \sinh^2 2x)$$

$$= \frac{2\sinh^2 2x - 2\cosh 2x - 2\cosh^2 2x}{\sinh^2 2x}$$

$$= \frac{2\sinh^2 2x - 2\cosh^2 2x - 2\cosh 2x}{\sinh^2 2x}$$

$$= \frac{-2(\cosh^2 2x - \sinh^2 2x) - 2\cosh 2x}{\sinh^2 2x}$$

$$= \frac{-2(1) - 2\cosh 2x}{\sinh^2 2x} \qquad \text{(Note that } \cosh^2 2x - \sinh^2 2x = 1)$$

$$\frac{dy}{dx} = \frac{-2(1 + \cosh 2x)}{\sinh^2 2x}$$

Or, $\dfrac{dy}{dx} = \dfrac{-2(1 + \cosh 2x)}{\cosh^2 2x - 1}$ (Note that since $\cosh^2 2x - \sinh^2 2x = 1$, then $\cosh^2 2x - 1 = \sinh^2 2x$)

$$= \frac{-2(1 + \cosh 2x)}{(\cosh 2x + 1)(\cosh 2x - 1)}$$

Note that from difference of two squares we have: $a^2 - b^2 = (a + b)(a - b)$.

Hence, $\cosh^2 2x - 1$ is also a difference of two squares since 1 is also 1^2. Therefore, $\cosh^2 2x - 1 = (\cosh2x + 1)(\cosh2x - 1)$ as represented above. Hence the expression above becomes:

$$\frac{dy}{dx} = \frac{-2}{(\cosh 2x - 1)}$$

10. Differentiate $y = \dfrac{\sinh^2 x}{x}$

276

Solution

$$y = \frac{\sinh^2 x}{x}$$

We use quotient rule as follows:

Therefore, $u = \sinh^2 x$

$\quad\quad v = x$

Hence, $\dfrac{du}{dx} = 2\sinh x \cosh x$

Since, $v = x$

Then, $\dfrac{dv}{dx} = 1$

Hence, $\dfrac{dy}{dx} = \dfrac{v\frac{du}{dx} - u\frac{dv}{dx}}{v^2}$

$$= \frac{x(2\sinh x \cosh x) - \sinh^2 x(1)}{x^2}$$

$$= \frac{2x\sinh x \cosh x - \sin^2 x}{x^2}$$

$$= \frac{\sinh x(2x\cosh x - \sinh x)}{x^2}$$

Exercise 26

1. If $y = \sinh 3x - 2\cosh x$, find $\dfrac{dy}{dx}$.

2. Find the derivative of $5x^2 \operatorname{sech} x$

3. If $y = \dfrac{\sinh 2x}{3x^2}$ find $\dfrac{dy}{dx}$.

4. Find the derivative of $\cosh^2 3x$

5. Find the derivative of $\cosh^3 4x^5$

6. Find the derivative of $\coth 2x^5$

7. Find the derivative of $(\operatorname{sech} 2x^2)^3$

8. Find the derivative of $x^2 \cosh 5x$

9. If $y = \dfrac{\tanh 5x}{\coth 3x}$ find $\dfrac{dy}{dx}$.

10. Differentiate $y = \dfrac{2\cosh^3 x}{3x}$

CHAPTER 27
DERIVATIVE OF LOGARITHMIC FUNCTIONS

If $y = \log_a x$, then $\dfrac{dy}{dx} = \dfrac{1}{x} \log_a e$, where a is any base.

If $y = \log_e x$, then $\dfrac{dy}{dx} = \dfrac{1}{x}$

Note that $\log_e x$ can also be represented as $\ln x$ and the value of e is 2.718 (to 3 decimal places)

Examples

1. Find the derivative of $\log_a 2x$.

Solution

$$y = \log_a 2x$$

We will use chain rule since we have a function that is not just x but $2x$.

Let $u = 2x$

Therefore, $y = \log_a u$

$$\frac{du}{dx} = 2$$

$$\frac{dy}{du} = \frac{1}{u} \log_a e$$

$$\frac{dy}{dx} = \frac{dy}{du} \times \frac{du}{dx}$$

$$= \frac{1}{u} \log_a e \times 2$$

$$= \frac{2}{u} \log_a e$$

$$= \frac{2}{2x} \log_a e \qquad \text{(since } u = 2x\text{)}$$

$$\frac{dy}{dx} = \frac{1}{x} \log_a e$$

2. Find the derivative of $\log_a(5x - 1)$

Solution

$$y = \log_a(5x - 1)$$

Let $u = 5x - 1$

Therefore, $y = \log_a u$

$$\frac{du}{dx} = 5$$

$$\frac{dy}{du} = \frac{1}{u} \log_a e$$

$$\frac{dy}{dx} = \frac{dy}{du} \times \frac{du}{dx}$$

$$= \frac{1}{u} \log_a e \ \times \ 5$$

$$= \frac{5}{u} \log_a e$$

$$\frac{dy}{dx} = \frac{5}{5x - 1} \log_a e \qquad \text{(since u = 5}x - 1\text{)}$$

3. Differentiate $\log_a(4x - 3)^2$ with respect to x

Solution

$$y = \log_a(4x - 3)^2$$

Let $u = (4x - 3)^2$

Therefore, $y = \log_a u$

$$\frac{du}{dx} = 2(4x - 3)^{2-1} \ \times \ 4 \qquad \text{(Note that 4 is from the derivative of } 4x - 3\text{)}$$

$$\frac{du}{dx} = 8(4x - 3)$$

$$\frac{dy}{du} = \frac{1}{u} \log_a e$$

$$\frac{dy}{dx} = \frac{dy}{du} \ \times \ \frac{du}{dx}$$

$$= \frac{1}{u} \log_a e \ \times \ 8(4x - 3)$$

$$= \frac{8(4x-3)}{u} \log_a e$$

$$= \frac{8(4x-3)}{(4x - 3)^2} \log_a e \qquad [\text{since u = } (4x - 3)^2]$$

$$\frac{dy}{dx} = \frac{8}{4x - 3} \log_a e \qquad \text{(One } 4x - 3 \text{ cancels out)}$$

4. If $y = \log_a\sqrt{1 + 2x}$, find $\frac{dy}{dx}$

Solution

$$y = \log_a\sqrt{1 + 2x}$$

Let $u = \sqrt{1 + 2x} \ = (1 + 2x)^{\frac{1}{2}}$

Therefore, $y = \log_a u$

$$\frac{du}{dx} = \frac{1}{2}(1 + 2x)^{\frac{1}{2}-1} \ \times \ 2$$

$$= (1 + 2x)^{-\frac{1}{2}} \qquad \text{(Note that } \frac{1}{2} \times 2 = 1\text{)}$$

$$\frac{du}{dx} = \frac{1}{(1+2x)^{\frac{1}{2}}}$$

$$\frac{dy}{du} = \frac{1}{u} \log_a e$$

$$\frac{dy}{dx} = \frac{dy}{du} \times \frac{du}{dx}$$

$$= \frac{1}{u} \log_a e \times \frac{1}{(1+2x)^{\frac{1}{2}}}$$

$$= \frac{1}{(1+2x)^{\frac{1}{2}}} \log_a e \times \frac{1}{(1+2x)^{\frac{1}{2}}} \qquad \text{[Note that u has been replaced with } (1+2x)^{\frac{1}{2}}\text{]}$$

$$\frac{dy}{dx} = \frac{1}{1+2x} \log_a e \qquad \text{[Note that } (1+2x)^{\frac{1}{2}} \times (1+2x)^{\frac{1}{2}} = (1+2x)^{1} \text{ by adding exponents]}$$

5. Find $\frac{dy}{dx}$ given that $y = \log_a\frac{1-3x}{1+3x}$

Solution

$$y = \log_a\frac{1-3x}{1+3x}$$

Or, $y = \log_a(1-3x) - \log_a(1+3x)$ $\qquad$ (Recall that $\log_x\left(\frac{a}{b}\right) = \log_x a - \log_x b$)

$$\frac{dy}{dx} = \frac{-3}{1-3x} \log_a e - \frac{3}{1+3x} \log_a e$$

$$= -\log_a e \left(\frac{3}{1-3x} + \frac{3}{1+3x}\right)$$

$$= -\log_a e \left[\frac{3(1+3x) + 3(1-3x)}{(1-3x)(1+3x)}\right]$$

$$= -\log_a e \left[\frac{3 + 9x + 3 - 9x}{1 - 9x^2}\right] \qquad \text{[Note that } (1-3x)(1+3x) = 1 - 9x^2\text{]}$$

$$= -\log_a e \left(\frac{6}{1 - 9x^2}\right)$$

$$\frac{dy}{dx} = \frac{-6}{1 - 9x^2} \log_a e$$

6. If $y = \log_{10}(x^2 - 2)$, find $\frac{dy}{dx}$.

Solution

$$y = \log_{10}(x^2 - 2)$$

The value 10 represents a in other examples. So we are going to differentiate y like the examples above except that we will write 10 wherever 'a' should be.

$$y = \log_{10}(x^2 - 2)$$

$$u = x^2 - 2$$

Hence, $y = \log_{10}u$

$$\frac{du}{dx} = 2x$$

$$\frac{dy}{du} = \frac{1}{u} \log_{10}e \qquad \text{(Just like differentiating } \log_a u)$$

$$\frac{dy}{dx} = \frac{dy}{du} \times \frac{du}{dx}$$

$$= \frac{1}{u} \log_{10}e \ \times \ 2x$$

$$= \frac{2x}{u} \log_{10}e$$

$$\frac{dy}{dx} = \frac{2x}{x^2 - 2} \log_{10}e$$

7. Find $\frac{dy}{dx}$ if $y = \log_{10} \frac{1}{x}$

Solution

$$y = \log_{10} \frac{1}{x}$$

Let us solve this question directly without using $u = \frac{1}{x}$ as follows:

$$\frac{dy}{dx} = \frac{\frac{d(\frac{1}{x})}{dx}}{\frac{1}{x}} \ \times \ \log_{10}e$$

$$= \frac{\frac{d(x^{-1})}{dx}}{x^{-1}} \ \times \ \log_{10}e$$

$$= \frac{-x^{-2}}{x^{-1}} \ \times \ \log_{10}e \qquad \text{(Note that the derivative of } x^{-1} \text{ is } -x^{-2})$$

$$= -x^{-2} \ \times \ x^1 \ \times \ \log_{10}e \qquad \text{(Since } \frac{1}{x^{-1}} = x^1)$$

$$= -x^{-1} \ \times \ \log_{10}e \qquad \text{(After adding the exponents of } x)$$

$$\frac{dy}{dx} = -\frac{1}{x} \log_{10}e$$

8. Find the derivative of $\log_e(2 - x^3)$

Solution

$$y = \log_e(2 - x^3)$$

Note that the base here is 'e' and not 'a'.

Let $u = 2 - x^3$

Hence, $y = \log_e u$

$$\frac{du}{dx} = -3x^2$$

$$\frac{dy}{du} = \frac{1}{u} \qquad \text{(Recall that the derivative of } \log_e x = \frac{1}{x})$$

$$\frac{dy}{dx} = \frac{dy}{du} \ \times \ \frac{du}{dx}$$

$$= \frac{1}{u} \ \times \ -3x^2$$

$$= \frac{-3x^2}{u}$$

$$\frac{dy}{dx} = \frac{-3x^2}{2 - x^3}$$

9. Find the derivative of $(\log_e 5x)^2$

<u>Solution</u>

$$y = (\log_e 5x)^2$$

Let $u = \log_e 5x$

Hence, $y = u^2$

$$\frac{du}{dx} = \frac{\frac{d(5x)}{dx}}{5x}$$

$$= \frac{5}{5x}$$

$$\frac{du}{dx} = \frac{1}{x}$$

$$\frac{dy}{du} = 2u$$

$$\frac{dy}{dx} = \frac{dy}{du} \times \frac{du}{dx}$$

$$= 2u \times \frac{1}{x}$$

$$= \frac{2u}{x}$$

$$\frac{dy}{dx} = \frac{2}{x}\log_e 5x \quad \text{(Since } u = \log_e 5x\text{)}$$

10. If $y = \ln\sqrt{3x^2 - 4}$, find $\frac{dy}{dx}$.

<u>Solution</u>

$$y = \ln\sqrt{3x^2 - 4}$$

Note that $\ln\sqrt{3x^2 - 4}$ is also the same as $\log_e\sqrt{3x^2 - 4}$

Hence, $y = \log_e\sqrt{3x^2 - 4}$

Or, $y = \log_e(3x^2 - 4)^{\frac{1}{2}}$

Let $u = (3x^2 - 4)^{\frac{1}{2}}$

Hence, $y = \log_e u$

$$\frac{du}{dx} = \frac{1}{2}(3x^2 - 4)^{\frac{1}{2} - 1} \times 6x \quad \text{(Note that } 6x \text{ is from the derivative of } 3x^2 - 4\text{)}$$

$$= \frac{6x}{2}(3x^2 - 4)^{-\frac{1}{2}}$$

$$\frac{du}{dx} = \frac{3x}{(3x^2 - 4)^{\frac{1}{2}}}$$

$$\frac{dy}{du} = \frac{1}{u}$$

$$\frac{dy}{dx} = \frac{dy}{du} \times \frac{du}{dx}$$

$$= \frac{1}{u} \times \frac{3x}{(3x^2 - 4)^{\frac{1}{2}}}$$

$$= \frac{1}{(3x^2 - 4)^{\frac{1}{2}}} \times \frac{3x}{(3x^2 - 4)^{\frac{1}{2}}}$$

$$\frac{dy}{dx} = \frac{3x}{3x^2 - 4}$$ (Note that $(3x^2 - 4)^{\frac{1}{2}} \times (3x^2 - 4)^{\frac{1}{2}} = 3x^2 - 4$, after adding their exponents)

11. Find $\frac{dy}{dx}$ if $y = \sqrt[3]{x}\ \text{In}2x$.

Solution

$y = \sqrt[3]{x}\ \text{In}2x$

We are going to apply product rule here.

$u = \sqrt[3]{x} = x^{\frac{1}{3}}$

$v = \text{In}2x$ (Note that In2x is the same as $\log_e 2x$)

$$\frac{du}{dx} = \frac{1}{3}x^{\frac{1}{3} - 1}$$

$$= \frac{1}{3}x^{-\frac{2}{3}}$$

$$\frac{dv}{dx} = \frac{2}{2x}$$

$$= \frac{1}{x}$$

Hence, $\frac{dy}{dx} = u\frac{dv}{dx} + v\frac{du}{dx}$ (Product rule)

$$= x^{\frac{1}{3}}\left(\frac{1}{x}\right) + \text{In}2x\left(\frac{1}{3}x^{-\frac{2}{3}}\right)$$

$$= x^{\frac{1}{3}}(x^{-1}) + \frac{1}{3}x^{-\frac{2}{3}}\ \text{In}2x$$

$$= x^{-\frac{2}{3}}\left(1 + \frac{\text{In}2x}{3}\right)$$

$$= x^{-\frac{2}{3}}\left(\frac{3 + \text{In}2x}{3}\right)$$

$$\frac{dy}{dx} = \frac{1}{x^{\frac{2}{3}}}\left(\frac{3 + \text{In}2x}{3}\right)$$

Or, $\frac{dy}{dx} = \frac{1}{\sqrt[3]{x^2}}\left(\frac{3 + \text{In}2x}{3}\right)$

12. Find $\frac{dy}{dx}$ given that $y = \text{In}(1 + 2x)^2$

Solution

$y = \text{In}(1 + 2x)^2$ [Also $y = \log_e(1 + 2x)^2$]

283

Let $u = (1 + 2x)^2$

Hence, $y = \ln u$

$$\frac{du}{dx} = 2(1 + 2x) \times \frac{d(2x)}{dx}$$

$$= 2(1 + 2x) \times 2x$$

$$\frac{du}{dx} = 4(1 + 2x)$$

$$\frac{dy}{du} = \frac{1}{u}$$

$$\frac{dy}{dx} = \frac{1}{u} \times 4(1 + 2x)$$

$$= \frac{4(1+2x)}{u}$$

$$= \frac{4(1+2x)}{(1+2x)^2}$$

$$= \frac{4}{1+2x}$$

13. Differentiate with respect to x: $5x\ln(3x^2 - 2)$

Solution

$$y = 5x\ln(3x^2 - 2)$$

By using product rule:

$$u = 5x$$

$$v = \ln(3x^2 - 2)$$

$$\frac{du}{dx} = 5$$

$$\frac{dv}{dx} = \frac{6x}{3x^2 - 2}$$

Hence, $\dfrac{dy}{dx} = u\dfrac{dv}{dx} + v\dfrac{du}{dx}$

$$= 5x\left(\frac{6x}{3x^2 - 2}\right) + \ln(3x^2 - 2)(5)$$

$$\frac{dy}{dx} = \frac{30x^2}{3x^2 - 2} + 5\ln(3x^2 - 2)$$

14. Find the derivative of $\dfrac{\ln x}{x}$

Solution

$$y = \frac{\ln x}{x}$$

We are going to use quotient rule as follows:

$$u = \ln x$$

$$v = x$$

$$\frac{du}{dx} = \frac{1}{x}$$

$$\frac{dv}{dx} = 1$$

Hence, $\dfrac{dy}{dx} = \dfrac{v\dfrac{du}{dx} - u\dfrac{dv}{dx}}{v^2}$

$$= \frac{x\left(\frac{1}{x}\right) - \ln x\,(1)}{x^2}$$

$$\frac{dy}{dx} = \frac{1 - \ln x}{x^2}$$

Exercise 27

1. Find the derivative of $\log_a 7x$.

2. Find the derivative of $\log_a(x^3 + 5)$

3. Differentiate $\log_a(2x^3 - 5)^6$ with respect to x

4. If $y = \log_a x^{\frac{2}{3}}$, find $\dfrac{dy}{dx}$

5. Find $\dfrac{dy}{dx}$ given that $y = \log_a\dfrac{1 + x^2}{1 - x^2}$

6. If $y = \log_5(5x^3 - 1)$, find $\dfrac{dy}{dx}$.

7. Find $\dfrac{dy}{dx}$ if $y = \log_2\dfrac{1}{x^2}$

8. Find the derivative of $\log_e(x^2 + 3)$

9. Find the derivative of $(\log_e x)^3$

10. If $y = \ln\sqrt{1 - 2x^5}$, find $\dfrac{dy}{dx}$.

11. Find $\dfrac{dy}{dx}$ if $y = x^4\ln x^2$

12. Find $\dfrac{dy}{dx}$ given that $y = \ln(3 - 7x)^4$

13. Differentiate with respect to x: $3x^2\ln(4x^3 + 1)$

14. Find the derivative of $\dfrac{\ln x^2}{x^2}$

15. Find the derivative of $\log_e(1 - 5x^2)^3$

CHAPTER 28
DERIVATIVE OF EXPONENTIAL FUNCTIONS

If $y = a^x$, then $\frac{dy}{dx} = a^x \log_e a$, where 'a' is any number.

If $y = e^x$, then $\frac{dy}{dx} = e^x$

Examples

1. Differentiate with respect to x: a^{5x}

Solution

$\quad y = a^{5x}$

Let $u = 5x$

$\quad\quad y = a^u$

$\quad\frac{du}{dx} = 5$

$\quad\frac{dy}{du} = a^u \log_e a$

Hence, $\frac{dy}{dx} = \frac{dy}{du} \times \frac{du}{dx}$

$\quad\quad = a^u \log_e a \times 5$

$\quad\quad = 5a^u \log_e a$

$\frac{dy}{dx} = 5a^{5x} \log_e a \quad$ (since u = 5x)

2. Find the derivative of $a^{x^2 - 3x + 4}$

Solution

$\quad y = a^{x^2 - 3x + 4}$

Let us solve this example directly without using $u = x^2 - 3x + 4$

$\quad\frac{dy}{dx} = \frac{d(x^2 - 3x + 4)}{dx} \times a^{x^2 - 3x + 4} \times \log_e a$

$\quad\quad = (2x - 3)(a^{x^2 - 3x + 4} \times \log_e a)$

3. If $y = 3x^2 a^{5x}$, find $\frac{dy}{dx}$.

Solution

$\quad y = 3x^2 a^{5x}$

We apply product rule as follows:

$\quad u = 3x^2$

$\quad v = a^{5x}$

$$\frac{du}{dx} = 6x$$

$$\frac{dv}{dx} = 5a^{5x}\log_e a$$

Hence, $\dfrac{dy}{dx} = u\dfrac{dv}{dx} + v\dfrac{du}{dx}$

$$= 3x^2(5a^{5x}\log_e a) + a^{5x}(6x)$$

$$= 15x^2 a^{5x}\log_e a + 6xa^{5x}$$

Factorizing the expression above gives:

$$\frac{dy}{dx} = 3xa^{5x}(5x\log_e a + 2)$$

4. Find the derivative of $\dfrac{e^x + e^{-x}}{e^x - e^{-x}}$

Solution

$$y = \frac{e^x + e^{-x}}{e^x - e^{-x}}$$

$$u = e^x + e^{-x}$$

$$v = e^x - e^{-x}$$

$$\frac{du}{dx} = e^x - e^{-x} \quad \text{(The derivative of } e^{-x} = -1 \times e^{-x} = -e^{-x}. \text{ The value } -1 \text{ is from the derivative of } -x)$$

$$\frac{dv}{dx} = e^x + e^{-x}$$

Hence, $\dfrac{dy}{dx} = \dfrac{v\dfrac{du}{dx} - u\dfrac{dv}{dx}}{v^2}$ (Quotient rule)

$$= \frac{(e^x - e^{-x})(e^x - e^{-x}) - (e^x + e^{-x})(e^x + e^{-x})}{(e^x - e^{-x})^2}$$

Expanding the numerator and denominator gives:

$$\frac{dy}{dx} = \frac{(e^x)(e^x) - (e^x)(e^{-x}) - (e^{-x})(e^x) + (e^{-x})(e^{-x}) - [(e^x)(e^x) + (e^x)(e^{-x}) + (e^{-x})(e^x) + (e^{-x})(e^{-x})]}{(e^x - e^{-x})^2}$$

$$= \frac{e^{2x} - 1 - 1 + e^{-2x} - (e^{2x} + 1 + 1 + e^{-2x})}{(e^x - e^{-x})^2} \quad \text{(Note that exponents have been added, and } e^0 = 1)$$

$$= \frac{e^{2x} - 2 + e^{-2x} - e^{2x} - 1 - 1 - e^{-2x})}{(e^x - e^{-x})^2}$$

$$\frac{dy}{dx} = \frac{-4}{(e^x - e^{-x})^2} \quad (e^{2x} \text{ and } e^{-2x} \text{ cancel out})$$

5. If $y = e^{\sqrt{x}}$ find $\dfrac{dy}{dx}$.

Solution

$$y = e^{\sqrt{x}} = e^{x^{\frac{1}{2}}}$$

$$\frac{dy}{dx} = \frac{d(x)^{\frac{1}{2}}}{dx} \times e^{x^{\frac{1}{2}}}$$

$$= \frac{1}{2}x^{-\frac{1}{2}} \times e^{x^{\frac{1}{2}}}$$

$$= \frac{1}{2}\left(\frac{1}{x^{\frac{1}{2}}}\right) \times e^{x^{\frac{1}{2}}}$$

$$= \frac{e^{x^{\frac{1}{2}}}}{2x^{\frac{1}{2}}}$$

$$\frac{dy}{dx} = \frac{e^{\sqrt{x}}}{2\sqrt{x}}$$

6. Find $\frac{dy}{dx}$ if $y = x^{\frac{1}{2}} e^{x^{\frac{1}{2}}}$

Solution

$$y = x^{\frac{1}{2}} e^{x^{\frac{1}{2}}}$$

From product rule:

$$\frac{dy}{dx} = x^{\frac{1}{2}}\frac{d(e^{x^{\frac{1}{2}}})}{dx} + e^{x^{\frac{1}{2}}}\frac{d(x^{\frac{1}{2}})}{dx}$$

$$= x^{\frac{1}{2}}\left(\frac{e^{x^{\frac{1}{2}}}}{2x^{\frac{1}{2}}}\right) + e^{x^{\frac{1}{2}}}\left(\frac{1}{2}x^{-\frac{1}{2}}\right) \quad \text{(Note that the derivative of } e^{x^{\frac{1}{2}}} \text{ is } \frac{e^{x^{\frac{1}{2}}}}{2x^{\frac{1}{2}}} \text{ from example 5)}$$

$$= \frac{e^{x^{\frac{1}{2}}}}{2} + \frac{e^{x^{\frac{1}{2}}}}{2x^{\frac{1}{2}}}$$

$$= \frac{x^{\frac{1}{2}}e^{x^{\frac{1}{2}}} + e^{x^{\frac{1}{2}}}}{2x^{\frac{1}{2}}}$$

$$\frac{dy}{dx} = \frac{e^{x^{\frac{1}{2}}}(x^{\frac{1}{2}} + 1)}{2x^{\frac{1}{2}}}$$

7. Find the derivative of $a^{2x} - a^{-2x}$

Solution

$$y = a^{2x} - a^{-2x}$$

$$\frac{dy}{dx} = \frac{d(2x)}{dx}(a^{2x}\log_e a) - \left[\frac{d(-2x)}{dx}(a^{-2x}\log_e a)\right]$$

$$= 2a^{2x}\log_e a - (-2a^{-2x}\log_e a)$$

$$= 2a^{2x}\log_e a + 2a^{-2x}\log_e a$$

$$= 2\log_e a(a^{2x} + a^{-2x})$$

8. Find the derivative of $e^{2x}\log_e 3x$

<u>Solution</u>

$$y = e^{2x}\log_e 3x$$

By product rule:

$$u = e^{2x}$$

$$v = \log_e 3x \quad (\text{or } v = \ln 3x)$$

$$\frac{du}{dx} = 2e^{2x}$$

$$\frac{dv}{dx} = \frac{3}{3x}$$

$$= \frac{1}{x}$$

$$\frac{dy}{dx} = e^{2x}\left(\frac{1}{x}\right) + \log_e 3x(2e^{2x})$$

$$= \frac{e^{2x}}{x} + 2e^{2x}\log_e 3x$$

$$\frac{dy}{dx} = e^{2x}\left(\frac{1}{x} + 2\log_e 3x\right)$$

9. Differentiate with respect to x: $2\sqrt{xe^x}$

<u>Solution</u>

$$y = 2\sqrt{xe^x}$$

Or, $y = 2\sqrt{x}\sqrt{e^x}$

From product rule:

$$u = 2\sqrt{x} = 2x^{\frac{1}{2}}$$

$$v = \sqrt{e^x} = (e^x)^{\frac{1}{2}}$$

$$v = e^{\frac{1}{2}x} \quad \text{(When the exponents are multiplied)}$$

$$\frac{du}{dx} = \frac{1}{2} \times 2x^{-\frac{1}{2}}$$

$$= x^{-\frac{1}{2}}$$

$$\frac{dv}{dx} = \frac{1}{2}e^{\frac{1}{2}x} \quad \text{(Note that } \frac{1}{2} \text{ is from the derivative of } \frac{1}{2}x\text{)}$$

$$\frac{dy}{dx} = u\frac{dv}{dx} + v\frac{du}{dx}$$

$$= 2x^{\frac{1}{2}}\left(\frac{1}{2}e^{\frac{1}{2}x}\right) + e^{\frac{1}{2}x}\left(x^{-\frac{1}{2}}\right)$$

$$= \left(\frac{1}{2} \times 2x^{\frac{1}{2}} \times e^{\frac{1}{2}x}\right) \times \frac{e^{\frac{1}{2}x}}{x^{\frac{1}{2}}}$$

$$= x^{\frac{1}{2}}e^{\frac{1}{2}x} + \frac{e^{\frac{1}{2}x}}{x^{\frac{1}{2}}}$$

$$= \frac{x\ e^{\frac{1}{2}x} + e^{\frac{1}{2}x}}{x^{\frac{1}{2}}}$$

$$= \frac{e^{\frac{1}{2}x}(x+1)}{x^{\frac{1}{2}}}$$

$$= \frac{e^{\frac{x}{2}}(x+1)}{x^{\frac{1}{2}}}$$

$$\frac{dy}{dx} = \frac{\sqrt{e^x}(x+1)}{\sqrt{x}}$$

10. Find the derivative of $2x\log_{10}x$

Solution

$y = 2x\log_{10}x$ (Note that this is similar to $2x\log_a x$)

$u = 2x$

$v = \log_{10}x$

$\frac{du}{dx} = 2$

$\frac{dv}{dx} = \frac{1}{x}\log_{10}e$ (Recall that the derivative of $\log_a x$ is $\frac{1}{x}\log_a e$)

$\frac{dy}{dx} = u\frac{dv}{dx} + v\frac{du}{dx}$

$\qquad = 2x\ (\frac{1}{x}\log_{10}e)\ +\ \log_{10}x(2)$

$\qquad = 2\log_{10}e + 2\log_{10}x$

$\frac{dy}{dx} = 2(\log_{10}e + \log_{10}x)$

Or, $\frac{dy}{dx} = 2\left(\frac{\log_e e}{\log_e 10} + \log_{10}x\right)$

Note that the rule of change of base in logarithm that has been applied above is given by:

If $\log_a b$ is to be converted to a new base, e, then: $\log_a b = \dfrac{\log_e b}{\log_e a}$

$\frac{dy}{dx} = 2\left(\frac{1}{\log_e 10} + \log_{10}x\right)$ (Recall that $\log_x x = 1$, or $\log_e e = 1$)

11. Find the derivative of 10^{2x}

Solution

$y = 10^{2x}$ (This is similar to a^{2x})

$\frac{dy}{dx} = \frac{d(2x)}{dx}\ \times\ 10^{2x}\log_e 10$

$$= 2 \times 10^{2x}\log_e 10$$

$$= (10^{2x})2\log_e 10$$

$$= 10^{2x}\log_e 10^2 \quad \text{(Recall that } a\log_x y = \log_x y^a)$$

$$\frac{dy}{dx} = 10^{2x}\log_e 100$$

12. If $y = e^{\ln x}$ find $\dfrac{dy}{dx}$

<u>Solution</u>

$$y = e^{\ln x} \quad \text{(Or } y = e^{\log_e x} \text{ since } \ln x = \log_e x)$$

Let, $u = \ln x$

$$y = e^u$$

$$\frac{du}{dx} = \frac{1}{x}$$

$$\frac{dy}{du} = e^u$$

$$\frac{dy}{dx} = e^u \times \frac{1}{x}$$

$$= \frac{e^u}{x}$$

$$= \frac{e^{\ln x}}{x} \quad \text{(Since } u = \ln x)$$

$$= \frac{e^{\log_e x}}{x} \quad \text{(Since } \ln x = \log_e x)$$

$$= \frac{x}{x} \quad \text{(Recall the identity that } m^{\log_m n} = n)$$

$$\frac{dy}{dx} = 1$$

This example shows that calculus works, since the derivative of x is 1, and the derivative of $e^{\log_e x} = 1$ because $e^{\log_e x} = x$. Hence, we could have solved this example by stating that:

$$\frac{d(e^{\ln x})}{dx} = \frac{d(x)}{dx} = 1$$

13. What is the derivative of $\dfrac{e^{\frac{1}{x}}}{x^2}$

<u>Solution</u>

$$y = \frac{e^{\frac{1}{x}}}{x^2}$$

From quotient rule:

$$u = e^{\frac{1}{x}}$$

$$v = x^2$$

$$\frac{du}{dx} = \frac{d\left(\frac{1}{x}\right)}{dx} \times e^{\frac{1}{x}}$$

$$= \frac{d(x^{-1})}{dx} \times e^{\frac{1}{x}}$$

$$= -1x^{-2} \times e^{\frac{1}{x}}$$

$$= \frac{-1}{x^2} \times e^{\frac{1}{x}}$$

$$\frac{du}{dx} = \frac{-e^{\frac{1}{x}}}{x^2}$$

$$\frac{du}{dx} = 2x$$

$$\frac{dy}{dx} = \frac{v\frac{du}{dx} - u\frac{dv}{dx}}{v^2} \qquad \text{(Quotient rule)}$$

$$= \frac{x^2\left(\frac{-e^{\frac{1}{x}}}{x^2}\right) - e^{\frac{1}{x}}(2x)}{(x^2)^2}$$

$$= \frac{-e^{\frac{1}{x}} - 2x\,e^{\frac{1}{x}}}{x^4} \qquad \text{(Note that } x^2 \text{ has cancelled out)}$$

14. Find the derivative of $e^{-5x} + 5e$

<u>Solution</u>

$$y = e^{-5x} + 5e$$

$$\frac{dy}{dx} = \frac{d(-5x)}{dx} \times e^{-5x} + \frac{d(5e)}{dx}$$

$$= -5e^{-5x} + 0 \qquad \text{(Note that 5e is a constant and the derivative of a constant is zero)}$$

$$\frac{dy}{dx} = -5e^{-5x}$$

15. If $y = e^{(2x+3)^2}$ find $\frac{dy}{dx}$.

<u>Solution</u>

$$y = e^{(2x+3)^2}$$

Let $u = (2x + 3)^2$

Hence, $y = e^u$

$$\frac{du}{dx} = 2(2x + 3)^{2-1} \times \frac{d(2x)}{dx} \qquad \text{(Using chain rule)}$$

$$= 2(2x + 3) \times 2$$

$$= 4(2x + 3)$$

$$\frac{dy}{du} = e^u$$

$$\frac{dy}{dx} = \frac{dy}{du} \times \frac{du}{dx}$$

$$= e^u \times 4(2x+3)$$

$$= e^{(2x+3)^2} \times 4(2x+3) \qquad \text{(Note that } u = (2x+3)^2)$$

$$\frac{dy}{dx} = 4(2x+3)e^{(2x+3)^2}$$

16. Differentiate with respect to θ, $5^{2\theta}$

Solution

$$y = 5^{2\theta} \qquad \text{(This is like } y = a^{2\theta})$$

$$\frac{dy}{d\theta} = \frac{d(2\theta)}{d\theta} \times 5^{2\theta}\log_e 5$$

$$= 2 \times 5^{2\theta}\log_e 5$$

$$= (5^{2\theta})2\log_e 5$$

$$= 5^{2\theta}\log_e 5^2 \qquad \text{(Note that } 2\log_e 5 = \log_e 5^2)$$

$$\frac{dy}{d\theta} = 5^{2\theta}\log_e 25$$

Exercise 28

1. Differentiate with respect to x: $5a^{2x}$

2. Find the derivative of $a^{5x^3 - x}$

3. If $y = x^4 a^{3x}$, find $\frac{dy}{dx}$.

4. Find the derivative of $\dfrac{3 + e^{-x}}{e^x}$

5. If $y = 2e^{\sqrt{3x}}$ find $\frac{dy}{dx}$.

6. Find $\frac{dy}{dx}$ if $y = 6x^3 \, e^{x^2}$

7. Find the derivative of $a^{-5x} + a^{5x}$

8. Find the derivative of $3e^{4x}\log_e 2x^2$

9. Differentiate with respect to x: $3x^2\sqrt{e^{5x}}$

10. Find the derivative of $x^4\log_{10}7x$

11. Find the derivative of 6^{3x}

12. Find the derivative of $e^{x^2 - 3x}$

13. What is the derivative of $\dfrac{2e^{\frac{3}{x}}}{10x^5}$

14. Find the derivative of $1 - 2e^{-3x}$

293

15. If $y = e^{(x^2+x)^3}$ find $\dfrac{dy}{dx}$.

16. Differentiate with respect to x, 2^{x^2}

17. Find the derivative of $\dfrac{a^x - a^{-x}}{a^{-x}}$

18. Find the derivative of $e^x \ln 10x^3$

19. Differentiate with respect to x: $e^{2x}\sqrt{5x}$

20. Find the derivative of $2x^5 \log_{10} 5x^4$

CHAPTER 29
LOGARITHMIC DIFFERENTIATION

A function of x could be raised to an exponent which is also a function of x. An example is x^x.
The differentiation of such a function is called logarithmic differentiation.

This type of function is differentiated by first taking the logarithm of both sides of the equation, and then differentiating implicitly before making $\dfrac{dy}{dx}$ the subject of the formula (i.e. solving for $\dfrac{dy}{dx}$). Finally, y (i.e. the function) is substituted into the final answer.

Examples

1. Find the derivative of x^x.

<u>Solution</u>

$$y = x^x$$

This is a function of x raised to an exponent which is also a function of x.

Hence, take the logarithm to base e of both sides as follows:

$$\log_e y = \log_e x^x$$

$$\log_e y = x\log_e x \qquad \text{(Note that } \log_m n^b = b\log_m n)$$

Now differentiate y implicitly and differentiate the right hand side using product rule. This gives:

$$\frac{1}{y}\frac{dy}{dx} = x\frac{1}{x} + \log_e x(1) \qquad \text{(Use of product rule on the right hand side, with u = } x \text{ and v} = \log_e x)$$

$$\frac{1}{y}\frac{dy}{dx} = 1 + \log_e x$$

Multiply both sides by y to make $\dfrac{dy}{dx}$ the subject of the formula as follows:

$$\frac{dy}{dx} = y(1 + \log_e x)$$

$$\frac{dy}{dx} = x^x(1 + \log_e x) \qquad \text{(Since y} = x^x)$$

2. Find the derivative of $x^{\ln x}$.

<u>Solution</u>

$$y = x^{\ln x}$$

Taking the logarithm of both sides gives:

$$\log_e y = \log_e x^{\ln x}$$

$$\log_e y = \ln x\log_e x$$

Differentiate y implicitly and treat the right hand side using product rule as follows:

$$\frac{1}{y}\frac{dy}{dx} = \ln x\left(\frac{1}{x}\right) + \log_e x\left(\frac{1}{x}\right)$$

$$\frac{1}{y}\frac{dy}{dx} = \left(\frac{\ln x}{x}\right) + \left(\frac{\ln x}{x}\right)$$ (Note that $\log_e x$ can also be written as $\ln x$)

$$\frac{1}{y}\frac{dy}{dx} = \frac{2\ln x}{x}$$

Multiply both sides by y. This gives:

$$\frac{dy}{dx} = y\left(\frac{2\ln x}{x}\right)$$

$$\frac{dy}{dx} = x^{\ln x}\left(\frac{2\ln x}{x}\right)$$ (Since $y = x^{\ln x}$)

Or, $\dfrac{dy}{dx} = \dfrac{2}{x}x^{\ln x}\ln x$

3. If $y = \ln(2 + x)^x$ find $\dfrac{dy}{dx}$.

Solution

$$y = \ln(2 + x)^x$$

In this case we are not going to take the logarithm of both sides since there is already logarithm on the right hand side. Hence, we proceed as follows:

$$y = \ln(2 + x)^x$$

$$y = x\log_e(2 + x)$$ [Note that $\ln(2 + x)$ can also be written as $\log_e(2 + x)$]

Using product rule gives:

$$\frac{dy}{dx} = x\left(\frac{1}{2 + x}\right) + \log_e(2 + x)(1)$$

$$\frac{dy}{dx} = \frac{x}{2 + x} + \log_e(2 + x)$$

4. Find $\dfrac{dy}{dx}$ if $y = (x^2 - 3)^{5x}$

Solution

$$y = (x^2 - 3)^{5x}$$

Taking the natural logarithm of both sides gives:

$$\log_e y = \log_e(x^2 - 3)^{5x}$$

$$\log_e y = 5x\log_e(x^2 - 3)$$

Using implicit differentiation for the left hand side and product rule for the right hand side gives:

$$\frac{1}{y}\frac{dy}{dx} = 5x\left(\frac{2x}{x^2 - 3}\right) + \log_e(x^2 - 3)(5)$$

$$\frac{1}{y}\frac{dy}{dx} = \frac{10x^2}{x^2 - 3} + 5\log_e(x^2 - 3)$$

Multiply both sides by y to obtain:

$$\frac{dy}{dx} = y\left[\frac{10x^2}{x^2 - 3} + 5\log_e(x^2 - 3)\right]$$

$$\frac{dy}{dx} = (x^2 - 3)^{5x}\left[\frac{10x^2}{x^2 - 3} + 5\log_e(x^2 - 3)\right] \qquad \text{[Note that } y = (x^2 - 3)^{5x}]$$

5. If $y = \dfrac{(x^2 - 5)(3x-1)^2}{x^7(2x^3-3)}$

Solution

This is not a function of x raised to an exponent which is also a function of x. This function can be differentiated by the use of product and quotient rule. However, the use of these rules will be a nightmare. Hence, we have to apply logarithmic differentiation. This is done as follows:

$$y = \frac{(x^2 - 5)(3x-1)^2}{x^7(2x^3-3)}$$

Taking the logarithm of both sides gives:

$$\log_e y = \log_e\left[\frac{(x^2 - 5)(3x-1)^2}{x^7(2x^3-3)}\right]$$

In order to continue, we have to recall the theory of logarithm as follows:

1. $\log_b(cd) = \log_b c + \log_b d$

2. $\log_b\left(\frac{c}{d}\right) = \log_b c - \log_b d$

3. $\log_b c^d = d\log_b c$

Therefore, we continue the logarithmic differentiation as follows:

$$\log_e y = \log_e\left[\frac{(x^2 - 5)(3x-1)^2}{x^7(2x^3-3)}\right]$$

Applying the theory above gives:

$$\log_e y = \log_e[(x^2 - 5)(3x - 1)^2] - \log_e[x^7(2x^3 - 3)]$$
$$\log_e y = \log_e(x^2 - 5) + \log_e(3x - 1)^2 - [\log_e x^7 + \log_e(2x^3 - 3)]$$
$$\log_e y = \log_e(x^2 - 5) + 2\log_e(3x - 1) - 7\log_e x - \log_e(2x^3 - 3)$$

Differentiating each term accordingly gives:

$$\frac{1}{y}\frac{dy}{dx} = \frac{2x}{x^2 - 5} + \frac{2 \times 3}{3x - 1} - \frac{7}{x} - \frac{6x}{2x^3 - 3}$$

Note that each function of x is differentiated first, and the value obtained is then divided by the original function. Hence, the expression above simplifies to gives:

$$\frac{1}{y}\frac{dy}{dx} = \frac{2x}{x^2 - 5} + \frac{6}{3x - 1} - \frac{7}{x} - \frac{6x}{2x^3 - 3}$$

Multiply both sides by y to obtain $\dfrac{dy}{dx}$ as follows:

$$\frac{dy}{dx} = y\left(\frac{2x}{x^2 - 5} + \frac{6}{3x - 1} - \frac{7}{x} - \frac{6x}{2x^3 - 3}\right)$$

Now replace y with its original value from the question as follows:

$$\frac{dy}{dx} = \left[\frac{(x^2 - 5)(3x-1)^2}{x^7(2x^3-3)}\right]\left(\frac{2x}{x^2 - 5} + \frac{6}{3x - 1} - \frac{7}{x} - \frac{6x}{2x^3 - 3}\right)$$

6. If $y = \dfrac{(2x-1)(x+4)(6x-5)}{(3-x)^2}$

Solution

$$y = \dfrac{(2x-1)(x+4)(6x-5)}{(3-x)^2}$$

This is a complex function where product rule and quotient rule would be difficult to apply.

Hence, we apply logarithmic differentiation as follows:

$$\log_e y = \log_e\left[\dfrac{(2x-1)(x+4)(6x-5)}{(3-x)^2}\right]$$

Applying the theory of logarithm gives:

$$\log_e y = \log_e[(2x-1)(x+4)(6x-5)] - \log_e(3-x)^2$$

$$\log_e y = \log_e(2x-1) + \log_e(x+4) + \log_e(6x-5) - 2\log_e(3-x)$$

Differentiating accordingly gives:

$$\dfrac{1}{y}\dfrac{dy}{dx} = \dfrac{2}{2x-1} + \dfrac{1}{x+4} + \dfrac{6}{6x-5} - \dfrac{2(-1)}{3-x}$$

$$\dfrac{1}{y}\dfrac{dy}{dx} = \dfrac{2}{2x-1} + \dfrac{1}{x+4} + \dfrac{6}{6x-5} + \dfrac{2}{3-x}$$

Multiply both sides by y. This gives:

$$\dfrac{dy}{dx} = y\left(\dfrac{2}{2x-1} + \dfrac{1}{x+4} + \dfrac{6}{6x-5} + \dfrac{2}{3-x}\right)$$

Finally replace y with its original expression as follows:

$$\dfrac{dy}{dx} = \left[\dfrac{(2x-1)(x+4)(6x-5)}{(3-x)^2}\right]\left(\dfrac{2}{2x-1} + \dfrac{1}{x+4} + \dfrac{6}{6x-5} + \dfrac{2}{3-x}\right)$$

7. Find the derivative of x^{x^2}

Solution

$$y = x^{x^2}$$

$$\log_e y = \log_e x^{x^2}$$

$$\log_e y = x^2\log_e x$$

$$\dfrac{1}{y}\dfrac{dy}{dx} = x^2\left(\dfrac{1}{x}\right) + \log_e x(2x)$$

$$\dfrac{1}{y}\dfrac{dy}{dx} = x + 2x\log_e x$$

$$\dfrac{1}{y}\dfrac{dy}{dx} = x(1 + 2\log_e x)$$

$$\dfrac{dy}{dx} = yx(1 + 2\log_e x)$$

$$= x^{x^2}x(1 + 2\log_e x) \qquad \text{(y has been replaced with } x^{x^2})$$

$$\dfrac{dy}{dx} = x^{x^2+1}(1 + 2\log_e x)$$

Note that the exponents of x^{x^2} and x were added to obtain x^{x^2+1} since x is also x^1

298

8. If $y = e^{e^x}$ find $\dfrac{dy}{dx}$.

Solution

$y = e^{e^x}$

$\log_e y = \log_e e^{e^x}$

$\log_e y = e^x \log_e e$

$\log_e y = e^x$ (Recall that logarithm of a number to the same base is 1. Hence, $\log_e e = 1$)

$\dfrac{1}{y}\dfrac{dy}{dx} = e^x$ (Note that the derivative of e^x is e^x)

$\dfrac{dy}{dx} = y e^x$

$\quad = e^{e^x} e^x$

$\dfrac{dy}{dx} = e^{e^x + x}$ (Their exponents have been added)

9. Find the derivative of 5^{2^x}

Solution

$y = 5^{2^x}$

$\log_e y = \log_e 5^{2^x}$

$\log_e y = 2^x \log_e 5$

Differentiate the left hand side implicitly. On the right hand side, differentiate 2^x and take $\log_e 5$ as a constant. This gives:

$\dfrac{1}{y}\dfrac{dy}{dx} = 2^x \log_e 2 \log_e 5$

Note that 2^x is treated in a similar way as a^x, and recall that the derivative of a^x is $a^x \log_e a$. Hence the derivative of 2^x is $2^x \log_e 2$, and $\log_e 5$ multiplies it since it is a constant. Therefore:

$\dfrac{dy}{dx} = y(2^x \log_e 2 \log_e 5)$

$\quad = 5^{2^x}(2^x \log_e 2 \log_e 5)$

$\dfrac{dy}{dx} = 5^{2^x}(2^x \log_e 2)(\log_e 5)$

10. If $y = x^{\ln(2x^2 - 5)}$, find $\dfrac{dy}{dx}$.

Solution

$y = x^{\ln(2x^2 - 5)}$

$\log_e y = \log_e x^{\ln(2x^2 - 5)}$

$\log_e y = \ln(2x^2 - 5)\log_e x$

$\dfrac{1}{y}\dfrac{dy}{dx} = \ln(2x^2 - 5)\left(\dfrac{1}{x}\right) + \log_e x\left(\dfrac{4x}{2x^2 - 5}\right)$ (Use of product rule)

$$\frac{1}{y}\frac{dy}{dx} = \frac{1}{x}\ln(2x^2 - 5) + \left(\frac{4x}{2x^2 - 5}\right)\log_e x$$

$$\frac{dy}{dx} = y\left[\frac{1}{x}\ln(2x^2 - 5) + \left(\frac{4x}{2x^2 - 5}\right)\log_e x\right]$$

$$\frac{dy}{dx} = x^{\ln(2x^2 - 5)}\left[\frac{\ln(2x^2 - 5)}{x} + \left(\frac{4x}{2x^2 - 5}\right)\log_e x\right]$$

Exercise 29

1. Find the derivative of x^{2x}

2. Find the derivative of $3x^{\ln 2x}$.

3. If $y = \ln(1 - 4x^2)^x$ find $\frac{dy}{dx}$.

4. Find $\frac{dy}{dx}$ if $y = (6x^2 - 5)^{3x}$

5. If $y = \frac{(2x^3 + 1)(x - 2)^3}{3x^2(x^3 - 1)^2}$

6. If $y = \frac{(2x - 1)(x^2 - 2)}{(1 - x)(x - 3)^2}$

7. Find the derivative of $5x^{3x^4}$

8. If $y = e^{e^{3x}}$ find $\frac{dy}{dx}$.

9. Find the derivative of 2^{e^x}

10. If $y = x^{\ln(x^2 - 4x)}$, find $\frac{dy}{dx}$.

11. If $y = \ln(10 + 2x^2)^x$ find $\frac{dy}{dx}$.

12. Find $\frac{dy}{dx}$ if $y = (3x^3 - 8)^{2x}$

13. If $y = \frac{(x + 3)^2(x - 3)^2}{(x^3 - 1)}$

14. If $y = 5x^{(\ln x)^2}$, find $\frac{dy}{dx}$.

15. Find the derivative of $10x^{3x^2}$

CHAPTER 30
DERIVATIVE OF ONE FUNCTION WITH RESPECT TO ANOTHER

We can differentiate one function with respect to another as illustrated by the examples shown below.

Examples

1. Differentiate x^{12} with respect to x^7.

Solution

Let $u = x^{12}$

And $v = x^7$

$$\frac{du}{dx} = 12x^{11}$$

$$\frac{dv}{dx} = 7x^6$$

Differentiating x^{12} with respect to x^7 means differentiating u (i.e. x^{12}) with respect to v (i.e. x^7).

This means $\frac{du}{dv}$. This follows the parametric equation rule given by:

$$\frac{du}{dv} = \frac{\frac{du}{dx}}{\frac{dv}{dx}}$$

$$= \frac{12x^{11}}{7x^6}$$

$$= \frac{12}{7}x^5$$

2. Differentiate $2e^x$ with respect to $\ln 2x$.

Solution

The differentiation of $2e^x$ with respect to $\ln 2x$ simply means:

$$\frac{\text{The derivative of } 2e^x}{\text{The derivative of } \ln 2x}$$

Hence, $\frac{d(2e^x)}{dx} = 2e^x$

And, $\frac{d(\ln 2x)}{dx} = \frac{2}{2x} = \frac{1}{x}$

Therefore, the differentiation of $2e^x$ with respect to $\ln 2x$ is given by:

$$\frac{2e^x}{\frac{1}{x}}$$

$$= 2xe^x$$

3. Differntiate $\sin 5x$ with respect to $\cos x$.

Solution

Hence, $\dfrac{d(\sin 5x)}{dx} = 5\cos 5x$

And, $\dfrac{d(\cos x)}{dx} = -\sin x$

Therefore, the differentiation of $\sin 5x$ with respect to $\cos x$ is given by:

$$\dfrac{\text{The derivative of } \sin 5x}{\text{The derivative of } \cos x}$$

$$= \dfrac{5\cos 5x}{-\sin x}$$

$$= -\dfrac{5\cos 5x}{\sin x}$$

4. Find the derivative of $2x^2 - 5$ with respect to $4x - 1$

Solution

This is obtained as follows:

$$\dfrac{\dfrac{d(2x^2 - 5)}{dx}}{\dfrac{d(4x - 1)}{dx}}$$

$$= \dfrac{4x}{4}$$

$$= x$$

Exercise 30

1. Differentiate x^3 with respect to x^5.
2. Differentiate e^{5x} with respect to $\ln x^2$
3. Differentiate $\cos x^2$ with respect to $\sin x$.
4. Find the derivative of $4x^3 - 5x^2$ with respect to $(x - 1)^2$
5. Differentiate $7x^4$ with respect to $3x^2$.
6. Differentiate $2a^x$ with respect to e^x.
7. Differentiate $\tan^2 x$ with respect to $\cos 5x$.
8. Find the derivative of $\ln(x - 1)$ with respect to e^{x-1}
9. Differentiate $\log_e x^5$ with respect to e^{5x}.
10. Differentiate $\ln(e^x - e^{-x})$ with respect to $\ln x^x$

CHAPTER 31
HIGHER DERIVATIVES (SUCCESSIVE DIFFERENTIATION)

If f(x) is differentiated it gives the first derivative denoted by f'(x) or $\frac{dy}{dx}$. If we differentiate f'(x),

it gives the second derivative denoted by f''(x) or $\frac{d^2y}{dx^2}$ (read as, dee two y dee x squared). Other

higher derivatives such as $\frac{d^3y}{dx^3}, \frac{d^4y}{dx^4}$ etc can also be obtained depending on the function.

Examples

1. Find the first, second and third derivatives of $2x^5 - 3x^4 + 5x^2 - 6$

Solution

$$y = 2x^5 - 3x^4 + 5x^2 - 6$$

$$\frac{dy}{dx} = 10x^4 - 12x^3 + 10x$$

$\frac{d^2y}{dx^2}$ is obtained by differentiating $\frac{dy}{dx}$, which means to differentiate $10x^4 - 12x^3 + 10x$.

Hence, $\frac{d^2y}{dx^2} = 40x^3 - 36x^2 + 10$

$\frac{d^3y}{dx^3}$ is obtained by differentiating $40x^3 - 36x^2 + 10$ as follows:

$$\frac{d^3y}{dx^3} = 120x^2 - 72x$$

In summary, the first derivative $\left(\frac{dy}{dx}\right)$ is $10x^4 - 12x^3 + 10x$, the second derivative $\left(\frac{d^2y}{dx^2}\right)$ is $40x^3 -$

$36x^2 + 10$, while the third derivative $\left(\frac{d^3y}{dx^3}\right)$ is $120x^2 - 72x$.

2. If $y = \ln x^2$, find $\frac{d^2y}{dx^2}$.

Solution

$$y = \ln x^2$$

$$\frac{dy}{dx} = \frac{2x}{x^2}$$

$$= \frac{2}{x}$$

$$\frac{d^2y}{dx^2} = \frac{d\left(\frac{2}{x}\right)}{dx}$$

$$= \frac{d(2x^{-1})}{dx}$$

$$= -2x^{-2}$$

$$\frac{d^2y}{dx^2} = \frac{-2}{x^2}$$

3. Find $\frac{d^3y}{dx^3}$ given that $y = e^{x^3}$

<u>Solution</u>

$$y = e^{x^3}$$

$$\frac{dy}{dx} = 3x^2 e^{x^3} \qquad \text{(Note that } 3x^2 \text{ is from the derivative of } x^3\text{)}$$

We now use product rule to obtain $\frac{d^2y}{dx^2}$ as follows:

$$\frac{d^2y}{dx^2} = 3x^2 \left[\frac{d(e^{x^3})}{dx} \right] + e^{x^3} \left[\frac{d(3x^2)}{dx} \right]$$

$$= 3x^2 (3x^2 e^{x^3}) + e^{x^3} (6x)$$

$$= 9x^4 e^{x^3} + 6x e^{x^3}$$

$$\frac{d^2y}{dx^2} = e^{x^3} (9x^4 + 6x)$$

Finally, let us also use product rule to obtain $\frac{d^3y}{dx^3}$ as follows:

$$\frac{d^3y}{dx^3} = e^{x^3} \left[\frac{d(9x^4 + 6x)}{dx} \right] + (9x^4 + 6x) \left[\frac{d(e^{x^3})}{dx} \right]$$

$$= e^{x^3} (36x^3 + 6) + (9x^4 + 6x)(3x^2 e^{x^3})$$

$$= 36x^3 e^{x^3} + 6e^{x^3} + 27x^6 e^{x^3} + 18x^3 e^{x^3}$$

$$= 27x^6 e^{x^3} + 36x^3 e^{x^3} + 18x^3 e^{x^3} + 6e^{x^3}$$

$$= 27x^6 e^{x^3} + 54x^3 e^{x^3} + 6e^{x^3}$$

$$\frac{d^3y}{dx^3} = 3e^{x^3} (9x^6 + 18x^3 + 2)$$

4. If $y = \sin 2x^3$, find the third derivative of y.

<u>Solution</u>

$$y = \sin 2x^3$$

$$\frac{dy}{dx} = \cos 2x^3 \frac{d(2x^3)}{dx} \qquad \text{(Recall that the derivative of } \sin x \text{ is } \cos x\text{)}$$

$$= \cos 2x^3 (6x^2)$$

$$\frac{dy}{dx} = 6x^2 \cos 2x^3$$

We now use product rule to obtain $\frac{d^2y}{dx^2}$ as follows:

$$\frac{d^2y}{dx^2} = 6x^2 \left[\frac{d(\cos 2x^3)}{dx} \right] + \cos 2x^3 \left[\frac{d(6x^2)}{dx} \right]$$

$$= 6x^2\left[-\sin 2x^3 \frac{d(2x^3)}{dx}\right] + \cos 2x^3(12x)$$

$$= 6x^2\left[-\sin 2x^3(6x^2)\right] + 12x\cos 2x^3$$

$$\frac{d^2y}{dx^2} = -36x^4\sin 2x^3 + 12x\cos 2x^3$$

Again, the use of product rule gives us $\frac{d^3y}{dx^3}$ as follows:

$$\frac{d^3y}{dx^3} = -36x^4\left[\frac{d(\sin 2x^3)}{dx}\right] + \sin 2x^3\left[\frac{d(-36x^4)}{dx}\right] + 12x\left[\frac{d(\cos 2x^3)}{dx}\right] + \cos 2x^3\left[\frac{d(12x)}{dx}\right]$$

Note that the derivative of $\sin 2x^3$ is $6x^2\cos 2x^3$ as obtained from $\frac{dy}{dx}$. Similarly, the differentiation of $\cos 2x^3$ will give us $-6x^2\sin 2x^3$ (since the derivative of $\sin x$ and $\cos x$ differs only by sign and the interchanging of sin with cos or cos with sin).

We now replace the derivative of $\sin 2x^3$ and $\cos 2x^3$ with $6x^2\cos 2x^3$ and $-6x^2\sin 2x^3$ respectively in the expression above. This gives:

$$\frac{d^3y}{dx^3} = -36x^4(6x^2\cos 2x^3) + \sin 2x^3(-144x^3) + 12x(-6x^2\sin 2x^3) + \cos 2x^3(12)$$

$$= -216x^6\cos 2x^3 - 144x^3\sin 2x^3 - 72x^3\sin 2x^3 + 12\cos 2x^3$$

$$\frac{d^3y}{dx^3} = -216x^6\cos 2x^3 - 216x^3\sin 2x^3 + 12\cos 2x^3$$

Or, $\frac{d^3y}{dx^3} = -12(18x^6\cos 2x^3 + 18x^3\sin 2x^3 - \cos 2x^3)$

5. Find $\frac{d^2y}{dx^2}$ given that $y = a^{x^2} - 5$

Solution

$$y = a^{x^2} - 5$$

$$\frac{dy}{dx} = a^{x^2}\left[\frac{d(x^2)}{dx}\right]\log_e a$$

$$= a^{x^2}(2x)\log_e a$$

$$\frac{dy}{dx} = 2x\, a^{x^2}\log_e a$$

We now use product rule to find $\frac{d^2y}{dx^2}$ as follows:

$$\frac{d^2y}{dx^2} = 2x\left[\frac{d\left(a^{x^2}\log_e a\right)}{dx}\right] + a^{x^2}\log_e a\left[\frac{d(2x)}{dx}\right]$$

$$= 2x\left[\frac{\log_e a\, d\left(a^{x^2}\right)}{dx}\right] + a^{x^2}\log_e a(2)$$

$$= 2x[\log_e a(2x a^{x^2}\log_e a)] + 2a^{x^2}\log_e a$$

$$= 4x^2 a^{x^2}(\log_e a)^2 + 2a^{x^2}\log_e a$$

$$\frac{d^2y}{dx^2} = 2a^{x^2}\log_e a(2x^2\log_e a + 1)$$

6. Find the second derivative of $y = \dfrac{\cos x}{x^3}$

Solution

$$y = \frac{\cos x}{x^3}$$

We use quotient rule as follows:

$$\frac{dy}{dx} = \frac{x^3\frac{d(\cos x)}{dx} - \cos x\frac{d(x^3)}{dx}}{(x^3)^2}$$

$$= \frac{x^3(-\sin x) - \cos x(3x^2)}{x^6}$$

$$\frac{dy}{dx} = \frac{-x^3\sin x - 3x^2\cos x}{x^6}$$

Divide each part by x^6 to separate into fractions as follows:

$$\frac{dy}{dx} = \frac{-x^3\sin x}{x^6} - \frac{3x^2\cos x}{x^6}$$

$$\frac{dy}{dx} = \frac{-\sin x}{x^3} - \frac{3\cos x}{x^4}$$

We now apply quotient rule again as follows:

$$\frac{d^2y}{dx^2} = \frac{x^3\frac{d(-\sin x)}{dx} - (-\sin x)\frac{d(x^3)}{dx}}{(x^3)^2} - \left[\frac{x^4\frac{d(3\cos x)}{dx} - 3\cos x\frac{d(x^4)}{dx}}{(x^4)^2}\right]$$

$$= \frac{x^3(-\cos x) + \sin x\,(3x^2)}{x^6} - \left[\frac{x^4(-3\sin x) - 3\cos x\,(4x^3)}{x^8}\right]$$

$$= \frac{-x^3\cos x + 3x^2\sin x}{x^6} + \frac{3x^4\sin x + 12x^3\cos x}{x^8}$$

Separating into fractions gives:

$$\frac{d^2y}{dx^2} = \frac{-x^3\cos x}{x^6} + \frac{3x^2\sin x}{x^6} + \frac{3x^4\sin x}{x^8} + \frac{12x^3\cos x}{x^8}$$

$$= \frac{-\cos x}{x^3} + \frac{3\sin x}{x^4} + \frac{3\sin x}{x^4} + \frac{12\cos x}{x^5}$$

Combining the fractions again by using x^5 as LCM gives:

$$\frac{d^2y}{dx^2} = \frac{-x^2\cos x + 3x\sin x + 3x\sin x + 12\cos x}{x^5}$$

$$\frac{d^2y}{dx^2} = \frac{6x\sin x + 12\cos x - x^2\cos x}{x^5}$$

7. Find the third derivative of $y = \log_e(1 + 2x)^2$

Solution

$$y = \log_e(1 + 2x)^2$$

$$\frac{dy}{dx} = \frac{2(1+2x)^{2-1} \times \frac{d(2x)}{dx}}{(1+2x)^2}$$ [Note the use of chain rule in finding the derivative of $(1 + 2x)^2$]

$$= \frac{2(1+2x) \times 2}{(1+2x)^2}$$

$$= \frac{4(1+2x)}{(1+2x)^2}$$

$$= \frac{4}{1+2x}$$ (1 + 2x cancels out)

$$\frac{dy}{dx} = 4(1 + 2x)^{-1}$$

Using chain rule gives $\frac{d^2y}{dx^2}$ as follows:

$$\frac{d^2y}{dx^2} = -1 \times 4(1 + 2x)^{-1-1} \times \frac{d(2x)}{dx}$$

$$= -4(1 + 2x)^{-2} \times 2$$

$$= -8(1 + 2x)^{-2}$$

Use of chain rule again gives $\frac{d^3y}{dx^3}$ as follows:

$$\frac{d^3y}{dx^3} = -2 \times -8(1 + 2x)^{-2-1} \times \frac{d(2x)}{dx}$$

$$= 16(1 + 2x)^{-3} \times 2$$

$$= 32(1 + 2x)^{-3}$$

$$\frac{d^3y}{dx^3} = \frac{32}{(1+2x)^3}$$

8. If y = sec2x, find $\frac{d^3y}{dx^3}$.

Solution

   y = sec2x

$$\frac{dy}{dx} = 2sec2x tan2x$$

We now use product rule to obtain $\frac{d^2y}{dx^2}$ as follows:

$$\frac{d^2y}{dx^2} = 2sec2x \frac{d(\tan 2x)}{dx} + \tan2x \frac{d(2sec\,2x)}{dx}$$

$$= 2sec2x(2sec^2 2x) + \tan2x(2 \times 2\,sec2x tan2x)$$

$$\frac{d^2y}{dx^2} = 4sec^3 2x + 4sec2x tan^2 2x$$

We now use chain rule for $4sec^3 2x$ and product rule for $4sec2x tan^2 2x$ to obtain $\frac{d^3y}{dx^3}$ as follows:

$$\frac{d^3y}{dx^3} = 3 \times 4\,sec^2 2x \frac{d(sec\,2x)}{dx} + 4sec2x \frac{d(\tan^2 2x)}{dx} + \tan^2 2x \frac{d(4sec\,2x)}{dx}$$

$$= 12sec^2 2x(2sec2x tan2x) + 4sec2x \left[2tan2x \frac{d(tan2x)}{dx}\right] + \tan^2 2x[4(2sec2x tan2x)]$$

307

$$= 24\sec^3 2x\tan 2x + 8\sec 2x\tan 2x(2\sec^2 2x) + 8\sec 2x\tan^3 2x$$

$$= 24\sec^3 2x\tan 2x + 16\sec^3 2x\tan 2x + 8\sec 2x\tan^3 2x$$

$$\frac{d^3 y}{dx^3} = 40\sec^3 2x\tan 2x + 8\sec 2x\tan^3 2x$$

9. Given that y = 2cos5x + 3sin5x, prove that $\frac{d^2 y}{dx^2}$ + 25y = 0

Solution

$$y = 2\cos 5x + 3\sin 5x$$

$$\frac{dy}{dx} = 2 \times 5(-\sin 5x) + 3 \times 5(\cos 5x)$$

$$= -10\sin 5x + 15\cos 5x$$

$$\frac{d^2 y}{dx^2} = -10 \times 5(\cos 5x) + 15 \times 5(-\sin 5x)$$

$$= -50\cos 5x - 75\sin 5x$$

We now obtain 25y as follows:

$$y = 2\cos 5x + 3\sin 5x$$

$$25y = 25(2\cos 5x + 3\sin 5x)$$

$$25y = 50\cos 5x + 75\sin 5x$$

We now simplify $\frac{d^2 y}{dx^2}$ + 25y as follows:

$$\frac{d^2 y}{dx^2} + 25y = -50\cos 5x - 75\sin 5x + 50\cos 5x + 75\sin 5x$$

$$= 0 \quad \text{(Equal terms with opposite signs cancel out)}$$

Therefore, $\frac{d^2 y}{dx^2}$ + 25y = 0   (As proven above)

10. If y = $x + \sqrt{4 + x^2}$, show that $(4 + x^2)\frac{d^2 y}{dx^2} + x\frac{dy}{dx} - y = 0$

Solution

$$y = x + \sqrt{4 + x^2}$$

$$y = x + (4 + x^2)^{\frac{1}{2}}$$

$$\frac{dy}{dx} = 1 + \frac{1}{2}(4 + x^2)^{\frac{1}{2} - 1} \times \frac{d(x^2)}{dx}$$

$$= 1 + \frac{1}{2}(4 + x^2)^{-\frac{1}{2}} \times 2x$$

$$\frac{dy}{dx} = 1 + x(4 + x^2)^{-\frac{1}{2}}$$

Using product rule, we obtain $\frac{d^2 y}{dx^2}$ as follows:

$$\frac{d^2y}{dx^2} = 0 + x\left[\frac{d(4+x^2)^{-\frac{1}{2}}}{dx}\right] + (4+x^2)^{-\frac{1}{2}}\frac{d(x)}{dx}$$

$$= x\left[-\frac{1}{2}(4+x^2)^{-\frac{1}{2}-1}\frac{d(x^2)}{dx}\right] + (4+x^2)^{-\frac{1}{2}}(1)$$

$$= x\left[-\frac{1}{2}(4+x^2)^{-\frac{3}{2}} \times 2x\right] + (4+x^2)^{-\frac{1}{2}}$$

$$= -x^2(4+x^2)^{-\frac{3}{2}} + (4+x^2)^{-\frac{1}{2}}$$

$$\frac{d^2y}{dx^2} = \frac{-x^2}{(4+x^2)^{\frac{3}{2}}} + \frac{1}{(4+x^2)^{\frac{1}{2}}}$$

Let us now simplify $(4+x^2)\dfrac{d^2y}{dx^2} + x\dfrac{dy}{dx} - y$ as follows:

$$(4+x^2)\left[\frac{-x^2}{(4+x^2)^{\frac{3}{2}}} + \frac{1}{(4+x^2)^{\frac{1}{2}}}\right] + x[1 + x(4+x^2)^{-\frac{1}{2}}] - [x + (4+x^2)^{\frac{1}{2}}]$$

Expanding the brackets gives:

$$-x^2(4+x^2)^{1-\frac{3}{2}} + (4+x^2)^{1-\frac{1}{2}} + x + x^2(4+x^2)^{-\frac{1}{2}} - x - (4+x^2)^{\frac{1}{2}}$$

$$= -x^2(4+x^2)^{-\frac{1}{2}} + (4+x^2)^{\frac{1}{2}} + x + x^2(4+x^2)^{-\frac{1}{2}} - x - (4+x^2)^{\frac{1}{2}}$$

$$= 0 \quad \text{(Note that equal terms with opposite signs cancel out one another to give zero)}$$

Therefore, $(4+x^2)\dfrac{d^2y}{dx^2} + x\dfrac{dy}{dx} - y = 0$ (As proven above)

Exercise 31

1. Find the first, second and third derivatives of $x^4 - 2x^3 - 7x^2 + 1$

2. If $y = \ln 2x^3$, find $\dfrac{d^2y}{dx^2}$.

3. Find $\dfrac{d^2y}{dx^2}$ given that $y = e^{5x^4}$

4. If $y = \sin^2 x^5$, find the second derivative of y.

5. Find $\dfrac{d^3y}{dx^3}$ given that $y = \ln(4x - 10)$

6. Find the second derivative of $y = \dfrac{\sin x}{x}$

7. Find the third derivative of $y = (\ln x)^2$

8. If $y = \cot x$, find $\dfrac{d^3y}{dx^3}$.

9. Given that $y = \sin^2 x + \cos 2x$, find $\dfrac{d^2y}{dx^2} - y$

10. If $y = 2e^{2x} + 5e^{-x}$ evaluate $\dfrac{d^2y}{dx^2} - \dfrac{dy}{dx} - 2y$

11. Given that $y = \cos^2 x + 2\sin x^2$ find $\dfrac{d^2y}{dx^2} - \dfrac{dy}{dx}$

12. Find the second derivative of $y = \dfrac{\tan x}{3x}$

13. If $y = e^{-x}\cos 2x$, find the second derivative of y.

14. If $y = 2\cos 3x + 5\sin 3x$, evaluate $\dfrac{d^2y}{dx^2} + 9y$

15. Find the second derivative of $y = \dfrac{\sin 3x}{3x}$

CHAPTER 32
MISCELLANEOUS PROBLEMS ON DIFFERENTIAL CALCULUS

This chapter covers worked examples on general problems involving combination of topics treated in the various previous chapters. More challenging problems would also be covered here.

Examples

1. If $y = \dfrac{1}{(2x^3 - 5)^4}$ find $\dfrac{dy}{dx}$.

Solution

$$y = \dfrac{1}{(2x^3 - 5)^4}$$

This can be written as:

$$y = (2x^3 - 5)^{-4}$$

We now use chain rule to differentiate it as follows:

$$\dfrac{dy}{dx} = -4(2x^3 - 5)^{-4-1} \times \dfrac{d(2x^3 - 5)}{dx}$$

$$= -4(2x^3 - 5)^{-5} \times 6x^2$$

$$= -24x^2(2x^3 - 5)^{-5}$$

$$\dfrac{dy}{dx} = \dfrac{-24x^2}{(2x^3 - 5)^5}$$

2. Given that $f(x) = 5x^4 - 3x^3 - 7x^2 + 9$, find $f'(2)$

Solution

$$f(x) = 5x^4 - 3x^3 - 7x^2 + 9$$

We differentiate $f(x)$ to obtain $f'(x)$ as follows:

$$f'(x) = 20x^3 - 9x^2 - 14x$$

In order to find $f'(2)$, we simply substitute 2 for x in $f'(x)$ as follows:

$$f'(x) = 20x^3 - 9x^2 - 14x$$

$$f'(2) = 20(2^3) - 9(2^2) - 14(2)$$

$$= 20(8) - 9(4) - 28$$

$$= 160 - 36 - 28$$

$$f'(2) = 96$$

3. A function $f(x)$ is given by $f(x) = \dfrac{\sqrt{4 + 3x^2}}{x^3}$. Find:

(a) the derivative of $f(x)$

(b) the gradient of $f(x)$ at the point $(2, \dfrac{1}{2})$

Solution

(a) $f(x) = \dfrac{\sqrt{4 + 3x^2}}{x^3}$

We apply quotient rule to differentiate f(x) as follows:

$$f'(x) = \frac{x^3\left[\dfrac{d(\sqrt{4+3x^2})}{dx}\right] - \sqrt{4+3x^2}\left[\dfrac{d(x^3)}{dx}\right]}{(x^3)^2}$$

$$= \frac{x^3\left[\dfrac{d(4+3x^2)^{\frac{1}{2}}}{dx}\right] - (4+3x^2)^{\frac{1}{2}}(3x^2)}{x^6}$$

$$= \frac{x^3\left[\dfrac{1}{2}(4+3x^2)^{\frac{1}{2}-1} \times \dfrac{d(3x^2)}{dx}\right] - 3x^2(4+3x^2)^{\frac{1}{2}}}{x^6}$$

$$= \frac{x^3\left[\dfrac{1}{2}(4+3x^2)^{-\frac{1}{2}} \times 6x\right] - 3x^2(4+3x^2)^{\frac{1}{2}}}{x^6}$$

$$= \frac{x^3\left[3x(4+3x^2)^{-\frac{1}{2}}\right] - 3x^2(4+3x^2)^{\frac{1}{2}}}{x^6}$$

$$= \frac{3x^4(4+3x^2)^{-\frac{1}{2}} - 3x^2(4+3x^2)^{\frac{1}{2}}}{x^6}$$

Let us factorize the expression above by taking $3x^2(4 + 3x^2)^{-\frac{1}{2}}$ as the common factor. This gives:

$$f'(x) = \frac{3x^2(4+3x^2)^{-\frac{1}{2}}[x^2 - (4+3x^2)]}{x^6}$$

Note that $(4 + 3x^2)^{\frac{1}{2}} \div (4 + 3x^2)^{-\frac{1}{2}} = (4 + 3x^2)^{\frac{1}{2}-(-\frac{1}{2})} = 4 + 3x^2$ as the exponent becomes 1. Recall that exponents are subtracted during division.

$$= \frac{3x^2(4+3x^2)^{-\frac{1}{2}}(x^2 - 4 - 3x^2)}{x^6}$$

$$= \frac{3x^2(4+3x^2)^{-\frac{1}{2}}(-2x^2 - 4)}{x^6}$$

$$= \frac{-3x^2(4+3x^2)^{-\frac{1}{2}}(2x^2 + 4)}{x^6}$$

$$= \frac{-3(4+3x^2)^{-\frac{1}{2}}(2x^2 + 4)}{x^4} \qquad \text{(Note that } x^2 \text{ has cancelled out of } x^6\text{)}$$

$$f'(x) = \frac{-3(2x^2 + 4)}{x^4(4+3x^2)^{\frac{1}{2}}}$$

313

(b) The gradient of f(x) at the point $(2, \frac{1}{2})$ is obtained by substituting 2 for x in f'(x). Note that the point $(2, \frac{1}{2})$ is at $x = 2$ and $y = \frac{1}{2}$. We ignore $y = \frac{1}{2}$ since y is not in the expression for f'(x).

Hence, $f'(x) = \dfrac{-3(2x^2 + 4)}{x^4(4 + 3x^2)^{\frac{1}{2}}}$

At $(2, \frac{1}{2})$, $f'(2) = \dfrac{-3(2(2)^2 + 4)}{(2)^4(4 + 3(2)^2)^{\frac{1}{2}}}$

$= \dfrac{-3(8 + 4)}{16(4 + 12)^{\frac{1}{2}}}$

$= \dfrac{-3(12)}{16(16)^{\frac{1}{2}}}$

$= \dfrac{-36}{16 \times 4}$ (Note that $(16)^{\frac{1}{2}} = \sqrt{16} = 4$)

$f'(2) = \dfrac{-9}{16}$ (In its lowest term)

Therefore the gradient of f(x) at the point $(2, \frac{1}{2})$ is $\dfrac{-9}{16}$

4. If $x^2y^2 - 3xy + 4xy^3 = 4$, find:

(a) the derivative of the expression

(b) the gradient at $(-1, 2)$

Solution

(a) $x^2y^2 - 3xy + 4xy^3 = 4$

The use of implicit differentiation combined with product rule gives us the derivative as follows:

$$x^2\dfrac{d(y^2)}{dx} + y^2\dfrac{d(x^2)}{dx} - \left[3x\dfrac{d(y)}{dx} + y\dfrac{d(3x)}{dx}\right] + 4x\dfrac{d(y^3)}{dx} + y^3(4) = \dfrac{d(4)}{dx}$$

$$x^2\left(2y\dfrac{dy}{dx}\right) + y^2(2x) - \left[3x\dfrac{dy}{dx} + y(3)\right] + 4x\left(3y^2\dfrac{dy}{dx}\right) + 4y^3 = 0$$

$$2x^2y\dfrac{dy}{dx} + 2xy^2 - 3x\dfrac{dy}{dx} - 3y + 12xy^2\dfrac{dy}{dx} + 4y^3 = 0$$

Collect terms in $\dfrac{dy}{dx}$ on one side of the equation. This gives:

$$2x^2y\dfrac{dy}{dx} - 3x\dfrac{dy}{dx} + 12xy^2\dfrac{dy}{dx} = 3y - 2xy^2 - 4y^3$$

Factorizing the left hand side gives:

$$\dfrac{dy}{dx}(2x^2y - 3x + 12xy^2) = 3y - 2xy^2 - 4y^3$$

Divide both sides by the terms in the bracket. This gives:

$$\dfrac{dy}{dx} = \dfrac{3y - 2xy^2 - 4y^3}{12x^2y - 3x + 12xy^2}$$

(b) At the point $(-1, 2)$ the gradient of the expression is obtained by simply substituting -1 for x and 2 for y in the expression for the derivative. This is done as follows:

$$\frac{dy}{dx} = \frac{3y - 2xy^2 - 4y^3}{12x^2y - 3x + 12xy^2}$$

$$= \frac{3(2) - 2(-1)(2)^2 - 4(2)^3}{12(-1)^2(2) - 3(-1) + 12(-1)(2)^2}$$

$$= \frac{6 + 8 - 32}{24 + 3 - 48}$$

$$= \frac{-18}{-21}$$

$$= \frac{6}{7}$$

5. Find the derivative of $(x + 3y^2)^3 = 7$

Solution

$(x + 3y^2)^3 = 7$

We differentiate implicitly and apply chain rule as follows:

$$3(x + 3y^2)^{3-1} \times \frac{d(x+3y^2)}{dx} = 0$$

$$3(x + 3y^2)^2\left(1 + 6y\frac{dy}{dx}\right) = 0$$

Dividing both sides by $3(x + 3y^2)^2$ gives:

$$1 + 6y\frac{dy}{dx} = 0 \qquad \left(\text{Note that } \frac{0}{3(x + 3y^2)^2} = 0\right)$$

$$6y\frac{dy}{dx} = -1$$

$$\frac{dy}{dx} = -\frac{1}{6y}$$

6. If $y = (3x^2 - 2x + 5)(2x - 3)$, find $\frac{dy}{dx}$.

Solution

$y = (3x^2 - 2x + 5)(2x - 3)$

We differentiate the expression by applying product rule as follows:

$$\frac{dy}{dx} = (3x^2 - 2x + 5)(2) + (2x - 3)(6x - 2)$$

$$= 6x^2 - 4x + 10 + 12x^2 - 4x - 18x + 6$$

$$\frac{dy}{dx} = 18x^2 - 26x + 16$$

7. Find the derivative of $a^{1 + \tan x}$

Solution

$y = a^{1 + \tan x}$

Let $u = 1 + \tan x$

Hence, $y = a^u$

$$\frac{du}{dx} = \sec^2 x$$

$$\frac{dy}{du} = a^u \log_e a$$

Hence, $\dfrac{dy}{dx} = \dfrac{dy}{du} \times \dfrac{du}{dx}$

$$= a^u \log_e a \times \sec^2 x$$

$$\frac{dy}{dx} = a^{1+\tan x} \sec^2 x \log_e a$$

8. Find the derivative of $\dfrac{\log_e x}{1+\cos x}$

Solution

$$y = \frac{\log_e x}{1+\cos x}$$

We apply product rule as follows:

$$\frac{dy}{dx} = \frac{(1 + \cos x)\frac{d(\log_e x)}{dx} - \log_e x \frac{d(1 + \cos x)}{dx}}{(1+\cos x)^2}$$

$$= \frac{(1 + \cos x)\frac{1}{x} - \log_e x(-\sin x)}{(1+\cos x)^2}$$

$$= \frac{\frac{1 + \cos x}{x} + \sin x \log_e x}{(1+\cos x)^2}$$

$$= \frac{\frac{1 + \cos x + x \sin x \log_e x}{x}}{(1+\cos x)^2}$$

$$\frac{dy}{dx} = \frac{1 + \cos x + x \sin x \log_e x}{x(1+\cos x)^2}$$

9. Differentiate with respect to x: $\ln(\cos x + \sin x)$

Solution

$$y = \ln(\cos x + \sin x)$$

$$\frac{dy}{dx} = \frac{\frac{d(\cos x + \sin x)}{dx}}{\cos x + \sin x}$$

$$= \frac{-\sin x + \cos x}{\cos x + \sin x}$$

$$\frac{dy}{dx} = \frac{\cos x - \sin x}{\cos x + \sin x}$$

10. If $y = e^{\sin 2x + \cos x}$ find $\frac{dy}{dx}$.

Solution

$$y = e^{\sin 2x + \cos x}$$

$$\frac{dy}{dx} = \frac{d(\sin 2x + \cos x)}{dx} \times e^{\sin 2x + \cos x}$$

$$= (2\cos 2x - \sin x)(e^{\sin 2x + \cos x})$$

11. Given that $y = e^x - e^{-x}$ show that $\frac{d^3y}{dx^3} + \frac{d^2y}{dx^2} + \frac{dy}{dx} + y = 4e^x$

Solution

$$y = e^x - e^{-x}$$

$$\frac{dy}{dx} = e^x - (-e^{-x}) \quad \text{(Note that } \frac{d(e^{-x})}{dx} = \frac{d(-x)}{dx} \times e^{-x} = -1 \times e^{-x} = -e^{-x})$$

$$= e^x + e^{-x}$$

$$\frac{d^2y}{dx^2} = e^x - e^{-x}$$

$$\frac{d^3y}{dx^3} = e^x + e^{-x}$$

Let us now substitute corresponding term into $\frac{d^3y}{dx^3} + \frac{d^2y}{dx^2} + \frac{dy}{dx} + y$ as follows:

$$(e^x + e^{-x}) + (e^x - e^{-x}) + (e^x + e^{-x}) + (e^x - e^{-x})$$

Collecting like terms together gives:

$$e^x + e^x + e^x + e^x + e^{-x} - e^{-x} + e^{-x} - e^{-x}$$

$$= 4e^x \quad \text{(Note that } e^{-x} \text{ cancels out each other)}$$

Hence, $\frac{d^3y}{dx^3} + \frac{d^2y}{dx^2} + \frac{dy}{dx} + y = 4e^x \quad$ (As proven above)

12. Differentiate with respect to x: $\ln\left(\frac{1 - 3x^2}{1 + 3x^2}\right)^{\frac{1}{2}}$

Solution

$$y = \ln\left(\frac{1 - 3x^2}{1 + 3x^2}\right)^{\frac{1}{2}}$$

$$= \log_e\left(\frac{1 - 3x^2}{1 + 3x^2}\right)^{\frac{1}{2}} \quad \text{(Note that "ln" is } \log_e)$$

$$= \log_e\left[\frac{(1 - 3x^2)^{\frac{1}{2}}}{(1 + 3x^2)^{\frac{1}{2}}}\right]$$

317

$$= \log_e(1 - 3x^2)^{\frac{1}{2}} - \log_e(1 + 3x^2)^{\frac{1}{2}}$$

$$y = \frac{1}{2}\log_e(1 - 3x^2) - \frac{1}{2}\log_e(1 + 3x^2)$$

$$\frac{dy}{dx} = \frac{1}{2}\frac{\frac{d(1-3x^2)}{dx}}{1-3x^2} - \frac{1}{2}\frac{\frac{d(1+3x^2)}{dx}}{1+3x^2}$$

$$= \frac{1}{2}\frac{-6x}{1-3x^2} - \frac{1}{2}\frac{6x}{1+3x^2}$$

$$= \frac{-3x}{1-3x^2} - \frac{3x}{1+3x^2} \qquad \text{(Note that } \frac{1}{2} \text{ reduces } 6x \text{ to } 3x\text{)}$$

$$= \frac{-3x(1+3x^2) - 3x(1-3x^2)}{(1-3x^2)(1+3x^2)}$$

$$= \frac{-3x - 9x^3 - 3x + 9x^3)}{(1-3x^2)(1+3x^2)}$$

$$= \frac{-6x}{1 + 3x^2 - 3x^2 - 9x^4}$$

$$\frac{dy}{dx} = \frac{-6x}{1 - 9x^4}$$

13. If $y = \dfrac{x}{\sqrt{9 - x^2}}$ show that: $(9 - x^2)\dfrac{d^2y}{dx^2} = 3x\dfrac{dy}{dx}$

Solution

$$y = \frac{x}{\sqrt{9 - x^2}}$$

Applying quotient rule gives $\dfrac{dy}{dx}$ as follows:

$$\frac{dy}{dx} = \frac{\sqrt{9 - x^2}(1) - x\frac{d(\sqrt{9 - x^2})}{dx}}{(\sqrt{9 - x^2})^2}$$

$$= \frac{(9 - x^2)^{\frac{1}{2}}(1) - x\frac{d[(9 - x^2)^{\frac{1}{2}}]}{dx}}{[(9 - x^2)^{\frac{1}{2}}]^2}$$

$$= \frac{(9 - x^2)^{\frac{1}{2}} - x\left[\frac{1}{2}(9 - x^2)^{\frac{1}{2} - 1}\right] \times (-2x)}{9 - x^2} \qquad \text{(Note that } -2x \text{ is from the derivative of } -x^2\text{)}$$

$$= \frac{(9 - x^2)^{\frac{1}{2}} - x\left[-x(9 - x^2)^{-\frac{1}{2}}\right]}{9 - x^2}$$

$$= \frac{(9 - x^2)^{\frac{1}{2}} + x^2(9 - x^2)^{-\frac{1}{2}}}{9 - x^2}$$

$$= \frac{(9 - x^2)^{\frac{1}{2}} + \frac{x^2}{(9 - x^2)^{\frac{1}{2}}}}{9 - x^2}$$

$$= \frac{\frac{9 - x^2 + x^2}{(9 - x^2)^{\frac{1}{2}}}}{9 - x^2}$$

$$= \frac{9}{(9 - x^2)^{\frac{1}{2}}(9 - x^2)}$$

$$= \frac{9}{(9 - x^2)^{\frac{3}{2}}}$$

$$\frac{dy}{dx} = 9(9 - x^2)^{-\frac{3}{2}}$$

Applying chain rule gives $\frac{d^2y}{dx^2}$ as follows:

$$\frac{d^2y}{dx^2} = \frac{-3}{2} \times 9(9 - x^2)^{-\frac{3}{2} - 1} \times -2x \qquad \text{(Note that } -2x \text{ is from the derivative of } -x^2\text{)}$$

$$= \frac{-27}{2} \times -2x(9 - x^2)^{-\frac{5}{2}}$$

$$= 27x(9 - x^2)^{-\frac{5}{2}}$$

$$\frac{d^2y}{dx^2} = \frac{27x}{(9 - x^2)^{\frac{5}{2}}}$$

Let us now show that $(9 - x^2)\dfrac{d^2y}{dx^2} = 3x\dfrac{dy}{dx}$

We simplify $(9 - x^2)\dfrac{d^2y}{dx^2}$ as follows:

$$(9 - x^2)\frac{27x}{(9 - x^2)^{\frac{5}{2}}}$$

$$= \frac{27x(9 - x^2)}{(9 - x^2)^{\frac{5}{2}}}$$

$$= 27x(9 - x^2)^{1 - \frac{5}{2}}$$

$$= 27x(9 - x^2)^{-\frac{3}{2}}$$

$$= \frac{27x}{(9 - x^2)^{\frac{3}{2}}}$$

Hence $(9 - x^2)\dfrac{d^2y}{dx^2}$ gives us $\dfrac{27x}{(9 - x^2)^{\frac{3}{2}}}$

Let us now simplify $3x\dfrac{dy}{dx}$ as follows:

$$3x[9(9 - x^2)^{-\frac{3}{2}}] \qquad \text{(Note that } \frac{dy}{dx} = 9(9 - x^2)^{-\frac{3}{2}} \text{ as obtained above)}$$

$$= 27x(9 - x^2)^{-\frac{3}{2}}$$

$$= \frac{27x}{(9 - x^2)^{\frac{3}{2}}}$$

319

Hence $3x\dfrac{dy}{dx}$ also gives us $\dfrac{27x}{(9-x^2)^{\frac{3}{2}}}$

Therefore, $(9-x^2)\dfrac{d^2y}{dx^2} = 3x\dfrac{dy}{dx}$ as both sides give $\dfrac{27x}{(9-x^2)^{\frac{3}{2}}}$

14. Given that $y = \dfrac{x}{x-1}$, show that: $(x-1)\dfrac{d^2y}{dx^2} + 2\dfrac{dy}{dx} = 0$

Solution

$$y = \dfrac{x}{x-1}$$

Using quotient rule gives $\dfrac{dy}{dx}$ as follows:

$$\dfrac{dy}{dx} = \dfrac{(x-1)(1) - x(1)}{(x-1)^2}$$

$$= \dfrac{x-1-x}{(x-1)^2}$$

$$\dfrac{dy}{dx} = \dfrac{-1}{(x-1)^2}$$

Using chain rule gives $\dfrac{d^2y}{dx^2}$ as follows:

$$\dfrac{dy}{dx} = \dfrac{-1}{(x-1)^2}$$

$$= -1(x-1)^{-2}$$

$$\dfrac{d^2y}{dx^2} = -2 \times -1(x-1)^{-2-1} \times \dfrac{d(x)}{dx}$$

$$= 2(x-1)^{-3} \times 1$$

$$\dfrac{d^2y}{dx^2} = \dfrac{2}{(x-1)^3}$$

From the question, let us now simplify $(x-1)\dfrac{d^2y}{dx^2} + 2\dfrac{dy}{dx}$ as follows:

$$(x-1)\dfrac{d^2y}{dx^2} + 2\dfrac{dy}{dx}$$

$$= (x-1)\dfrac{2}{(x-1)^3} + (2)\dfrac{-1}{(x-1)^2}$$

$$= \dfrac{2}{(x-1)^2} - \dfrac{2}{(x-1)^2}$$

$$= 0$$

This shows that $(x-1)\dfrac{d^2y}{dx^2} + 2\dfrac{dy}{dx}$ is equal to zero.

15. Find, with respect to x, the derivative of $\left(x - \dfrac{5}{x}\right)^3$

Solution

$$y = \left(x - \frac{5}{x}\right)^3$$

This can also be written as:

$$y = (x - 5x^{-1})^3$$

$$\frac{dy}{dx} = 3(x - 5x^{-1})^{3-1} \times \frac{d(x - 5x^{-1})}{dx} \qquad \text{(By use of chain rule)}$$

$$= 3(x - 5x^{-1})^2 \times 1 - (-1 \times 5x^{-1-1})$$

$$= 3(x - 5x^{-1})^2 \times 1 - (-5x^{-2})$$

$$= 3(x - 5x^{-1})^2 \times (1 + 5x^{-2})$$

$$= 3(x - 5x^{-1})^2(1 + 5x^{-2})$$

$$\frac{dy}{dx} = 3\left(x - \frac{5}{x}\right)^2\left(1 + \frac{5}{x^2}\right)$$

16. Given that $\exp(2x^2 + 2y^2 - 16) = x + y$, find:

(a) $\dfrac{dy}{dx}$

(b) $\dfrac{dy}{dx}$ at $\left(\dfrac{1}{2} \ \dfrac{1}{2}\right)$

Solution

$$\exp(2x^2 + 2y^2 - 16) = x + y$$

This can also be written as

$$e^{2x^2 + 2y^2 - 16} = x + y \qquad \text{(Note that } \exp x = e^x)$$

We now differentiate implicitly as follows:

$$\frac{d(2x^2 + 2y^2 - 16)}{dx} \times (e^{2x^2 + 2y^2 - 16}) = \frac{d(x)}{dx} + \frac{d(y)}{dx}$$

$$\left(4x + 4y\frac{dy}{dx}\right)(e^{2x^2 + 2y^2 - 16}) = 1 + \frac{dy}{dx}$$

Expanding the bracket gives:

$$4x(e^{2x^2 + 2y^2 - 16}) + 4y\frac{dy}{dx}(e^{2x^2 + 2y^2 - 16}) = 1 + \frac{dy}{dx}$$

Collecting terms in $\dfrac{dy}{dx}$ on the left hand side gives:

$$4y\frac{dy}{dx}(e^{2x^2 + 2y^2 - 16}) - \frac{dy}{dx} = 1 - 4x(e^{2x^2 + 2y^2 - 16})$$

Factorizing the left hand side gives:

$$\frac{dy}{dx}[4y(e^{2x^2 + 2y^2 - 16}) - 1] = 1 - 4x(e^{2x^2 + 2y^2 - 16})$$

Hence, $\dfrac{dy}{dx} = \dfrac{1 - 4x(e^{2x^2 + 2y^2 - 16})}{4y(e^{2x^2 + 2y^2 - 16}) - 1}$

Or, $\dfrac{dy}{dx} = \dfrac{-[4x(e^{2x^2 + 2y^2 - 16}) - 1]}{4y(e^{2x^2 + 2y^2 - 16}) - 1}$

(b) In order to find $\dfrac{dy}{dx}$ at $\left(\dfrac{1}{2}\ \dfrac{1}{2}\right)$, we simply substitute $x = \dfrac{1}{2}$ and $y = \dfrac{1}{2}$ into the expression for $\dfrac{dy}{dx}$ as follows:

$$\frac{dy}{dx} = \frac{-[4x(e^{2x^2+2y^2-16}) - 1]}{4y(e^{2x^2+2y^2-16}) - 1}$$

Since $x = \dfrac{1}{2}$ and $y = \dfrac{1}{2}$, this simplifies to give:

$$\frac{dy}{dx} \text{ at } \left(\frac{1}{2}\ \frac{1}{2}\right) = \frac{-[4\left(\frac{1}{2}\right)(e^{2x^2+2y^2-16}) - 1]}{4\left(\frac{1}{2}\right)(e^{2x^2+2y^2-16}) - 1}$$

$$= \frac{-[2(e^{2x^2+2y^2-16}) - 1]}{2(e^{2x^2+2y^2-16}) - 1}$$

The numerator cancels out the denominator to give:

$$\frac{dy}{dx} = \frac{-1}{1}$$

$$\frac{dy}{dx} = -1$$

17. If $y = x^3 - 2x^2 + 5$, show that: $x\dfrac{dy}{dx} - 3y - 2x^2 + 15 = 0$

Solution

$$y = x^3 - 2x^2 + 5$$

$$\frac{dy}{dx} = 3x^2 - 4x$$

Let us now simplify $x\dfrac{dy}{dx} - 3y - 2x^2 + 15$ as follows:

$x\dfrac{dy}{dx} - 3y - 2x^2 + 15$

$= x(3x^2 - 4x) - 3y - 2x^2 + 15$

$= x(3x^2 - 4x) - 3(x^3 - 2x^2 + 5) - 2x^2 + 15$ (Note that $x^3 - 2x^2 + 5$ has been substituted for y)

$= 3x^3 - 4x^2 - 3x^3 + 6x^2 - 15 - 2x^2 + 15$

$= 0$

Therefore, $x\dfrac{dy}{dx} - 3y - 2x^2 + 15 = 0$ as proven above.

18. Determine $\dfrac{d^2}{dx^2}\left(x\sin\dfrac{1}{x}\right)$

Solution

This means the second derivative of $x\sin\dfrac{1}{x}$

Let $y = x\sin\dfrac{1}{x}$

Or, $y = x\sin x^{-1}$

We now apply product rule to obtain $\dfrac{dy}{dx}$ as follows:

$$\dfrac{dy}{dx} = x\left[\dfrac{d(x^{-1})}{dx}\cos x^{-1}\right] + \sin x^{-1}\left[\dfrac{d(x)}{dx}\right]$$

$$= x(-x^{-2}\cos x^{-1}) + \sin x^{-1}(1)$$

$$= x\left(\dfrac{-1}{x^2}\cos x^{-1}\right) + \sin x^{-1}$$

$$= \dfrac{-x}{x^2}\cos x^{-1} + \sin x^{-1}$$

$$= \dfrac{-\cos x^{-1}}{x} + \sin x^{-1}$$

$$\dfrac{dy}{dx} = -\dfrac{1}{x}\cos\dfrac{1}{x} + \sin\dfrac{1}{x}$$

Or, $\dfrac{dy}{dx} = -x^{-1}\cos x^{-1} + \sin x^{-1}$

We now obtain $\dfrac{d^2y}{dx^2}$ as follows:

$$\dfrac{d^2y}{dx^2} = -x^{-1}\left[\dfrac{d(\cos x^{-1})}{dx}\right] + \cos x^{-1}\left[\dfrac{d(-x^{-1})}{dx}\right] + \dfrac{d(\sin x^{-1})}{dx}$$

$$= -x^{-1}\left[\dfrac{d(x^{-1})}{dx}(-\sin x^{-1})\right] + \cos x^{-1}(x^{-2}) + \left[\dfrac{d(x^{-1})}{dx}\cos x^{-1}\right]$$

$$= -x^{-1}[-x^{-2}(-\sin x^{-1})] + \cos x^{-1}(x^{-2}) - x^{-2}(\cos x^{-1})$$

$$= -x^{-1}(x^{-2}\sin x^{-1}) + x^{-2}\cos x^{-1} - x^{-2}\cos x^{-1}$$

$$= -x^{-1}(x^{-2}\sin x^{-1}) \quad \text{(Note that } x^{-2}\cos x^{-1}\text{ cancels out)}$$

$$= -x^{-3}\sin x^{-1}$$

$$\dfrac{d^2y}{dx^2} = -\dfrac{1}{x^3}\sin\dfrac{1}{x}$$

19. Given that $y = \dfrac{e^x + e^{-x}}{e^x - e^{-x}}$

(a) find $\dfrac{d^2y}{dx^2}$

(b) show that $\dfrac{d^2y}{dx^2} + 2y\dfrac{dy}{dx} = 0$

Solution

(a) $y = \dfrac{e^x + e^{-x}}{e^x - e^{-x}}$

By using quotient rule we obtain $\dfrac{dy}{dx}$ as follows:

$$\dfrac{dy}{dx} = \dfrac{e^x - e^{-x}\left[\dfrac{d(e^x + e^{-x})}{dx}\right] - \left(e^x + e^{-x}\left[\dfrac{d(e^x - e^{-x})}{dx}\right]\right)}{(e^x - e^{-x})^2}$$

$$= \dfrac{(e^x - e^{-x})(e^x - e^{-x}) - (e^x + e^{-x})(e^x + e^{-x})}{(e^x - e^{-x})^2}$$

$$= \frac{(e^x - e^{-x})^2 - (e^x + e^{-x})^2}{(e^x - e^{-x})^2}$$

Separating into fractions gives:

$$= \frac{(e^x - e^{-x})^2}{(e^x - e^{-x})^2} - \frac{(e^x + e^{-x})^2}{(e^x - e^{-x})^2}$$

$$\frac{dy}{dx} = 1 - \frac{(e^x + e^{-x})^2}{(e^x - e^{-x})^2}$$

We now apply quotient and chain rules to obtain $\dfrac{d^2y}{dx^2}$ as follows:

$$\frac{d^2y}{dx^2} = 0 - \frac{(e^x - e^{-x})^2[2(e^x + e^{-x})(e^x - e^{-x})] - (e^x + e^{-x})^2[2(e^x - e^{-x})(e^x + e^{-x})]}{[(e^x - e^{-x})^2]^2}$$

$$= -\left[\frac{2(e^x - e^{-x})^3(e^x + e^{-x}) - 2(e^x + e^{-x})^3(e^x - e^{-x})}{(e^x - e^{-x})^4}\right]$$

$$= \frac{-2(e^x - e^{-x})^3(e^x + e^{-x}) + 2(e^x + e^{-x})^3(e^x - e^{-x})}{(e^x - e^{-x})^4}$$

$$= \frac{2(e^x + e^{-x})^3(e^x - e^{-x}) - 2(e^x - e^{-x})^3(e^x + e^{-x})}{(e^x - e^{-x})^4} \qquad \text{(After rearranging the numerator)}$$

Separating into fractions gives:

$$= \frac{2(e^x + e^{-x})^3(e^x - e^{-x})}{(e^x - e^{-x})^4} - \frac{2(e^x - e^{-x})^3(e^x + e^{-x})}{(e^x - e^{-x})^4}$$

$$\frac{d^2y}{dx^2} = \frac{2(e^x + e^{-x})^3}{(e^x - e^{-x})^3} - \frac{2(e^x + e^{-x})}{e^x - e^{-x}}$$

(b) Let us simplify $\dfrac{d^2y}{dx^2} + 2y\dfrac{dy}{dx}$ as follows:

$$\frac{d^2y}{dx^2} + 2y\frac{dy}{dx}$$

$$= \frac{2(e^x + e^{-x})^3}{(e^x - e^{-x})^3} - \frac{2(e^x + e^{-x})}{e^x - e^{-x}} + 2\left(\frac{e^x + e^{-x}}{e^x - e^{-x}}\right)\left[1 - \frac{(e^x + e^{-x})^2}{(e^x - e^{-x})^2}\right]$$

Expanding bracket gives:

$$= \frac{2(e^x + e^{-x})^3}{(e^x - e^{-x})^3} - \frac{2(e^x + e^{-x})}{e^x - e^{-x}} + 2\left(\frac{e^x + e^{-x}}{e^x - e^{-x}}\right) - \frac{2(e^x + e^{-x})^3}{(e^x - e^{-x})^3}$$

$$= 0 \qquad \text{(Since equal terms with opposite signs cancel out each other)}$$

Therefore, $\dfrac{d^2y}{dx^2} + 2y\dfrac{dy}{dx}$ gives zero as proven above.

Or, $\dfrac{d^2y}{dx^2} + 2y\dfrac{dy}{dx} = 0$

20. Find the derivative of $\ln(\tan 2x)$

Solution

$$y = \ln(\tan 2x)$$

324

$$\frac{dy}{dx} = \frac{\frac{d(\tan 2x)}{dx}}{\tan 2x}$$

$$= \frac{2\sec^2 2x}{\tan 2x}$$

Or, $\dfrac{dy}{dx} = \dfrac{2\left(\frac{1}{\cos 2x}\right)\left(\frac{1}{\cos 2x}\right)}{\left(\frac{\sin 2x}{\cos 2x}\right)}$ (Note that $\sec 2x = \dfrac{1}{\cos 2x}$ and $\tan 2x = \dfrac{\sin 2x}{\cos 2x}$)

$$= \left(\frac{2}{\cos 2x}\right)\left(\frac{1}{\cos 2x}\right) \times \frac{\cos 2x}{\sin 2x}$$

$$\frac{dy}{dx} = \frac{2}{\sin 2x \cos 2x}$$ (Since $\cos 2x$ cancels $\cos 2x$)

21. Given that $y = x^3 + 3x^2$, determine $2\dfrac{dy}{dx} - x\dfrac{d^2y}{dx^2}$

Solution

$$y = x^3 + 3x^2$$

$$\frac{dy}{dx} = 3x^2 + 6x$$

$$\frac{d^2y}{dx^2} = 6x + 6$$

Hence, $2\dfrac{dy}{dx} - x\dfrac{d^2y}{dx^2}$ is simplified as follows:

$$2(3x^2 + 6x) - x(6x + 6)$$

$$= 6x^2 + 12x - 6x^2 - 6x$$

$$= 6x$$

Therefore, $2\dfrac{dy}{dx} - x\dfrac{d^2y}{dx^2} = 6x$

22. Find $\dfrac{dy}{dx}$ if $y = 3(3x + \sqrt{x})^2$

Solution

$$y = 3(3x + \sqrt{x})^2$$

Applying chain rule gives:

$$\frac{dy}{dx} = 3 \times 2(3x + x^{\frac{1}{2}})^{2-1} \times \frac{d(3x + x^{\frac{1}{2}})}{dx}$$ (Note that $\sqrt{x} = x^{\frac{1}{2}}$)

$$= 6(3x + x^{\frac{1}{2}}) \times (3 + \frac{1}{2}x^{-\frac{1}{2}})$$

$$= 6(3x + x^{\frac{1}{2}})\left(3 + \frac{1}{2x^{\frac{1}{2}}}\right)$$

$$= 6(3x + \sqrt{x})\left(3 + \frac{1}{2\sqrt{x}}\right)$$

Expanding the bracket gives:

$$= 6\left(9x + \frac{3x}{2\sqrt{x}} + 3\sqrt{x} + \frac{1}{2}\right)$$

$$= 6\left(9x + \frac{3}{2}\sqrt{x} + 3\sqrt{x} + \frac{1}{2}\right) \qquad \left(\text{Note that } \frac{3x}{2\sqrt{x}} = \frac{3}{2}x^{1-\frac{1}{2}} = \frac{3}{2}x^{\frac{1}{2}} = 3\sqrt{x}\right)$$

$$= 6\left(9x + \frac{9}{2}\sqrt{x} + \frac{1}{2}\right)$$

$$\frac{dy}{dx} = 54x + 27\sqrt{x} + 3$$

23. If $y = \dfrac{5x^2 + 7}{x^4}$,

(a) find $\dfrac{d^2y}{dx^2}$

(b) show that $x^2\dfrac{d^2y}{dx^2} + 7x\dfrac{dy}{dx} + 8y = 0$

Solution

(a) $y = \dfrac{5x^2 + 7}{x^4}$

$$= \frac{5x^2}{x^4} + \frac{7}{x^4}$$

$$= \frac{5}{x^2} + \frac{7}{x^4}$$

$$y = 5x^{-2} + 7x^{-4}$$

$$\frac{dy}{dx} = -10x^{-3} - 28x^{-5}$$

Similarly,

$$\frac{d^2y}{dx^2} = 30x^{-4} + 140x^{-6}$$

(b) Let us now simplify $x^2\dfrac{d^2y}{dx^2} + 7x\dfrac{dy}{dx} + 8y$ by substituting appropriately as follows:

$$x^2(30x^{-4} + 140x^{-6}) + 7x(-10x^{-3} - 28x^{-5}) + 8(5x^{-2} + 7x^{-4})$$

$$= 30x^{-2} + 140x^{-4} - 70x^{-2} - 196x^{-4} + 40x^{-2} + 56x^{-4}$$

$$= 30x^{-2} + 40x^{-2} - 70x^{-2} + 140x^{-4} + 56x^{-4} - 196x^{-4}$$

$$= 0$$

Therefore $x^2\dfrac{d^2y}{dx^2} + 7x\dfrac{dy}{dx} + 8y$ gives zero as obtained above.

Or, $x^2\dfrac{d^2y}{dx^2} + 7x\dfrac{dy}{dx} + 8y = 0$

24. If $y = (2x + 5)^4 + \dfrac{x-1}{2x-1}$ find $\dfrac{dy}{dx}$.

Solution

$$(2x + 5)^4 + \frac{x-1}{2x-1}$$

We now use chain rule and quotient rule as follows:

$$\frac{dy}{dx} = 4(2x+5)^{4-1} \times \frac{d(2x+5)}{dx} + \frac{(2x-1)\frac{d(x-1)}{dx} - \left[(x-1)\frac{d(2x-1)}{dx}\right]}{(2x-1)^2}$$

$$= 4(2x+5)^3(2) + \frac{(2x-1)(1) - [(x-1)(2)]}{(2x-1)^2}$$

$$= 8(2x+5)^3 + \frac{(2x-1) - 2(x-1)}{(2x-1)^2}$$

$$= 8(2x+5)^3 + \frac{2x-1-2x+2}{(2x-1)^2}$$

$$\frac{dy}{dx} = 8(2x+5)^3 + \frac{1}{(2x-1)^2}$$

Exercise 32

1. If $y = x^2(3x^4 - 5)^3$ find $\frac{dy}{dx}$.

2. Given that $f(x) = 2x^5 - x^4 + 2x^3 + x^2 - 3x + 4$ find $f'(-2)$

3. A function $f(x)$ is given by $f(x) = \frac{\sqrt{(3x^2-1)^3}}{x^2}$. Find:

(a) the derivative of $f(x)$

(b) the gradient of $f(x)$ at the point $(1, -2)$

4. If $3xy^3 - y^2 - 4x^3 = 5y$, find:

(a) the derivative of the expression

(b) the gradient at $(-1, 1)$

5. Find the derivative of $(2x^2 + y^2)^2 = 0$

6. If $y = (x^2 + 5)^3(x^3 - 1)$, find $\frac{dy}{dx}$.

7. Find the derivative of $e^{\sin x + \tan x}$

8. Find the derivative of $\frac{x^2}{\sin^2 x}$

9. Differentiate with respect to x: $\ln(\sin^2 x + \cos 3x)$

10. If $y = a^{\cos 5x}$ find $\frac{dy}{dx}$.

11. Given that $y = x^2 e^{-3x}$ evaluate $\frac{d^2y}{dx^2} + \frac{dy}{dx} - y$

12. Differentiate with respect to x: $\sin\left(\frac{1-2x^3}{x^2}\right)$

13. If $y = \frac{2x-1}{x^2}$, find $x^4\frac{d^2y}{dx^2} - 3(2x-1)\frac{dy}{dx}$

14. Given that $y = 5e^{-2x} + 3e^x$, evaluate $\dfrac{d^2y}{dx^2} + \dfrac{dy}{dx} - 2y$

15. Find the derivative of $\left(\dfrac{1}{x^2} - \dfrac{2}{x}\right)^5$

16. Given that $e^{x^3 - y^3} = xy$, find:

(a) $\dfrac{dy}{dx}$

(b) $\dfrac{dy}{dx}$ at $(1, -1)$

17. If $y = \sin(\sin x)$, evaluate $\dfrac{d^2y}{dx^2} + \tan x \dfrac{dy}{dx} + y\cos^2 x$

18. Determine $\dfrac{d^2}{dx^2}\left(x^2\cos\dfrac{1}{x^2}\right)$

19. Given that $y = \dfrac{1 + e^{-x}}{1 - e^{-x}}$

(a) find $\dfrac{d^2y}{dx^2}$

(b) evaluate $\dfrac{d^2y}{dx^2} - \dfrac{dy}{dx} + y$

20. Find the derivative of $\ln(\sec^2 x)$

21. Given that $y = x + \tan x$, determine $\cos^2 x \dfrac{d^2y}{dx^2} - 2y + 2x$

22. Find $\dfrac{dy}{dx}$ if $y = \left(3x + \dfrac{\sqrt{5x}}{2}\right)^3$

23. If $y = \dfrac{x^3 + 2x^2 - 5x - 3}{x^2}$, find $\dfrac{d^2y}{dx^2}$

24. If $y = (x^3 + 1)^3 + \dfrac{5x - 2}{x^2 + 1}$ find $\dfrac{dy}{dx}$.

25. Given that $y = 2\cos 5x + 7\sin 5x$, determine $\dfrac{d^2y}{dx^2} + 25y$

CHAPTER 33
COLLECTION AND TABULATION OF DATA

When a large volume of data is obtained, it is necessary to present such data in frequency table. Sometimes the tally system which involves the use of vertical and horizontal strokes is applied.

Examples

1. A die is rolled 50 times and the following data is obtained. Represent the data in a frequency table.

4	6	4	3	5	3	1	4	6	5	6	4	2
6	4	5	6	2	1	6	4	3	4	6	1	5
1	3	6	2	2	4	3	4	5	3	4	1	2
3	1	2	1	5	3	4	3	4	2	5		

Solutions

The data which ranges from 1 to 6 is summarized as shown on the table below.

Number on die	Frequency
1	7
2	7
3	9
4	12
5	7
6	8

The data can also be represented on a horizontal table as shown below.

Number in die	1	2	3	4	5	6
Frequency	7	7	9	12	7	8

2. The scores of 40 students in a physics test are presented below. Prepare a frequency distribution table for the data.

64	66	68	63	70	63	67	64	70	69
66	64	65	70	62	70	66	69	67	64
61	63	67	62	68	64	63	69	70	63
63	61	68	67	68	63	61	67	69	68

<u>Solution</u>

The data which ranges from 61 to 70 is summarized as shown on the table below.

Score	61	62	63	64	65	66	67	68	69	70
Frequency	3	2	7	5	1	3	5	5	4	5

Exercise 33

1.   The marks obtained in an examination by 40 students in a class are as shown below. Represent the data in a frequency table.

71	74	74	70	70	72	74	74	65	69
66	68	65	73	66	72	69	69	67	65
71	73	67	68	68	69	70	69	71	65
67	67	68	72	74	73	67	67	69	68

2.   The score of 20 students in a test are as shown below. Represent the score in a frequency table.

6	6	8	9	5	6	7	5	6	9
8	6	7	5	9	5	9	6	6	5

3.   The number of seeds in a sample of 40 cocoa pods are as shown below. Represent the information using a frequency table.

28	22	28	28	27	29	20	20	20	24
21	25	25	20	22	20	26	29	30	24
21	23	27	22	28	30	23	29	20	23
23	21	28	27	28	23	21	27	30	28

4.   The ages of 30 students in a senior high school is represented below. Show the data using a frequency table.

14	15	12	13	10	13	11	14	10	12
15	11	15	10	12	10	11	13	14	14
15	13	12	11	14	14	13	12	10	15

CHAPTER 34
MEAN, MEDIAN AND MODE OF UNGROUPED DATA

The mean, median and mode are averages of sets of statistical data. They are called "measures of central tendency".

Mean

The mean of a set of data is obtained by adding all the data and then dividing the result by the total number in the data set.

For an ungrouped data given in a frequency table, the mean can be calculated by using the formula:

$$\bar{x} = \frac{\sum fx}{\sum f}$$

Where $\bar{x}$ is the mean, $\sum$ is a symbol representing summation, x is each number in the data, and f is the frequency.

Median

The median of a data is the middle number when the data is arranged in an increasing or decreasing order of size. For an odd number of data, the position of the middle number is obtained by the expression:

Median = number in the $(\frac{N+1}{2})$th position

where N is the total number of data.

If there is an even number of data, the median is the average of the two middle numbers. In such a case the positions of the two middle numbers is obtained by the expression:

$$\text{Median} = \frac{\text{Number in the } \left(\frac{N}{2}\right)\text{th position} + \text{Number in the } \left(\frac{N+2}{2}\right)\text{th position}}{2}$$

However, for large data in a frequency table which has an even number of data, the median is given by:

$$\text{Median} = \frac{\text{Number in the } \left(\frac{\sum f}{2}\right)\text{th position} + \text{Number in the } \left(\frac{\sum f+2}{2}\right)\text{th position}}{2}$$

Where $\sum f$ is the total frequency of the data.

Note that for an odd number of data, only one number will be at the middle. However, for an even number of data, two numbers will be at the middle. The average of the two numbers gives the median of the data.

Mode

The mode is the most occurring number in a set of data. It is the number with the highest frequency. If a set of data has two modes, we say it is bimodal.

Examples

1. Find the mean, median and mode of the data below:

   2, 5, 0, 3, 1, 6, 9, 7, 3

Solution

There are 9 numbers in the data. So, the mean is obtained as follows:

$$\text{Mean} = \frac{2 + 5 + 0 + 3 + 1 + 6 + 9 + 7 + 3}{9}$$

$$= \frac{36}{9}$$

$$= 4$$

∴   The mean is 4

In order to calculate the median, first arrange the numbers in ascending order as follows:

   0, 1, 2, 3, 3, 5, 6, 7, 9

By inspection, the number that is at the middle of the data is 3

∴   The median is 3

Or, since there are 9 numbers in the data and 9 is an odd number, then the position of the middle number is obtained as follows:

$$\text{Median} = \text{number in the } (\frac{N+1}{2})\text{th position}$$

$$= \text{number in the } (\frac{9+1}{2})\text{th position}$$

$$= \text{number in the } (\frac{10}{2})\text{th position}$$

$$= \text{number in the } 5^{th} \text{ position.}$$

$$= 3 \text{ (Since 3 is in the } 5^{th} \text{ position in the data arranged above)}$$

∴ The median is 3

The most occurring number in the data is 3 since it appears twice while every other number appears once.

∴ The mode is 3.

2. Find: a. the mean;

          b. the median;

          c. the mode of the data below.

   6, 7, 10, 5, 11, 5, 9, 7, 10, 13, 5, 8, 7, 5, 12

Solutions

a. There are 15 numbers in the data. So, the mean is obtained as follows:

$$\text{Mean} = \frac{6+7+10+5+11+5+9+7+10+13+5+8+7+5+12}{15}$$

$$= \frac{120}{15}$$

$$= 8$$

∴   The mean is 8

b. In order to calculate the median, first arrange the numbers in ascending order as follows:

5, 5, 5, 5, 6, 7, 7, 7, 8, 9, 10, 10, 11, 12, 13

By inspection, the number that is at the middle of the data is 7

∴   The median is 7

Or, since there are 15 numbers in the data and 15 is an odd number, then the position of the middle number is obtained as follows:

$$\text{Median} = \text{number in the } (\frac{N+1}{2})\text{th position}$$

$$= \text{number in the } (\frac{15+1}{2})\text{th position}$$

$$= \text{number in the } (\frac{16}{2})\text{th position}$$

$$= \text{number in the } 8^{th} \text{ position.}$$

$$= 7 \quad (\text{Since 7 is in the } 8^{th} \text{ position of the data})$$

∴   The median is 7

c. The most occurring number in the data is 5. It occurs four times.

∴   The mode is 5.

3. Find the mean, median and mode of the data below:

11, 14, 10, 16, 18, 12, 11, 15, 10, 11, 15, 13

Solutions

There are 12 numbers in the data. So, the mean is obtained as follows:

$$\text{Mean} = \frac{11+14+10+16+18+12+11+15+10+11+15+13}{12}$$

$$= \frac{156}{12}$$

$$= 13$$

∴ The mean is 13

In order to calculate the median, first arrange the numbers in ascending order as follows:

10, 10, 11, 11, 11, 12, 13, 14, 15, 15, 16, 18

By inspection, the two numbers that are at the middle of the data are 12 and 13. So, we take their average.

$$\text{median} = \frac{12+13}{2}$$

$$= \frac{25}{2}$$

$$= 12.5$$

∴ The median is 12.5

Or, since there are 12 numbers in the data and 12 is an even number, then the positions of the two middle numbers and their average are obtained as follows:

$$\text{Median} = \frac{\text{Number in the }\left(\frac{N}{2}\right)\text{th position} + \text{Number in the }\left(\frac{N+2}{2}\right)\text{th position}}{2}$$

$$= \frac{\text{Number in the }\left(\frac{12}{2}\right)\text{th position} + \text{Number in the }\left(\frac{12+2}{2}\right)\text{th position}}{2}$$

$$= \frac{\text{Number in the 6th position} + \text{Number in the }\left(\frac{14}{2}\right)\text{th position}}{2}$$

$$= \frac{12 + 13}{2} \quad \text{(Note that 12 is in the 6}^{\text{th}}\text{ position while 13 is in the 7}^{\text{th}}\text{ position)}$$

$$= \frac{25}{2}$$

$$= 12.5$$

∴ The median is 12.5

The most occurring number in the data is 11 since it appears three times.

∴ The mode is 11.

4. Find the mean, median and mode of the data below:

   50, 56, 58, 52, 55, 59, 51, 55

Solutions

There are 8 numbers in the data. So, the mean is obtained as follows:

$$\text{Mean} = \frac{50 + 56 + 58 + 52 + 55 + 59 + 51 + 55}{8}$$

$$= \frac{436}{8}$$

$$= 54.5$$

∴ The mean is 54.5

In order to calculate the median, first arrange the numbers in ascending order as follows:

50, 51, 52, 55, 55, 56, 58, 59.

By inspection, the two numbers that are at the middle of the data are 55 and 55. So, we take their average.

$$\text{median} = \frac{55+55}{2}$$
$$= \frac{110}{2}$$
$$= 55$$

∴　　The median is 55

Or, since there are 8 numbers in the data and 8 is an even number, then the positions of the two middle numbers and their average are obtained as follows:

$$\text{Median} = \frac{\text{Number in the } \left(\frac{N}{2}\right)\text{th position} + \text{Number in the } \left(\frac{N+2}{2}\right)\text{th position}}{2}$$

$$= \frac{\text{Number in the } \left(\frac{8}{2}\right)\text{th position} + \text{Number in the } \left(\frac{8+2}{2}\right)\text{th position}}{2}$$

$$= \frac{\text{Number in the 4th position} + \text{Number in the } \left(\frac{10}{2}\right)\text{th position}}{2}$$

$$= \frac{\text{Number in the 4th position} + \text{Number in the 5th position}}{2}$$

$$= \frac{55+55}{2} \quad \text{(Note that 55 is in the 4}^{th} \text{ and the 5}^{th} \text{ position)}$$

$$= \frac{110}{2}$$

$$= 55$$

∴　　The median is 55

The most occurring number in the data is 55.

∴　　The mode is 55.

5. The table below shows the marks of 50 students in a test.

Mark	1	2	3	4	5	6	7
No of Student	8	16	10	5	3	6	2

Calculate: a. the mean b. the median c. the mode of the marks

Solutions

a. Note that the number of students is also the frequency. Presenting the table as shown below allows for easy calculation of the mean.

Mark (x)	No of student (f)	Fx
1	8	8
2	16	32
3	10	30
4	5	20
5	3	15
6	6	36
7	2	14
Total:	$\sum f = 50$	$\sum fx = 155$

Note that the column fx is obtained by multiplying the values of numbers in the column f by numbers in the column x.

Mean, $\bar{x} = \dfrac{\sum fx}{\sum f}$

$= \dfrac{155}{50}$

$= 3.1$

b. Since there are 50 students, i.e. the total frequency is 50, and 50 is an even number, then the positions of the two middle marks and their average are obtained as follows:

$$\text{Median} = \frac{\text{Number in the } \left(\frac{\sum f}{2}\right)\text{th position} + \text{Number in the } \left(\frac{\sum f + 2}{2}\right)\text{th position}}{2}$$

$$= \frac{\text{Number in the } \left(\frac{50}{2}\right)\text{th position} + \text{Number in the } \left(\frac{50+2}{2}\right)\text{th position}}{2}$$

$$= \frac{\text{Number in the 25th position} + \text{Number in the } \left(\frac{52}{2}\right)\text{th position}}{2}$$

$$= \frac{\text{Number in the 25th position} + \text{Number in the 26th position}}{2}$$

$= \dfrac{3+3}{2}$ (Note that 3 is the mark in the 25th position and in the 26th position)

$= \dfrac{6}{2}$

$= 3$

∴   The median is 3

Use the frequency (number of students) to locate the marks in the 25th and 26th position as follows:

The first frequency of 8 shows that mark 1 occupies the position of 1st to 8th. Adding the second frequency of 16 to the first frequency gives, 8 + 16 = 24. This shows that after the 8th position occupied by the mark 1, the positions 9th to 24th is occupied by the mark 2. Adding the third frequency of 10 to the previous sum of frequencies gives, 10 + 24 = 34. This shows that after the 24th position occupied by the mark 2, the positions 25th to 34th is occupied by the mark 3. Hence, 3 is the mark in the 25th and 26th position which are at the middle of the data.

c. The mode is the mark that has the highest frequency. From the table, the mark 2 has the highest frequency of 16. So, the mode is 2.

∴ The mode is 2.

Note that the mode is not the frequency itself, but that particular mark that has the highest frequency. Avoid the mistake of taking 16 (frequency) as the mode.

6. The table below shows the ages of 30 students in a school.

Age	10	11	12	13	14	15
No of Student	1	4	3	7	9	6

Calculate: a. the mean b. the median c. the mode of the ages

Solutions

a. Using the number of students as the frequency, the table can be presented for easy calculation of the mean as follows:

Age (x)	No of student (f)	Fx
10	1	10
11	4	44
12	3	36
13	7	91
14	9	126
15	6	90
Total:	$\sum f = 30$	$\sum fx = 397$

Note that the column fx is obtained by multiplying the values of numbers in the column f by numbers in the column x.

Mean, $\bar{x} = \dfrac{\Sigma fx}{\Sigma f}$

$= \dfrac{397}{30}$

$= 13.3$

b. Since there are 30 students, i.e. the total frequency is 30, and 30 is an even number, then the positions of the two middle ages and their average are obtained as follows:

Median = $\dfrac{\text{Age in the } \left(\frac{\Sigma f}{2}\right)\text{th position} + \text{Age in the } \left(\frac{\Sigma f + 2}{2}\right)\text{th position}}{2}$

$= \dfrac{\text{Age in the } \left(\frac{30}{2}\right)\text{th po sition} + \text{Age in the } \left(\frac{30+2}{2}\right)\text{th position}}{2}$

$= \dfrac{\text{Age in the 15th position} + \text{Age in the } \left(\frac{32}{2}\right)\text{th position}}{2}$

$= \dfrac{\text{Age in the 15th position} + \text{Age in the 16th position}}{2}$

$= \dfrac{13 + 14}{2}$

$= \dfrac{27}{2}$

$= 13.5$

∴   The median is 13.5

Note that 13 is the age in the 15th position while 14 is the age in the 16th position.

The frequency (number of students) was used to locate the ages in the 15th and 16th position as follows:

The first frequency of 1 shows that age 10 occupies the 1st position. Adding the second frequency of 4 to the first frequency gives, 1 + 4 = 5. This shows that after the 1th position occupied by the age 10, the positions 2nd to 5th is occupied by the age 11. Adding the third frequency of 3 to the previous sum of frequencies gives, 3 + 5 = 8. This shows that after the 5th position occupied by the age 11, the positions 6th to 8th is occupied by the age 12. Adding the fourth frequency of 7 to the previous sum of frequencies gives, 7 + 8 = 15. This shows that after the 8th position occupied by the age 12, the positions 9th to 15th is occupied by the age 13. Adding the fifth frequency of 9 to the previous sum of frequencies gives, 9 + 15 = 24. This shows that after the 15th position occupied by the age 13, the positions 16th to 24th is occupied by the age 14. Hence, 13 is the age in the 15th position while 14 is the age in the 16th position.

c.   The mode is the age that has the highest frequency. From the table, the age 14 has the highest frequency of 9.

∴ The mode is 14.

Note that 9 is the frequency. It should not be taken as the mode.

Range

Range is the difference between the highest and lowest values in a given set of data. It is a measure of dispersion or variation.

Examples

1. Find the range of the following set of numbers: 4, 8, 2, 5, 8, 3, 6, 4, 9, 2, 5

Solution

The highest number in the data set is 9, while the lowest number is 2.

∴   Range = Highest number – Lowest number

$\qquad$ = 9 – 2 = 7

$\quad$ Range = 7

2. The monthly salaries of five workers in a company are: $845, $1205, $694, $626 and $864. What is the range of the salaries?

Solution

Range = Highest salary – Lowest salary

$\qquad$ = 1205 – 626

$\quad$ Range = $579

Exercise 34

1. Find the mean, median and mode of the data below:

$\qquad$ 1, 6, 10, 4, 1, 2, 5, 2, 3, 2, 8

2. Find: a. the mean;

$\qquad$ b. the median;

$\qquad$ c. the mode of the data below.

$\quad$ 20, 24, 21, 25, 22, 25, 28, 26, 20, 23, 25, 27 and 26

3. Find the mean, median and mode of the data below:

$\quad$ 101, 105, 120, 116, 109, 112, 118, 115, 105 and 111

4. Find the mean, median and mode of the data below:

0, 6, 8, 2, 5, 9, 1, 5, 4, 7, 5, 2, 3, 3

5. The table below shows the marks of 50 students in a test.

Mark	3	4	5	6	7	8	9
No of Student	8	16	10	5	3	6	2

Calculate: a. the mean b. the median c. the mode of the marks

6. The table below shows the ages of 30 students in a school.

Age	10	11	12	13	14	15
No of Student	1	4	3	7	9	6

Calculate: a. the mean b. the median c. the mode of the ages

7. Find the range of the following set of data

a. 12, 17, 21, 15, 19, 13, 11, 16, 22, 12, 13

b. 231kg, 258kg, 213kg, 243kg, 216kg, 271kg, 262kg, 219kg, 238kg, 231kg.

CHAPTER 35
COLLECTION AND TABULATION OF GROUPED DATA

Statistical data containing numerous values is easier to work with when the values are grouped into class intervals.

Examples

1. The data below gives the marks of 30 students in an exam.

43	45	50	47	51	58	52	47	42	54
61	50	45	55	57	41	46	49	51	50
59	44	53	57	49	40	48	52	51	58

Taking class intervals 40 – 44, 45 – 49, ..., construct a frequency distribution for the data.

Solution

The data is summarized as shown on the table below. Note that the highest value in the given data falls within the range 60 – 64.

Class interval	40 – 44	45 - 49	50 - 54	55 - 59	60 - 64
Frequency	5	8	10	6	1

2. The data below gives the ages of lecturers in a university.

34	62	54	41	51	63	31	44	48	50
33	45	59	55	47	31	39	55	60	40
63	45	53	55	36	58	61	34	34	43
47	35	43	51	35	48	42	51	36	31

Taking class intervals 31 – 35, 36 – 40, ..., construct a frequency table for the data.

Solution

The data is summarized as shown on the table below. Note that the highest value in the given data falls within the range 61 – 65.

Age	31 – 35	36 - 40	41 - 45	46 - 50	51 - 55	56 - 60	61 - 65
Frequency	9	4	7	5	8	3	4

Terms Used in Grouped Data

The table below will be used to explain the terms used in grouped data.

Class interval	Frequency
8 – 14	3
15 – 21	5
22 – 28	8
29 – 35	18

1. Class limit: The end numbers in each class interval are called the class limits. 8 is the lower class limit, while 14 is the upper class limit of the first class interval.

2. Class boundaries: The class boundary for the second class interval is 14.5 – 21.5. The lower class boundary is 14.5 which is obtained by subtracting 0.5 from 15 (the lower class limit). The upper class boundary is 21.5 which is obtained by adding 0.5 to 21 (the upper class limit). Other class boundaries are obtained in a similar way.

3. Class width: For each class interval the difference between the upper class boundary and the lower class boundary gives the class width. From the table above the class width for the third class interval is 28.5 – 21.5 = 7

4. Class mid-value: This is half of the sum of the lower and upper class limit of a given class interval. The class-value of the first class interval is given by: $\frac{8+14}{2} = \frac{22}{2} = 11$.

Examples

1. Copy and complete the table below.

Class interval	Frequency	Class boundary	Class width	Class mid-value
55 – 59	3			
60 – 64	2			
65 – 69	5			
70 – 74	4			
75 – 79	1			

<u>Solution</u>

The completed table is as shown below

Class interval	Frequency	Class boundary	Class width	Class mid-value
55 – 59	3	54.5 – 59.5	5	57
60 – 64	2	59.5 – 64.5	5	62
65 – 69	5	64.5 – 69.5	5	67
70 – 74	4	69.5 – 74.5	5	72
75 – 79	1	74.5 – 79.5	5	77

Note that the class boundaries are obtained by subtracting and adding 0.5 to the lower and upper class limits respectively. This 0.5 is obtained by finding the difference between the lower class limit of one class and the upper class limit of the previous class and dividing the result by 2. This gives, for example, $\frac{60-59}{2} = \frac{1}{2} = 0.5$.

The class width is the difference between the upper and lower class boundaries. The class mid-values are obtained by taking the mean of the upper and lower class limits.

2. Copy and complete the table below.

Class interval	Frequency	Class boundary	Class width	Class mid-value
0 – 19	2			
20 – 39	8			
40 – 59	3			
60 – 79	1			
80 – 99	4			

Solution

The completed table is as shown below

Class interval	Frequency	Class boundary	Class width	Class mid-value
0 – 19	2	-0.5 – 19.5	20	9.5
20 – 39	8	19.5 – 39.5	20	29.5
40 – 59	3	39.5 – 59.5	20	49.5
60 – 79	1	59.5 – 79.5	20	69.5
80 – 99	4	79.5 – 99.5	20	89.5

Exercise 35

1. The data below gives the scores of 50 students in an exam.

43	45	50	47	51	58	52	47	42	54
61	50	45	55	57	41	46	49	51	50
59	44	53	57	49	40	48	52	51	58
48	54	43	54	61	60	49	57	45	42
56	45	57	61	54	62	44	47	46	62

Taking class intervals 40 – 44, 45 – 49, ..., construct a frequency distribution for the data.

2. The data below shows the weights in kg of students in a school.

24	32	44	51	31	23	51	34	48	40
53	45	29	35	27	51	29	35	50	30
43	55	53	35	26	28	41	44	54	43
27	45	33	41	55	28	32	51	26	39

Taking class intervals 21 – 25, 26 – 30, ..., construct a frequency table for the data.

3. Copy and complete the table below.

Class interval	Frequency	Class boundary	Class width	Class mid-value
5 – 9	2			
10 – 14	5			
15 – 19	5			
20 – 24	7			
25 – 29	1			

4. Copy and complete the table below.

Class interval	Frequency	Class boundary	Class width	Class mid-value
1 – 20	1			
21 – 40	4			
41 – 60	7			
61 – 80	3			
81 – 100	5			

5. Copy and complete the table below.

Class interval	Frequency	Class boundary	Class width	Class mid-value
0 – 90	2			
100 – 190	4			
200 – 290	1			
300 – 390	7			
400 – 490	1			

CHAPTER 36
MEAN, MEDIAN AND MODE OF GROUPED DATA

Mean

The mean of a grouped data can be calculated by substituting the class mid value as the values of x in the formula given by:

$$\text{Mean } \bar{x} = \frac{\sum fx}{\sum f}$$

Median

The median of a grouped data can be estimated by:

$$\text{Median} = L + C\left(\frac{\frac{\sum f}{2} - CF_{bm}}{F_m}\right)$$

Where, $\frac{\sum f}{2}$ determines the median class

L = Lower class boundary of the median class

CF_{bm} = Cumulative frequency before the median class

F_m = Frequency of the median class

C = Class width

Mode

The mode of a grouped data can be calculated as follows:

$$\text{Mode} = L + C\left(\frac{\Delta_1}{\Delta_1 + \Delta_2}\right)$$

Where, L = Lower class boundary of modal class

C = Class width

Δ_1 = Difference between the frequency of the modal class and the frequency before it

Δ_2 = Difference between the frequency of the modal class and the frequency after it

Examples

1. The following table shows the frequency distribution of ages, in years of 50 people at a bus stop.

Ages	10 - 19	20 - 29	30 - 39	40 - 49	50 - 59	60 - 69
Number of people	6	12	16	9	5	2

Calculate: a. the mean

b. the median

c. the mode of the distribution

<u>Solution</u>

Ages	Number of people (f)	Cumulative frequency	Class mid-value(x)	fx	Class boundary	Class width
10 – 19	6	6	14.5	87	9.5-19.5	10
20 – 29	12	6+12=18	24.5	294	19.5-29.5	10
30 – 39	16	18+16=34	34.5	552	29.5-39.5	10
40 – 49	9	34+9=43	44.5	400.5	39.5-49.5	10
50 – 59	5	43+5=48	54.5	272.5	49.5-59.5	10
60 – 69	2	48+2=50	64.5	129	59.5-69.5	10
	$\sum f = 50$			$\sum f = 1735$		

a. Mean $\bar{x} = \dfrac{\sum fx}{\sum f}$

$= \dfrac{1735}{50} = 34.7$

b. Median $= L + C\left(\dfrac{\frac{\sum f}{2} - CF_{bm}}{F_m}\right)$

$\dfrac{\sum f}{2} = \dfrac{50}{2} = 25$

This shows that the median class falls in the 25th position

∴ Median class $= 30 – 39$

Lower class boundary of median class, $L = 29.5$

Class width, $C = 10$

Cumulative frequency before the median class, $CF_{bm} = 18$

Frequency of the median class, $F_m = 16$

∴ Median $= L + C\left(\dfrac{\frac{\sum f}{2} - CF_{bm}}{F_m}\right)$

$= 29.5 + 10\left(\dfrac{25 - 18}{16}\right)$

$= 29.5 + 10\left(\dfrac{7}{16}\right)$

$= 29.5 + 4.375 = 33.875$

∴ Median $= 33.9$ (To 1 d.p)

c. Mode $= L + C\left(\dfrac{\Delta_1}{\Delta_1 + \Delta_2}\right)$

Modal class = 30 – 39

Lower class boundary of modal class, L = 29.5

Class width = 10

Δ_1 = Modal class frequency – frequency before it

= 16 - 12 = 4

Δ_2 = Modal class frequency – frequency after it

= 16 - 9 = 7

∴ Mode = L + C$\left(\dfrac{\Delta_1}{\Delta_1 + \Delta_2}\right)$

= 29.5 + 10$\left(\dfrac{4}{4 + 7}\right)$

= 29.5 + 10$\left(\dfrac{4}{11}\right)$

= 29.5 + 3.64 = 33.14

∴ Mode = 33.1 (To 1 d.p)

2. The data below is the weight of students in a high school.

Weight	31 - 35	36 - 40	41 - 45	46 - 50	51 - 55	56 - 60	61 - 65
Number of student	2	9	7	5	8	3	6

Determine: a. the mean

b. the median

c. the mode of the weights

Solution

Weight	Number of student (f)	Cumulative frequency	Class mid-value(x)	fx	Class boundary	Class width
31 – 35	2	2	33	66	30.5-35.5	5
36 – 40	9	2+9=11	38	349	35.5-40.5	5
41 – 45	7	11+7=18	43	301	40.5-45.5	5
46 – 50	5	18+5=23	48	240	45.5-50.5	5
51 – 55	8	23+8=31	53	424	50.5-55.5	5
56 – 60	3	31+3=34	58	174	55.5-60.5	5
61 – 65	6	34+6=40	63	378	60.5-65.5	5
	Σf = 40			Σfx= 1932		

a. Mean $\bar{x}$ = $\dfrac{\Sigma fx}{\Sigma f}$

$$= \frac{1932}{40}$$

$$= 48.3$$

b. Median $= L + C\left(\dfrac{\frac{\Sigma f}{2} - CF_{bm}}{F_m}\right)$

$$\frac{\Sigma f}{2} = \frac{40}{2} = 20$$

This shows that the median class falls in the 20th position

$\therefore$ Median class $= 46 - 50$

Lower class boundary of median class, $L = 45.5$

Class width, $C = 5$

Cumulative frequency before the median class, $CF_{bm} = 18$

Frequency of the median class, $F_m = 5$

$\therefore$ Median $= L + C\left(\dfrac{\frac{\Sigma f}{2} - CF_{bm}}{F_m}\right)$

$$= 45.5 + 5\left(\frac{20 - 18}{5}\right)$$

$$= 45.5 + 5\left(\frac{2}{5}\right)$$

$$= 45.5 + 2 = 47.5$$

$\therefore$ Median $= 47.5$

c. Mode $= L + C\left(\dfrac{\Delta_1}{\Delta_1 + \Delta_2}\right)$

Modal class $= 36 - 40$

Lower class boundary of modal class, $L = 35.5$

Class width $= 5$

Δ_1 = Modal class frequency – frequency before it

$= 9 - 2 = 7$

Δ_2 = Modal class frequency – frequency after it

$= 9 - 7 = 2$

$\therefore$ Mode $= L + C\left(\dfrac{\Delta_1}{\Delta_1 + \Delta_2}\right)$

$$= 35.5 + 5\left(\frac{7}{7 + 2}\right)$$

$$= 35.5 + 5\left(\frac{7}{9}\right)$$

$$= 35.5 + 3.89 = 39.39$$

$\therefore$ Mode $= 39.4$ (To 1 d.p)

Exercise 36

1. The following table shows the weights of 30 people at a company.

Weight	60 - 64	65 - 69	70 - 74	75 - 79	80 - 84	85 - 89
Number of people	1	12	7	5	3	2

Calculate: a. the mean

b. the median

c. the mode of the distribution

2. The data below is the load distribution in tones, a chain can support.

Load	83 - 85	86 - 88	89 - 91	92 - 94	95 - 97
Number of chain	2	8	5	14	1

Determine: a. the mean

b. the median

c. the mode of the weights

3. The data below is the ages, in years, of 50 people at a party.

Ages	1 - 20	21 - 40	41 - 60	61 - 80	81-100
Number of people	3	21	17	7	2

Determine: a. the mean

b. the median

c. the mode of the weights

CHAPTER 37
MEAN DEVIATION

The mean deviation of a set of data is the mean of the absolute deviation of the values from the mean of the group. The mean deviation for data not given in a frequency table is given by:

$$\text{Mean deviation} = \frac{\sum |x - \bar{x}|}{N}$$ where x is each value in the data, $\bar{x}$, is the mean and N is the number of values in the data.

For data given in a frequency table, the mean deviation is given by:

$$\text{Mean deviation} = \frac{\sum f|x - \bar{x}|}{\sum f}$$

Examples

1. Calculate the mean deviation of the following data: 2, 4, 1, 3, 0

Solution

Let us first calculate the mean of the data.

$$\text{Mean, } \bar{x} = \frac{2 + 4 + 1 + 3 + 0}{5}$$
$$= \frac{10}{5}$$
$$= 2$$

$$\therefore \ \bar{x} = 2$$

The deviation from the mean $(x - \bar{x})$ is now tabulated as follows.

Data (x)	$x - \bar{x}$ ($\bar{x} = 2$)	$|x - \bar{x}|$
2	0	0
4	2	2
1	-1	1
3	1	1
0	-2	2
		$\sum|x - \bar{x}| = 6$

$$\therefore \quad \text{Mean deviation} = \frac{\sum |x - \bar{x}|}{N} = \frac{6}{5} = 1.2$$

2. Calculate the mean deviation of the following data: 6, 2, 5, 8, 3, 6, 4, 5, 7, 4.

Solution

Let us first calculate the mean of the data.

Mean, $\bar{X} = \dfrac{6 + 2 + 5 + 8 + 3 + 6 + 4 + 5 + 7 + 4}{10}$

$= \dfrac{50}{10}$

$= 5$

$\therefore\ \bar{x} = 5$

The deviation from the mean $(x - \bar{x})$ is now tabulated as follows.

Data (x)	$x - \bar{x}$ $(\bar{x} = 5)$	$\|x - \bar{x}\|$
6	1	1
2	-3	3
5	0	0
8	3	3
3	-2	2
6	1	1
4	-1	1
5	0	0
7	2	2
4	-1	1
		$\sum\|x - \bar{x}\| = 14$

$\therefore\quad$ Mean deviation $= \dfrac{\sum|x - \bar{x}|}{N} = \dfrac{14}{10} = 1.4$

3. The marks obtained by 40 students in a mathematics test are as shown below. Calculate the mean deviation of the data.

Marks	31 - 40	41 - 50	51 - 60	61 - 70	71 - 80	81 - 90	91- 100
Number of student	1	2	8	11	8	6	4

Solution

The table below summarizes the determination of the mean and the values needed for the mean deviation. Note that the mean used on the table has been calculated below the table.

Mark	mid-value x	$x - \bar{x}$ $\bar{x} = 69.75$	$\|x - \bar{x}\|$	No of student, f	fx	$f\|x - \bar{x}\|$
31 – 40	35.5	-34.25	34.25	1	35.5	34.25
41 – 50	45.5	-24.25	24.25	2	91	48.50
51 – 60	55.5	-14.24	14.24	8	444	114
61 – 70	65.5	-4.25	4.25	11	720.5	46.75
71 – 80	75.5	5.75	5.75	8	604	46
81 – 90	85.5	15.75	15.75	6	513	94.5
91 – 100	95.5	25.75	25.75	4	382	103
				$\sum f$= 40	$\sum fx$ =2790	

Mean, $\bar{x} = \dfrac{\sum fx}{\sum f} = \dfrac{2790}{40} = 69.75$

Using the values from the table above, $\sum f\|x - \bar{x}\|$ = 34.25 + 48.50 + 114 + 46.75 + 46 + 94.5 + 103 = 487

∴   Mean deviation = $\dfrac{\sum f|x - \bar{x}|}{\sum f}$

$= \dfrac{487}{40}$

∴   Mean deviation = 12.2

4. The ages of 50 people in a hospital are as shown below. Calculate the mean deviation of the ages.

Age	1 - 5	6 - 10	11 – 15	16 - 20	21 - 25	26 - 30
Number of people	6	9	14	10	4	7

Solution

The table below summarizes the determination of the mean and the values needed for the mean deviation. The mean has been calculated below the table.

Age	mid-value x	$x - \bar{x}$ $\bar{x} = 14.8$	$\|x - \bar{x}\|$	No of people, f	fx	$f\|x - \bar{x}\|$
1 – 5	3	-11.8	11.8	6	18	70.8
6 – 10	8	-6.8	6.8	9	72	61.2
11 – 15	13	-1.8	1.8	14	182	25.2
16 – 20	18	3.2	3.2	10	180	32
21 – 25	23	8.2	8.2	4	92	32.8
26 – 30	28	13.2	13.2	7	196	92.4
				$\sum f$= 50	$\sum fx$ =740	

Mean, $\bar{x} = \dfrac{\sum fx}{\sum f} = \dfrac{740}{50} = 14.8$

Using the values from the table above, $\sum f|x - \bar{x}| = 70.8 + 61.2 + 25.2 + 32 + 32.8 + 92.4 = 314.4$

$\therefore$ Mean deviation $= \dfrac{\sum f|x - \bar{x}|}{\sum f}$

$= \dfrac{314.4}{50}$

$\therefore$ Mean deviation $= 6.29$

5. The table below shows the number of cars owned by some political public office holders.

Number of cars	1	2	3	4	5	6
Number of politicians	9	15	11	7	3	5

Calculate the mean deviation of the data.

Solution

The table below summarises the calculations of the mean and the mean deviation.

Cars X	$x - \bar{x}$ $\bar{x} = 2.9$	$|x - \bar{x}|$	No of politicians, f	Fx	$f|x - \bar{x}|$
1	-1.9	1.9	9	9	17.1
2	-0.9	0.9	15	30	13.5
3	0.1	0.1	11	33	1.1
4	1.1	1.1	7	28	7.7
5	2.1	2.1	3	15	6.3
6	3.1	3.1	5	30	15.5
			$\sum f = 50$	$\sum fx = 145$	

Mean, $\bar{x} = \dfrac{\sum fx}{\sum f} = \dfrac{145}{50} = 2.9$

Using the values from the table above, $\sum f|x - \bar{x}| = 17.1 + 13.5 + 1.1 + 7.7 + 6.3 + 15.5 = 61.2$

$\therefore$ Mean deviation $= \dfrac{\sum f|x - \bar{x}|}{\sum f}$

$= \dfrac{61.2}{50}$

$\therefore$ Mean deviation $= 1.224$

Exercise 37

1. Calculate the mean deviation of the following data: 0, 5, 7, 4, 5, 3
2. Calculate the mean deviation of the following data: 4, 6, 5, 9, 9, 5, 2, 4, 8, 6, 8
3. Calculate the mean deviation of the following data: 1, 3, 1, 4, 6
4. The marks obtained by 30 students in a physics test are as shown below. Calculate the mean deviation of the data.

Marks	0 - 9	10 - 19	20 - 29	30 - 39	40 - 49	50 – 59	60- 69
Number of student	4	1	5	8	3	2	7

5. The ages of 100 people in a village are as shown below. Calculate the mean deviation of the ages.

Age	11 - 20	21 - 30	31 – 40	41 - 50	51 - 60	61 - 70
Number of people	12	9	15	24	29	11

6. The number of employees in 50 enterprises are as shown below. Calculate the mean deviation of the data.

Marks	0 - 4	5 - 9	10 - 14	15 - 19	20 - 24	25 – 29	30- 34
Number of student	2	11	15	3	4	2	13

7. The table below shows the number of farms owned by some people in a city.

Number of farms	2	4	6	8	10	12
Number of people	3	5	10	6	8	8

Calculate the mean deviation of the data.

8. A die is rolled 50 times and the following data is obtained.

```
2   5   4   3   5   3   1   4   6   5   6   4   2
6   1   5   6   2   1   6   4   3   4   3   1   6
1   3   6   4   2   4   3   4   5   3   4   1   2
3   1   2   2   5   6   4   3   4   6   5
```

a. Present the data in a frequency table
b. Calculate the mean deviation of the data.

CHAPTER 38
VARIANCE AND STANDARD DEVIATION

Variance is the mean of the squares of the deviations from the mean. Standard deviation is the positive square root of the variance.

Variance, standard deviation and mean deviation are also regarded as measures of dispersion or variation.

The variance of data not given on a frequency table is given by:

$$\text{Variance} = \frac{\sum(x - \bar{x})^2}{N}$$

For data given on a frequency table, the variance is given by:

$$\text{Variance} = \frac{\sum f(x - \bar{x})^2}{\sum f}$$

Standard deviation is the square root of variance.

Examples

1. Calculate the variance and standard deviation of the following data: 4, 2, 1, 5.

<u>Solution</u>

Let us first calculate the mean of the data.

$$\text{Mean, } \bar{x} = \frac{4 + 2 + 1 + 5}{4} = \frac{12}{4} = 3$$

We now present the deviation from the mean as follows.

Data x	$x - \bar{x}$ $(\bar{x} = 3)$	$(x - \bar{x})^2$
4	1	1
2	-1	1
1	-2	4
5	2	4
		$\sum(x - \bar{x})^2 = 10$

$$\text{Variance} = \frac{\sum(x - \bar{x})^2}{N} = = \frac{10}{4} = 2.5$$

∴   Standard deviation $= \sqrt{2.5} = 1.58$

2. Calculate the variance and standard deviation of the data below:

   2, 5, 3, 2, 6, 5, 7, 2.

Solution

Let us first calculate the mean of the data as follows:

$$\text{Mean, } \bar{x} = \frac{2 + 5 + 3 + 2 + 6 + 5 + 7 + 2}{8}$$

$$= \frac{32}{8}$$

$$= 4$$

The deviation from the mean is as presented below.

Data x	$x - \bar{x}$ $(\bar{x} = 4)$	$(x - \bar{x})^2$
2	-2	4
5	1	1
3	-1	1
2	-2	4
6	2	4
5	1	1
7	3	9
2	-2	4
		$\Sigma(x - \bar{x})^2 = 28$

$$\text{Variance} = \frac{\Sigma(x - \bar{x})^2}{N} = \frac{28}{8} = 3.5$$

$$\therefore \quad \text{Standard deviation} = \sqrt{3.5} = 1.87$$

3. The distances in Km, from school to the homes of 30 students are as shown below.

Calculate:

a. the variance

b. the standard deviation of the data

Distance (Km)	0 − 4	5 - 9	10 − 14	15 - 19	20 - 24	25 - 29
Number of students	2	10	8	6	3	1

Solutions

The working is set out as shown on the table below

Distance	mid-value x	No of student f	fx	$x - \bar{x}$ $\bar{x} = 12.2$	$(x - \bar{x})^2$	$f(x - \bar{x})^2$
0 – 4	2	2	4	-10.2	104.04	208.08
5 – 9	7	10	70	-5.2	27.04	270.4
10 – 14	12	8	96	-0.2	0.04	0.32
15 – 19	17	6	102	4.8	23.04	138.24
20 – 24	22	3	66	9.8	96.04	288.12
25 – 29	27	1	27	14.8	219.04	219.04
		$\sum f = 30$	$\sum fx = 365$			

Mean, $\bar{x} = \dfrac{\sum fx}{\sum f} = \dfrac{365}{30} = 12.2$

a. Using the values from the table above, $\sum f(x - \bar{x})^2 = 208.08 + 270.4 + 0.32 + 138.24 + 288.12 + 219.04 = 1124.2$

$\therefore$ Variance $= \dfrac{\sum f(x - \bar{x})^2}{\sum f}$

$= \dfrac{1124.2}{30}$

$\therefore$ Variance = 37.5

b. Standard deviation $= \sqrt{\text{Variance}}$

$= \sqrt{37.5}$

$\therefore$ Standard deviation = 6.1

4. The projected population in millions, of 20 states in a country are as shown below. Calculate:

a. the variance

b. the standard deviation of the data

Population	1 - 5	6 - 10	11 – 15	16 - 20	21 - 25	26 - 30
Number of state	1	8	5	3	2	1

Solutions

The working is set out as shown on the table below

Popula-tion	mid-value x	No of states f	fx	$x - \bar{x}$ $\bar{x} = 13$	$(x - \bar{x})^2$	$f(x - \bar{x})^2$
1 – 5	3	1	3	-10	100	100
6 – 10	8	8	64	-5	25	200
11 – 15	13	5	65	0	0	0
16 – 20	18	3	54	5	25	75
21 – 25	23	2	46	10	100	200
26 – 30	28	1	28	15	225	225
		$\sum f = 20$	$\sum fx = 260$			

Mean, $\bar{x} = \dfrac{\sum fx}{\sum f} = \dfrac{260}{20} = 13$

a. Using the values from the table above, $\sum f(x - \bar{x})^2 = 100 + 200 + 0 + 75 + 200 + 225 = 800$

$\therefore$ Variance $= \dfrac{\sum f(x - \bar{x})^2}{\sum f}$

$= \dfrac{800}{20}$

$\therefore$ Variance $= 40$

b. Standard deviation $= \sqrt{\text{Variance}}$

$= \sqrt{40}$

$\therefore$ Standard deviation $= 6.3$

5. The scores obtained by 100 students in a test are as shown below. Calculate:
a. the variance
b. the standard deviation of the scores

Scores	2	3	4	5	6	7
Number of student	10	22	18	30	12	8

Solutions
The working is set out as shown on the table below

Scores x	No of students f	fx	$x - \bar{x}$ $\bar{x} = 4.4$	$(x - \bar{x})^2$	$f(x - \bar{x})^2$
2	12	24	-2.4	5.76	69.12
3	18	54	-1.4	1.96	35.28
4	22	88	-0.4	0.16	3.52
5	24	120	0.6	0.36	8.64
6	14	84	1.6	2.56	35.84
7	10	70	2.6	6.76	67.6
	$\sum f = 100$	$\sum fx = 440$			

Mean, $\bar{x} = \dfrac{\sum fx}{\sum f} = \dfrac{440}{100} = 4.4$

a. Using the values from the table above, $\sum f(x - \bar{x})^2 = 69.12 + 35.28 + 3.52 + 8.64 + 35.84 + 67.6$
= 220

$\therefore$ Variance $= \dfrac{\sum f(x - \bar{x})^2}{\sum f}$

$= \dfrac{220}{100}$

$\therefore$ Variance = 2.2

b. Standard deviation $= \sqrt{\text{Variance}}$

$= \sqrt{2.2}$

$\therefore$ Standard deviation = 1.48

Exercise 38

1. Calculate the variance and standard deviation of the following data: 3, 5, 4, 7, 6.

2. Calculate the variance and standard deviation of the data below:

1, 0, 4, 3, 5, 8, 6, 4, 7, 2.

3. The scores of 50 students in a test are as shown below. Calculate:

a. the variance

b. the standard deviation of the data

Scores	0 – 9	10 - 19	20 – 29	30 - 39	40 – 49	50 - 59
Number of students	5	12	6	18	5	4

4. The ages of employees in an organization are as shown below. Calculate:

a. the variance

b. the standard deviation of the data

Age	20 - 24	25 - 29	30 - 34	35 - 39	40 - 44
Number of empoyees	8	6	3	1	2

5. The scores obtained by 40 students in a test are as shown below. Calculate:

a. the variance

b. the standard deviation of the scores

Scores	5	6	7	8	9	10
Number of student	1	2	4	12	20	1

CHAPTER 39
QUARTILES AND PERCENTILES BY INTERPOLATION METHOD

When a distribution is divide into four equal parts, it is called a quartile. When it is divided into hundred equal parts, such a division is called percentile.

The first quartile is also called lower quartile, and it is denoted by Q_1.

The second quartile is also called median, and it is denoted by Q_2.

The third quartile is also called upper quartile, and it is denoted by Q_3.

The lower quartile for a grouped data is calculated as follows:

$$Q_1 = L_1 + C\left(\frac{\frac{\Sigma f}{4} - CF_{bQ_1}}{F_{Q_1}}\right)$$

Where, $\frac{\Sigma f}{4}$ determines the lower quartile class

L_1 = Lower class boundary of the lower quartile class

CF_{bQ_1} = Cumulative frequency before the lower quartile class

F_{Q_1} = Frequency of the lower quartile class

C = Class width

The median is calculated as follows:

$$Q_2 = L_2 + C\left(\frac{\frac{\Sigma f}{2} - CF_{bm}}{F_m}\right)$$

Where, $\frac{\Sigma f}{2}$ determines the median class

L_2 = Lower class boundary of the median class

CF_{bm} = Cumulative frequency before the median class

F_m = Frequency of the median class

C = Class width

The upper quartile is calculated as follows:

$$Q_3 = L_3 + C\left(\frac{\frac{3\Sigma f}{4} - CF_{bQ_3}}{F_{Q_3}}\right)$$

Where, $\frac{3\Sigma f}{4}$ determines the upper quartile class

L_3 = Lower class boundary of the upper quartile class

CF_{bQ_3} = Cumulative frequency before the upper quartile class

F_{Q_3} = Frequency of the upper quartile class

C = Class width

The interquartile range is given by:

Interquartile range = $Q_3 - Q_1$

The semi-interquartile range is also called quartile deviation, and it is given by:

Semi-interquartile range = $\dfrac{Q_3 - Q_1}{2}$

The percentile is calculated as follows:

$$P_N = L_N + C\left(\dfrac{\frac{N\Sigma f}{100} - CF_{bP_N}}{F_{P_N}}\right)$$

Where P_N is the N percentile, and $\dfrac{N\Sigma f}{100}$ determines the N percentile class

L_N = Lower class boundary of the N percentile class

CF_{bP_N} = Cumulative frequency before the N percentile class

F_{P_N} = Frequency of the N percentile class

C = Class width

Examples

1. The following is the record of marks of 40 students in an examination:

64 84 91 58 43 86 73 33 76 80 57 33 53 29 40 27 72 19 51
67 37 14 18 92 13 45 61 39 23 22 22 41 27 51 63 47 19 35
39 76

Using class interval 11 – 20, 21 – 30, ..., prepare a frequency table for the distribution.
Hence calculate the:

a. median

b. lower quartile

c. upper quartile

d. interquartile range

e. quartile deviation/semi-interquartile range

f. 40th percentile

g. 85th percentile

Solutions

The frequency table is as shown below.

Class interval	Frequency
11 – 20	5
21 – 30	6
31 – 40	7
41 – 50	4
51 – 60	5
61 – 70	4
71 – 80	5
81 – 90	2
91 – 100	2

a. In order to calculate the median, a table of the class boundaries and cumulative frequency has to be drawn as shown below.

Class interval	Class boundary	Frequency	Cumulative frequency	Class width
11 – 20	10.5 – 20.5	5	5	10
21 – 30	20.5 – 30.5	6	11	10
31 – 40	30.5 – 40.5	7	18	10
41 – 50	40.5 – 50.5	4	22	10
51 – 60	50.5 – 60.5	5	27	10
61 – 70	60.5 – 70.5	4	31	10
71 – 80	70.5 – 80.5	5	36	10
81 – 90	80.5 – 90.5	2	38	10
91 - 100	90.5 – 100.5	2	40	10

The median is calculated as follows:

$$Q_2 = L_2 + C\left(\frac{\frac{\Sigma f}{2} - CF_{bm}}{F_m}\right)$$

$\frac{\Sigma f}{2} = \frac{40}{2} = 20$. This shows that the median class is at the 20th position. This is the class, 41 – 50. This position is obtained by counting the frequency to get to the 20th position. 5 + 6 + 7 = 18. This shows that the 18th position is occupied by the class 31 – 40. After this class, the next frequency is 4. When 4 positions are added to 18 positions, it gives 22. This means that these 4 positions are the 19th, 20th, 21st and 22nd positions. These 4 positions are occupied by the class 41 – 50 as shown on the table. Hence the class in the 20th position is 41 – 50. You can also look at the cumulative frequency and see where the 20th position class falls.

L_2 = Lower class boundary of the median class = 40.5

CF_{bm} = Cumulative frequency before the median class = 18

F_m = Frequency of the median class = 4

C = Class width = 10. The class limit is the difference between an upper class boundary and a lower class boundary. For example, 20.5 − 10.5 = 10.

$$\therefore \quad Q_2 = L_2 + C\left(\frac{\frac{\Sigma f}{2} - CF_{bm}}{F_m}\right)$$

$$= 40.5 + 10\left(\frac{\frac{40}{2} - 18}{4}\right)$$

$$= 40.5 + 10\left(\frac{20 - 18}{4}\right)$$

$$= 40.5 + 10\left(\frac{2}{4}\right)$$

$$= 40.5 + \left(\frac{10 \times 2}{4}\right)$$

$$= 40.5 + 5$$

$$Q_2 = 45.5$$

b. The lower quartile is calculated as follows:

$$Q_1 = L_1 + C\left(\frac{\frac{\Sigma f}{4} - CF_{bQ1}}{F_{Q1}}\right)$$

$\frac{\Sigma f}{4} = \frac{40}{4}$ = 10. Hence the lower quartile class is at the 10th position. This class is, 21 − 30.

L_1 = Lower class boundary of the lower quartile class = 20.5

CF_{bQ1} = Cumulative frequency before the lower quartile class = 5

F_{Q1} = Frequency of the lower quartile class = 6

C = Class width = 10

$$\therefore \quad Q_1 = L_1 + C\left(\frac{\frac{\Sigma f}{4} - CF_{bQ1}}{F_{Q1}}\right)$$

$$= 20.5 + 10\left(\frac{\frac{40}{4} - 5}{6}\right)$$

$$= 20.5 + 10\left(\frac{10 - 5}{6}\right)$$

$$= 20.5 + 10\left(\frac{5}{6}\right)$$

$$= 20.5 + \left(\frac{10 \times 5}{6}\right)$$

$$= 20.5 + 8.3$$

$$Q_1 = 28.8$$

c. The upper quartile is calculated as follows:

$$Q_3 = L_3 + C\left(\frac{\frac{3\Sigma f}{4} - CF_{bQ3}}{F_{Q3}}\right)$$

$\frac{3\Sigma f}{4} = \frac{3 \times 40}{4}$ = 30. Hence the upper quartile class is at the 30th position. This class is, 61 − 70.

L_3 = Lower class boundary of the upper quartile class = 60.5

CF_{bQ_3} = Cumulative frequency before the upper quartile class = 27

F_{Q_3} = Frequency of the upper quartile class = 4

C = Class width = 10

$$\therefore \quad Q_3 = L_3 + C\left(\frac{\frac{3\sum f}{4} - CF_{bQ_3}}{F_{Q_3}}\right)$$

$$= 60.5 + 10\left(\frac{\frac{3 \times 40}{4} - 27}{4}\right)$$

$$= 60.5 + 10\left(\frac{30 - 27}{4}\right)$$

$$= 60.5 + 10\left(\frac{3}{4}\right)$$

$$= 60.5 + \left(\frac{10 \times 3}{4}\right)$$

$$= 60.5 + 7.5$$

$$Q_3 = 68$$

d. Interquartile range = $Q_3 - Q_1$

$$= 68 - 28.8$$

$$= 39.2$$

e. Quartile deviation/semi-interquartile range is given by:

$$Q = \frac{Q_3 - Q_1}{2}$$

$$= \frac{68 - 28.8}{2}$$

$$= \frac{39.2}{2}$$

$$Q = 19.6$$

f. The 40th percentile is calculated as follows:

$$P_N = L_N + C\left(\frac{\frac{N\sum f}{100} - CF_{bP_N}}{F_{P_N}}\right)$$

$P_N = P_{40}$

$\frac{N\sum f}{100} = \frac{40 \times 40}{100}$ = 16. Hence the 40th percentile class is at the 16th position. This class is: 31 - 40

$L_N = L_{40}$ = Lower class boundary of the 40th percentile class = 30.5

$CF_{bP_N} = CF_{bP_{40}}$ = Cumulative frequency before the 40th percentile class = 11

$F_{P_N} = F_{P_{40}}$ = Frequency of the 40th percentile class = 7

C = Class width = 10

Hence, $P_{40} = L_{40} + C\left(\dfrac{\frac{40\sum f}{100} - CF_{bP_{40}}}{F_{P_{40}}}\right)$

$= 30.5 + 10\left(\dfrac{\frac{40 \times 40}{100} - 11}{7}\right)$

$= 30.5 + 10\left(\dfrac{16 - 11}{7}\right)$

$= 30.5 + 10\left(\dfrac{5}{7}\right)$

$= 30.5 + \left(\dfrac{10 \times 5}{7}\right)$

$= 30.5 + 7.1$

$P_{40} = 37.6$

g. The 85th percentile is calculated as follows:

$P_N = L_N + C\left(\dfrac{\frac{N\sum f}{100} - CF_{bP_N}}{F_{P_N}}\right)$

$P_N = P_{85}$

$\dfrac{N\sum f}{100} = \dfrac{85 \times 40}{100}$ = 34. Hence the 85th percentile class is at the 34th position. This class is: 71 - 80

$L_N = L_{85}$ = Lower class boundary of the 85th percentile class = 70.5

$CF_{bP_N} = CF_{bP_{85}}$ = Cumulative frequency before the 85th percentile class = 31

$F_{P_N} = F_{P_{85}}$ = Frequency of the 85th percentile class = 5

C = Class width = 10

Hence, $P_{85} = L_{85} + C\left(\dfrac{\frac{85 \times 40}{100} - CF_{bP_{85}}}{F_{P_{85}}}\right)$

$= 70.5 + 10\left(\dfrac{\frac{85 \times 40}{100} - 31}{5}\right)$

$= 70.5 + 10\left(\dfrac{34 - 31}{5}\right)$

$= 70.5 + 10\left(\dfrac{3}{5}\right)$

$= 70.5 + \left(\dfrac{10 \times 3}{5}\right)$

$= 70.5 + 6$

$P_{85} = 76.5$

2. The table below shows the distribution of marks scored by students in an examination.

Class interval	Frequency
60 – 64	2
65 – 69	4
70 – 74	7
75 – 79	13
80 – 84	10
85 – 89	8
90 – 94	5
95 – 99	1

From the data, calculate:

a. median

b. lower quartile

c. upper quartile

d. interquartile range

e. semi-interquartile range

f. 70th percentile

g. the pass mark if 25% of the students passed

h. the pass mark if it was later agreed that only 40% of the students should fail.

Solution

a. In order to calculate the median, a table of the class boundaries and cumulative frequency has to be drawn as shown below.

Class interval	Class boundary	Frequency	Cumulative frequency	Class width
60 – 64	59.5 – 64.5	2	2	5
65 – 69	64.5 – 69.5	4	6	5
70 – 74	69.5 – 74.5	7	13	5
75 – 79	74.5 – 79.5	13	26	5
80 – 84	79.5 – 84.5	10	36	5
85 – 89	84.5 – 89.5	8	44	5
90 – 94	89.5 – 94.5	5	49	5
95 – 99	94.5 – 99.5	1	50	5

The median is calculated as follows:

$$Q_2 = L_2 + C\left(\frac{\frac{\Sigma f}{2} - CF_{bm}}{F_m}\right)$$

$\frac{\Sigma f}{2} = \frac{50}{2} = 25$. This shows that the median class is at the 25th position. This is the class, 75 – 79.

This is obtained by looking at the cumulative frequency to see where the 25th position class falls.

L_2 = Lower class boundary of the median class = 74.5

CF_{bm} = Cumulative frequency before the median class = 13

F_m = Frequency of the median class. This is also 13

C = Class width = 5

$$\therefore \quad Q_2 = L_2 + C\left(\frac{\frac{\Sigma f}{2} - CF_{bm}}{F_m}\right)$$

$$= 74.5 + 5\left(\frac{\frac{50}{2} - 13}{13}\right)$$

$$= 74.5 + 5\left(\frac{25 - 13}{13}\right)$$

$$= 74.5 + 5\left(\frac{12}{13}\right)$$

$$= 74.5 + \left(\frac{5 \times 12}{13}\right)$$

$$= 74.5 + 4.6$$

$$Q_2 = 79.1$$

b. The lower quartile is calculated as follows:

$$Q_1 = L_1 + C\left(\frac{\frac{\Sigma f}{4} - CF_{bQ1}}{F_{Q1}}\right)$$

$\frac{\Sigma f}{4} = \frac{50}{4}$ = 12.5. Hence the lower quartile class is at the 12th or 13th position. This class is, 70 – 74.

L_1 = Lower class boundary of the lower quartile class = 69.5

CF_{bQ_1} = Cumulative frequency before the lower quartile class = 6

F_{Q_1} = Frequency of the lower quartile class = 7

C = Class width = 5

$$\therefore \quad Q_1 = L_1 + C\left(\frac{\frac{\Sigma f}{4} - CF_{bQ1}}{F_{Q1}}\right)$$

$$= 69.5 + 5\left(\frac{\frac{50}{4} - 6}{7}\right)$$

$$= 69.5 + 5\left(\frac{12.5 - 6}{7}\right)$$

$$= 69.5 + 5\left(\frac{6.5}{7}\right)$$

$$= 69.5 + \left(\frac{5 \times 6.5}{7}\right)$$

$$= 69.5 + 4.6$$

$$Q_1 = 74.1$$

c. The upper quartile is calculated as follows:

$$Q_3 = L_3 + C\left(\frac{\frac{3\sum f}{4} - CF_{bQ_3}}{F_{Q_3}}\right)$$

$\frac{3\sum f}{4} = \frac{3 \times 50}{4} = 37.5$. Hence the upper quartile class is at the 37th or 38th position. This class is, 85 – 89.

L_3 = Lower class boundary of the upper quartile class = 84.5

CF_{bQ_3} = Cumulative frequency before the upper quartile class = 36

F_{Q_3} = Frequency of the upper quartile class = 8

C = Class width = 5

$$\therefore \quad Q_3 = L_3 + C\left(\frac{\frac{3\sum f}{4} - CF_{bQ_3}}{F_{Q_3}}\right)$$

$$= 84.5 + 5\left(\frac{\frac{3 \times 50}{4} - 36}{8}\right)$$

$$= 84.5 + 5\left(\frac{37.5 - 36}{8}\right)$$

$$= 84.5 + 5\left(\frac{1.5}{8}\right)$$

$$= 84.5 + 0.9$$

$$Q_3 = 85.4$$

d. Interquartile range = $Q_3 - Q_1$

$$= 85.4 - 74.1$$

$$= 11.3$$

e. Semi-interquartile range, $Q = \frac{Q_3 - Q_1}{2}$

$$= \frac{85.4 - 74.1}{2}$$

$$= \frac{11.3}{2}$$

$$Q = 5.65$$

f. The 70th percentile is calculated as follows:

$$P_N = L_N + C\left(\frac{\frac{N\sum f}{100} - CF_{bP_N}}{F_{P_N}}\right)$$

$P_N = P_{70}$

$\frac{N\sum f}{100} = \frac{70 \times 50}{100} = 35$. Hence the 70th percentile class is at the 35th position. This class is: 80 - 84

$L_N = L_{70}$ = Lower class boundary of the 70th percentile class = 79.5

$CF_{bP_N} = CF_{bP_{70}}$ = Cumulative frequency before the 70th percentile class = 26

$F_{P_N} = F_{P_{70}}$ = Frequency of the 70th percentile class = 10

C = Class width = 5

Hence, $P_{70} = L_{70} + C(\dfrac{\frac{70\Sigma f}{100} - CF_{bP_{70}}}{F_{P_{70}}})$

$= 79.5 + 5(\dfrac{\frac{70 \times 50}{100} - 26}{10})$

$= 79.5 + 5(\dfrac{35 - 26}{10})$

$= 79.5 + 5(\dfrac{9}{10})$

$= 79.5 + 4.5$

$P_{70} = 84$

g. If 25% of the students passed, then the first 75% (i.e. 100 − 25) of the students failed. This means that the pass mark is at the 75th percentile.

Note that the pass mark is always at the failure percentile.

Hence the 75th percentile is calculated as follows:

$P_N = L_N + C(\dfrac{\frac{N\Sigma f}{100} - CF_{bP_N}}{F_{P_N}})$

$P_N = P_{75}$

$\dfrac{N\Sigma f}{100} = \dfrac{75 \times 50}{100} = 37.5$. Hence the 75th percentile class is at the 37.5th position. This class is: 85 - 89

$L_N = L_{75}$ = Lower class boundary of the 75th percentile class = 84.5

$CF_{bP_N} = CF_{bP_{75}}$ = Cumulative frequency before the 75th percentile class = 36

$F_{P_N} = F_{P_{75}}$ = Frequency of the 75th percentile class = 8

C = Class width = 5

Hence, $P_{75} = L_{75} + C(\dfrac{\frac{75\Sigma f}{100} - CF_{bP_{75}}}{F_{P_{75}}})$

$= 84.5 + 5(\dfrac{\frac{75 \times 50}{100} - 36}{8})$

$= 84.5 + 5(\dfrac{37.5 - 36}{8})$

$= 84.5 + 5(\dfrac{1.5}{8})$

$= 84.5 + 0.9$

$P_{75} = 85.4$

Hence the pass mark is 85.4

h. If 40% of the students should fail, then the pass mark is at the 40th percentile.

Hence the 40th percentile is calculated as follows:

$$P_N = L_N + C\left(\frac{\frac{N\Sigma f}{100} - CF_{bP_N}}{F_{P_N}}\right)$$

$P_N = P_{40}$

$\frac{N\Sigma f}{100} = \frac{40 \times 50}{100} = 20$. Hence the 40th percentile class is at the 20th position. This class is: 75 - 79

$L_N = L_{40} =$ Lower class boundary of the 40th percentile class = 74.5

$CF_{bP_N} = CF_{bP_{40}} =$ Cumulative frequency before the 40th percentile class = 13

$F_{P_N} = F_{P_{40}} =$ Frequency of the 40th percentile class = 13

C = Class width = 5

Hence, $P_{40} = L_{40} + C\left(\frac{\frac{40\Sigma f}{100} - CF_{bP_{40}}}{F_{P_{40}}}\right)$

$$= 74.5 + 5\left(\frac{\frac{40 \times 50}{100} - 13}{13}\right)$$

$$= 74.5 + 5\left(\frac{20 - 13}{13}\right)$$

$$= 74.5 + 5\left(\frac{7}{13}\right)$$

$$= 74.5 + 2.7$$

$P_{40} = 77.2$

Hence the pass mark is 77.2

3. The table below shows the masses of some items sold in a supermarket.

Mass	1.5 − 1.9	2.0 − 2.4	2.5 − 2.9	3.0 − 3.4	3.5 − 3.9	4.0 − 4.5
Number of Items	5	12	6	18	5	4

From the table given above, estimate:

a. median

b. lower quartile

c. upper quartile

d. the 55th percentile

<u>Solution</u>

a. In order to calculate the median, a table of the class boundaries and cumulative frequency has to be drawn as shown below.

Class interval	Class boundary	Frequency	Cumulative frequency	Class width
1.5 – 1.9	1.45 – 1.95	5	5	0.5
2.0 – 2.4	1.95 – 2.45	12	17	0.5
2.5 – 2.9	2.45 – 2.95	6	23	0.5
3.0 – 3.4	2.95 – 3.45	18	41	0.5
3.5 – 3.9	3.45 – 3.95	5	46	0.5
4.0 – 4.4	3.95 – 4.45	4	50	0.5

Note that in computing the class boundaries, a difference between an upper class limit and a lower class limit, such as, 2.0 – 1.9 = 0.1, is first determined and then divided by 2 to give, 0.1/2 = 0.05. It is this 0.05 that is added and subtracted from the class limit values to obtain the class boundary values. This is the method applied in obtaining the class boundaries of any given grouped data.

The median is calculated as follows:

$$Q_2 = L_2 + C\left(\frac{\frac{\Sigma f}{2} - CF_{bm}}{F_m}\right)$$

$\frac{\Sigma f}{2} = \frac{50}{2} = 25$. Hence the median class is at the 25th position. This is the class, 3.0 – 3.4. This is obtained by looking at the cumulative frequency to see where the 25th position class falls.

L_2 = Lower class boundary of the median class = 2.95

CF_{bm} = Cumulative frequency before the median class = 23

F_m = Frequency of the median class = 18

C = Class width = 0.5, i.e. 1.95 – 1.45 = 0.5.

$$\therefore \quad Q_2 = L_2 + C\left(\frac{\frac{\Sigma f}{2} - CF_{bm}}{F_m}\right)$$

$$= 2.95 + 0.5\left(\frac{\frac{50}{2} - 23}{18}\right)$$

$$= 2.95 + 0.5\left(\frac{25 - 23}{18}\right)$$

$$= 2.95 + 0.5\left(\frac{2}{18}\right)$$

$$= 2.95 + 0.06$$

$$Q_2 = 3.01$$

b. The lower quartile is calculated as follows:

$$Q_1 = L_1 + C\left(\frac{\frac{\Sigma f}{4} - CF_{bQ1}}{F_{Q1}}\right)$$

$\frac{\Sigma f}{4} = \frac{50}{4}$ = 12.5. This shows that the lower quartile class is at the 12th or 13th position. This class is, 2.0 – 2.4.

L_1 = Lower class boundary of the lower quartile class = 1.95

CF_{bQ1} = Cumulative frequency before the lower quartile class = 5

F_{Q1} = Frequency of the lower quartile class = 12

C = Class width = 0.5

$$\therefore \quad Q_1 = L_1 + C\left(\frac{\frac{\Sigma f}{4} - CF_{bQ1}}{F_{Q1}}\right)$$

$$= 1.95 + 0.5\left(\frac{\frac{50}{4} - 5}{12}\right)$$

$$= 1.95 + 0.5\left(\frac{12.5 - 5}{12}\right)$$

$$= 1.95 + 0.5\left(\frac{7.5}{12}\right)$$

$$= 1.95 + 0.31$$

$$Q_1 = 2.26$$

c. The upper quartile is calculated as follows:

$$Q_3 = L_3 + C\left(\frac{\frac{3\Sigma f}{4} - CF_{bQ3}}{F_{Q3}}\right)$$

$\frac{3\Sigma f}{4} = \frac{3 \times 50}{4}$ = 37.5. This shows that the upper quartile class is at the 37th or 38th position. This class is, 3.0 – 3.4.

L_3 = Lower class boundary of the upper quartile class = 2.95

CF_{bQ3} = Cumulative frequency before the upper quartile class = 23

F_{Q3} = Frequency of the upper quartile class = 18

C = Class width = 0.5

$$\therefore \quad Q_3 = L_3 + C\left(\frac{\frac{3\Sigma f}{4} - CF_{bQ3}}{F_{Q3}}\right)$$

$$= 2.95 + 0.5\left(\frac{\frac{3 \times 50}{4} - 23}{18}\right)$$

$$= 2.95 + 0.5\left(\frac{37.5 - 23}{18}\right)$$

$$= 2.95 + 0.5\left(\frac{14.5}{18}\right)$$

$$= 2.95 + 0.4$$

$Q_3 = 3.35$

d. The 55th percentile is calculated as follows:

$$P_N = L_N + C\left(\frac{\frac{N\Sigma f}{100} - CF_{bP_N}}{F_{P_N}}\right)$$

$P_N = P_{55}$

$\frac{N\Sigma f}{100} = \frac{55 \times 50}{100} = 27.5$. Hence the 55th percentile class is at the 27th and 28th position. This class is: $3.0 - 3.4$

$L_N = L_{55}$ = Lower class boundary of the 55th percentile class = 2.95

$CF_{bP_N} = CF_{bP_{55}}$ = Cumulative frequency before the 55th percentile class = 23

$F_{P_N} = F_{P_{55}}$ = Frequency of the 55th percentile class = 18

C = Class width = 0.5

Hence, $P_{55} = L_{55} + C\left(\frac{\frac{55\Sigma f}{100} - CF_{bP_{55}}}{F_{P_{55}}}\right)$

$$= 2.95 + 0.5\left(\frac{\frac{55 \times 50}{100} - 23}{18}\right)$$

$$= 2.95 + 0.5\left(\frac{27.5 - 23}{18}\right)$$

$$= 2.95 + 0.5\left(\frac{4.5}{18}\right)$$

$$= 2.95 + 0.13$$

$P_{55} = 3.08$

Exercise 39

1. The following is the record of marks of 40 students in an examination:

34 74 92 58 46 76 73 23 66 70 57 43 53 39
50 37 82 29 54 77 67 19 18 96 15 55 41 29
33 52 22 81 77 81 58 27 20 55 49 96

Using class interval $11 - 20$, $21 - 30$, ..., prepare a frequency table for the distribution.
Hence calculate the:

a. median

b. lower quartile

c. upper quartile

d. interquartile range

e. quartile deviation/semi-interquartile range

f. 30th percentile

g. 68th percentile

2. The table below shows the distribution of marks scored by students in an examination.

Class interval	Frequency
10 – 14	1
15 – 19	3
20 – 24	8
25 – 29	11
30 – 34	7
35 – 39	9
40 – 44	10
45 – 49	5

From the data, calculate:

a. median

b. lower quartile

c. upper quartile

d. interquartile range

e. semi-interquartile range

f. 80th percentile

g. the pass mark if 35% of the students passed

h. the pass mark if 15% the students should fail.

3. The table below shows the height of some flowers sold in a farm.

Mass	0.5 – 0.9	1.0 – 1.4	1.5 – 1.9	2.0 – 2.4	2.5 – 2.9	3.0 – 3.4
No of Items	4	15	12	9	7	3

From the table given above, estimate:

a. median

b. lower quartile

c. upper quartile

d. the 45th percentile

e. pass mark if 90% of the students passed

4. The table below shows the distribution of marks scored by students in an test.

Class interval	Frequency
0 − 4	1
5 − 9	4
10 − 14	7
15 − 19	5
20 − 24	9
25 − 29	1
30 − 34	2
35 − 39	1

From the data, calculate:

a. median

b. lower quartile

c. upper quartile

d. interquartile range

e. semi-interquartile range

f. 60th percentile

g. the pass mark if 10% of the students passed

5. The table below shows the weight in gram of some seeds found in some cocoa pods.

Mass	0 − 0.4	0.5 − 0.9	1.0 − 1.4	1.5 − 1.9	2.0 − 2.4	2.5 − 2.9
No of Items	9	21	16	22	28	4

From the table given above, estimate:

a. median

b. lower quartile

c. upper quartile

d. the 25th percentile

e. pass mark if 68% of the students passed

CHAPTER 40
THE BASIC THEORY OF PROBABILITY

Probability is the likelihood of an event happening. Mathematically probability is given by:

$$\text{Probability} = \frac{\text{number of required outcome}}{\text{number of total or possible outcome}}$$

If the probability of an event happening is x, then the probability that it will not happen will be given by: $1 - x$

Probability must lie between the values of 0 and 1. If an event cannot happen, then its probably is 0. If an event is certain to happen, then its probability is 1.

Mutually Exclusive Events

When there is no member/element common between two or more similar events, then we say they are mutually exclusive events. For example the event of odd numbers or even numbers are mutually exclusive. They are disjoint sets.

Addition Law of Probability

If two events are mutually exclusive, then the probability of one or the other happening is the sum of their individual probabilities.

Independent Events

When a die is thrown, and a coin is tossed, these two events have no effect on each other. Such events are called independent events

Product law of probability

If two events are independent, then the probability of both events happening is the product is the product (multiplication) of their individual probabilities.

CHAPTER 41
PROBABILITY ON SIMPLE EVENTS

Examples

1. The table below give the number of students in each age group in a class.

Age (Years)	12	13	14	15	16	17
number of students	6	3	10	4	2	5

If a student is chosen at random, find the probability that the student is:

(a) 13 years old

(b) 15 years old or less

(c) at least 16 years old

(d) most 13 years old

(e) not 17 years old

<u>Solution</u>

(a) Pr. (13 years old) $= \dfrac{\text{Number of students who are 13 years old}}{\text{Total number of students}}$

$\qquad = \dfrac{3}{30}$

$\qquad = \dfrac{1}{10}$ (when $\dfrac{3}{30}$ is express in its lowest term, it gives $\dfrac{1}{10}$)

(b) Pr. (15 years or less) $= \dfrac{\text{Students who are 15 years and below}}{\text{Total number of students}}$

$\qquad = \dfrac{4+10+3+6}{30}$

$\qquad = \dfrac{23}{30}$

(c) Pr. (At least 16 years old) $= \dfrac{\text{Students who are 16 years and above}}{\text{Total number of students}}$

$\qquad = \dfrac{2+5}{30}$

$\qquad = \dfrac{7}{30}$

(d) Pr. (At most 13 years) $= \dfrac{\text{Students who are 13 years and below}}{\text{Total number of students}}$

$\qquad = \dfrac{3+6}{30}$

$\qquad = \dfrac{9}{30}$

$= \dfrac{3}{10}$ (When expressed in its lowest term)

(e) Pr. (17 years old) $= \dfrac{\text{number of students who are 17 years old}}{\text{total number of students}}$

$= \dfrac{5}{30}$

$= \dfrac{1}{6}$

Therefore, Pr. (Not 17 years old) = 1 - Pr. (17 years old)

$= 1 - \dfrac{1}{6}$

$= \dfrac{5}{6}$

2. The probability that a seed will germinate is $\dfrac{2}{5}$. What is the probability that it will not germinate?

<u>Solution</u>

Pr. (It will germinate) $= \dfrac{2}{5}$

Pr. (It will not germinate) $= 1 - \dfrac{3}{5}$

$= \dfrac{2}{5}$

3. A letter is chosen at random from the alphabet. Find the probability that it is one of the letters of the word: PROBABILITY.

<u>Solution</u>

In this case a letter should not be counted more than once. Avoiding repetition, the word can now be written as:

PROBALITY (i.e. 9 letters). Note that there are 26 letters of the alphabet.

Therefore, Pr. (one letter from PROBABILITY) $= \dfrac{9}{26}$

4. The probability that a boy gains admission into a higher institution is $\dfrac{3}{7}$. What is the probability that he does not gain admission into the institution?

<u>Solution</u>

Pr. (He gains admission) $= \dfrac{3}{7}$

Pr. (He does not gain admissions) = $1 - \dfrac{3}{7}$

$$= \dfrac{4}{7}$$

5. Out of every 100 cars, 4 develop mechanical fault within 6 months of purchase. What is the probability of buying a car which will not develop a mechanical fault within 6 months of purchase?

Solution

Total number of cars is 100. Number of cars with fault within 6 months is 4. Number of cars without fault within an months of purchase is going is 96, (i.e. 100 - 4 = 96).

Therefore, Pr. (Buying a car that will not develop fault) = $\dfrac{\text{Number of cars without fault}}{\text{Total number of cars}}$

$$= \dfrac{96}{100}$$

$$= \dfrac{24}{25} \quad \text{(In its lowest term after equal division by 4)}$$

6. In Mr. Smith's extended family, the number of males is 16, while the number of females is 14. Find the probability that Mr. Smith has:

(a) a male child

(b) a female child

Solution

(a) Total number of family members = 16 + 14 = 30

Therefore, Pr. (a male child) = $\dfrac{\text{Family members who are males}}{\text{Total number of family members}}$

$$= \dfrac{16}{30}$$

$$= \dfrac{8}{15}$$

(b) Pr. (a female child) = $\dfrac{\text{Family members who are females}}{\text{Total number of family members}}$

$$= \dfrac{14}{30}$$

$$= \dfrac{7}{15}$$

7. A survey shows that 36% of all women take size 8 shoes. What is the probability that Khan's grandmother takes size 8 shoes?

Solution

Pr. (Khan's grandmother takes size 8 shoes) = $\dfrac{36}{100}$ (Note that the total percentage is always 100%)

$\qquad = \dfrac{9}{25}$ (In its lowest term)

8. In a secondary school, 46 out of every 50 students are at least 130cm tall. What is the probability that a student chosen at random from the school is less than 130cm tall?

Solution

Total number of students for the sample = 50

Number of students who are at least 130cm tall = 46

Number of students who are less than 130cm tall = 50 - 46 = 4

Therefore, Pr. (a student less than 130cm tall) = $\dfrac{\text{Number of students less than 130cm tall}}{\text{Total number of students in the sample}}$

$\qquad = \dfrac{4}{50}$

$\qquad = \dfrac{2}{25}$

9. A number is chosen at random between 1 and 16, both inclusive. What is the probability that it is:

(a) even

(b) prime

(c) odd or prime

(d) divisible by 4

(e) a perfect square or a perfect cube

Solution

(a) Total numbers in all from 1 to 16 = 16

The even numbers are 2, 4, 6, 8, 10, 12, 14, 16

Therefore the number of even numbers is 8

Hence Pr. (even number selected) = $\dfrac{\text{Number of even numbers}}{\text{Total numbers in all}}$

$\qquad = \dfrac{8}{16}$

$\qquad = \dfrac{1}{2}$

(b) The prime numbers are 2, 3, 5, 7, 11, 13

Therefore the number of prime numbers is 6

Hence Pr. (prime number selected) = $\dfrac{\text{Number of prime numbers}}{\text{Total numbers in all}}$

$= \dfrac{6}{16}$

$= \dfrac{3}{8}$

(c) The odd numbers are 1, 3, 5, 7, 9, 11, 13, 15

The prime numbers are 1, 3, 5, 7, 11, 13

Since OR in probability means addition, then we add all the odd and prime numbers together, but we must not count any number twice. This gives 1, 3, 5, 7, 9, 11, 13, 15, which is a total of 8 numbers.

Hence Pr. (odd or prime number selected) = $\dfrac{\text{Number of odd and even numbers}}{\text{Total numbers in all}}$

$= \dfrac{8}{16}$

$= \dfrac{1}{8}$

(d) The numbers divisible by 4 are 4, 8, 12, 16

This gives a total of 4 numbers

Hence Pr. (a number divisible by 4) = $\dfrac{\text{The four numbers divisible by 4}}{\text{Total numbe rs in all}}$

$= \dfrac{4}{16}$

$= \dfrac{1}{4}$

(e) The perfect square numbers are 1, 4, 9, 16

The perfect cube numbers are 1, 8

Since OR in probability means addition, then we add all the set of values above without counting any number twice. This gives 1, 4, 8, 9, 15, which is a total of 5 numbers.

Hence Pr. (perfect square or perfect cube selected) = $\dfrac{15}{16}$

10. A letter is chosen at random from the alphabet. Find the probability that it is:

(a) T

(b) E or P

(c) not B or G

(d) either D, J, N, U, W or Y

(e) one of the letters of the word REJECTED

Solution

(a) There are 26 letters of the alphabet, out of which there is 1 T.

Therefore, Pr. (T) = $\dfrac{\text{Number of Ts}}{\text{Total numbers of alphabets}}$

$= \dfrac{1}{26}$

(b) Pr. (E or P) = $\dfrac{\text{Number of Es and Ps}}{\text{Total numbers of alphabets}}$

$= \dfrac{6}{26}$

$= \dfrac{1}{13}$

(c) Pr. (B or G) = $\dfrac{2}{26} = \dfrac{1}{13}$

Therefore, Pr. (not B or G) = 1 - Pr. (B or G)

$= 1 - \left(\dfrac{6}{13}\right)$

$= \dfrac{12}{13}$

(d) The letters D, J, N, U, W and Y makes a total of 6 letters.

Pr. (D, J, N, U, W or Y) = $\dfrac{6}{26}$

$= \dfrac{3}{13}$

(e) Writing the letters of the word REJECTED without repeating a letter gives REJCTD. This gives a total of 6 letters

Therefore Pr. (one of the letters of REJECTED) = $\dfrac{6}{26}$

$= \dfrac{3}{13}$

11. A letter is selected at random from the word PROBABILITY. What is the probability of selecting the letter B.

Solution

In this case the total letters of the word PROBABILITY gives 11. The repeated letters should be counted more than once since this is not a case of letter from the alphabet. In the 26 alphabet each letter appears once, that is why they are counted once. But in PROBABILITY (or other words that might be given) some letters appear more than once, hence they should be counted as many times as they appear.

In PROBABILITY, B appears 2 times.

Therefore, Pr. (selecting B) = $\dfrac{2}{11}$

Exercice 41

1. The table below give the number of students in each mark group in a class.

Mark	5	6	7	8	9	10
Number of students	3	6	2	4	1	4

If a student is chosen at random, find the probability that the student scored:

(a) 7 marks

(b) 6 marks or less

(c) at least 9 marks

(d) at most 8 marks

(e) 5 or 8 maks

2. The probability that a seed will germinate is $\frac{3}{4}$. What is the probability that it will not germinate?

3. A letter is chosen at random from the alphabet. Find the probability that it is one of the letters of the word: MATHEMATICS.

4. The probability that a man wins an election is $\frac{3}{5}$. What is the probability that he does not win.

5. Out of every 10 bulbs, 2 do not last long. What is the probability that a bulb will last long when lit?

6. In family, the number of males is 3, while the number of females is 2. Find the probability that another child born into the family is:

(a) a male child

(b) a female child

7. A survey shows that 44% of all women take size 7 shoes. What is the probability that a mother of two takes size 7 shoes?

8. In a secondary school, 30 out of every 100 students are at least 160cm tall. What is the probability that a student chosen at random from the school is less than 160cm tall?

9. A number is chosen at random between 1 and 20, both inclusive. What is the probability that it is:

(a) prime

(b) odd

(c) even or prime

(d) divisible by 3

(e) a number less than 10 or a perfect cube

10. A letter is chosen at random from the alphabet. Find the probability that it is:

(a) F

(b) M or Q or Y

(c) in the word COME

(d) either in the word BUT or in REMOVE

(e) one of the letters of the word SURPRISED

11. A letter is selected at random from the word RESPIRATION. What is the probability of selecting the letter I.

CHAPTER 42
PROBABILITY ON PACK OF PLAYING CARDS

A pack of playing cards contains 52 cards of 4 types. There are 13 clubs, 13 diamonds, 13 hearts and 13 spades. Each of the set of 13 cards contains Ace (A), 2, 3, 4, 5, 6, 7, 8, 9, 10, Jack (J), Queen (Q), and King (K). This means that out of the 52 cards, each card is four in number, i.e. Aces are 4 in number, 1s are 4 in number, 2s are 4 in number, 3s are 4 in number, 4s are 4 in number, 5s are 4 in number, 6s are 4 in number, 7s are 4 in number, 8s are 4 in number, 9s are 4 in number, 10s are 4 in number, Jacks are 4 in number, Queens are 4 in number, and Kings are 4 in number. Clubs and spades are black, diamonds and hearts are red. This means that there are 26 black cards and 26 red cards. This also means that out of the 4 Aces cards, 2 are black and 2 are red. Out of the four cards that are 1, two are black and two are red, out of the four cards that are 2, two are black and two are red, and so on.

Examples

1. A card is picked at random from a pack of playing cards. Find the probability of picking a spade. <u>Solution</u>

There are 13 spades in a pack of playing cards.

Therefore, Pr. (picking a spade) $= \dfrac{\text{Number of Spades}}{\text{Total numbers of cards}}$

$$= \frac{13}{52}$$

$$= \frac{1}{4} \quad \text{(In its lowest term)}$$

2. A card is picked at random from a pack of playing cards. Find the probability of picking a red card.

<u>Solution</u>

There are 26 red cards in a pack of playing cards.

Therefore, Pr. (picking a red card) $= \dfrac{\text{Number of red cards}}{\text{Total numbers of cards}}$

$$= \frac{26}{52}$$

$$= \frac{1}{2} \quad \text{(In its lowest term)}$$

3. A card is picked at random from a pack of playing cards. Find the probability of picking a red 5.

Solution

There are two red 5 cards in a pack of playing cards.

Therefore, Pr. (picking a red 5) = $\dfrac{\text{Number of red 5}}{\text{Total numbers of cards}}$

$= \dfrac{2}{52}$

$= \dfrac{1}{26}$ (In its lowest term)

4. A card is picked at random from a pack of playing cards. Find the probability of picking a 3.

Solution

There are 4 cards that are 3 in a pack of playing cards.

Therefore, Pr. (picking a 3) = $\dfrac{\text{Number of cards that are 3}}{\text{Total numbers of cards}}$

$= \dfrac{4}{52}$

$= \dfrac{1}{13}$ (In its lowest term)

5. A card is picked at random from a pack of playing cards. Find the probability of picking a black Ace.

Solution

There are 2 cards that are black ace in a pack of playing cards.

Therefore, Pr. (picking a black ace) = $\dfrac{\text{Number of cards that are black ace}}{\text{Total numbers of cards}}$

$= \dfrac{2}{52}$

$= \dfrac{1}{26}$ (In its lowest term)

6. A card is picked at random from a pack of playing cards. Find the probability of picking a card that is not a Jack.

Solution

(a) There are 4 cards that are Jacks.

Therefore, Pr. (picking a Jack) = $\dfrac{\text{Number of jacks}}{\text{Total numbers of cards}}$

$$= \frac{4}{52}$$

$$= \frac{1}{13}$$

Hence, Pr. (picking a card that in not a Jack) = 1 - Pr. (picking a Jack)

$$= 1 - \frac{1}{13}$$

$$= \frac{12}{13}$$

7. A card is picked at random from a pack of playing cards. Find the probability of picking

(a) a black or red card

(b) a 2 or a 5

(c) either a heart or the king of spades

(d) a club or a red Queen

(e) a diamond or a 9

(f) a 6 or a black card

Solution

(a) There are 26 black cards and 26 red card

Since or in probability means plus, then we have to add the numbers. This gives a total of: 26 + 26 = 52

Therefore, Pr. (picking a black or red card) = $\dfrac{\text{Number of black and red cards}}{\text{Total numbers of cards}}$

$$= \frac{52}{52}$$

$$= 1$$

(b) There are 4 cards that are 2, and 4 cards that are 5. This gives a total of 8 cards.

Therefore, Pr. (picking a 2 or a 5) = $\dfrac{8}{52}$

$$= \frac{2}{13}$$

(c) There are 13 cards that are Hearts, and 1 king that is a spade. This gives a total of 14 cards.

Therefore, Pr. (picking either a heart or the king of spades) = $\dfrac{14}{52}$

$$= \frac{7}{26}$$

(d) There are 13 cards that are club, and 2 cards that are red Queen, (i.e. the Queen of hearts and the queen of diamond). This gives a total of 15 cards.

Therefore, Pr. (picking a club or a red Queen) = $\dfrac{15}{52}$

(e) There are 13 cards that are diamonds, and 4 cards that are 9. But one of the 9 is in diamond and has already been counted among the 13 diamonds. So it must not be counted twice. Hence we count the other three 9 (each from clubs, hearts and spades). This will give a total of 16 (13 + 3) cards.

Therefore, Pr. (picking a diamond or a 9) = $\dfrac{16}{52}$

$= \dfrac{4}{13}$

(f) There are 4 cards that are 6, and 26 cards that are black. But two of the 26 black cards are among the four cards that are 6, and these two black 6 cards have already been counted among the 26 black cards. So they must not be counted twice. Hence we count the other two 6 cards that are red. This will give a total of 28 (26 + 2) cards.

Therefore, Pr. (picking a 6 or a back card) = $\dfrac{28}{52}$

$= \dfrac{7}{13}$

8. A card is picked at random from a pack of playing cards and then replaced. A second card is picked. What is the probability of picking:

(a) a 3 and a 10

(b) a queen and an ace

(c) two kings

(d) two red cards

(e) two cards of different colours

(f) two cards of the same colour

Solution

In probability problems, when two items are selected, it is important to logically analyse the situation when solving the problem. This will help you to know if addition (use of OR) is involved or multiplication (use of AND) is involved. For example, for a queen and a king to be selected, it simply means that, either the queen is selected first and then the king, or the king is selected first and then the queen. When this logical analysis is understood, then most questions in probability become easy to solve.

(a) There are four cards that are 3, and four cards that are 10

Therefore, Pr. (picking a 3) = $\dfrac{4}{52}$

$= \dfrac{1}{13}$

Similarly, Pr. (picking a 10) = $\dfrac{4}{52}$

$$= \frac{1}{13}$$

Recall that "and" in probability means multiplication.

The probability of picking a 3 and a 10 means that:

Either the first is a 3 AND the second is a 10, OR the first is a 10 AND the second is a 3.

This can be calculated by putting x in place of AND and + in place of OR in the above statement as follows:

Pr. (picking a 3) x Pr. (picking a 10) + Pr. (picking a 10) x Pr. (picking a 3)

$$= (\frac{1}{13} \times \frac{1}{13}) + (\frac{1}{13} \times \frac{1}{13})$$

$$= \frac{1}{169} + \frac{1}{169}$$

$$= \frac{2}{169}$$

Therefore, Pr. (picking a 3 and a 10) = $\frac{2}{169}$

(b) There are 4 cards that are queen, and 4 cards that are ace

Therefore, Pr. (picking a queen) = $\frac{4}{52}$

$$= \frac{1}{13}$$

Similarly, Pr. (picking an ace) = $\frac{4}{52}$

$$= \frac{1}{13}$$

The probability of picking a queen and an ace means that:

Either you first pick a queen AND then an ace, OR you first pick an ace AND then a queen.

This can be calculated by putting x in place of AND and + in place of OR in the above statement as follows:

Pr. (picking a queen) x Pr. (picking an ace) + Pr. (picking an ace) x Pr. (picking a queen)

$$= (\frac{1}{13} \times \frac{1}{13}) + (\frac{1}{13} \times \frac{1}{13})$$

$$= \frac{1}{169} + \frac{1}{169}$$

$$= \frac{1}{169}$$

Therefore, Pr. (picking a queen and an ace) = $\frac{2}{169}$

(c) There are four cards that are King

Therefore, Pr. (picking a king) = $\frac{4}{52}$

$$= \frac{1}{13}$$

The probability of picking two kings means that:

The first is a king AND the second is a king

= Pr. (picking a king) x Pr. (picking a king)

$$= \frac{1}{13} \times \frac{1}{13}$$

$$= \frac{1}{169}$$

Therefore, Pr. (picking two kings) = $\frac{1}{169}$

(d) There are 26 cards that are red

Therefore, Pr. (picking a red card) = $\frac{26}{52}$

$$= \frac{1}{2}$$

The probability of picking two red cards means that:

The first is a red card AND the second is a red card

= Pr. (picking a red card) x Pr. (picking a red card)

$$= \frac{1}{2} \times \frac{1}{2}$$

$$= \frac{1}{4}$$

Therefore, Pr. (picking two red cards) = $\frac{1}{4}$

(e) There are two colours of cards, red and black.

Therefore, Pr. (picking a red card) = $\frac{1}{2}$ (i.e from $\frac{26}{52}$ since there are 26 red cards)

Similarly, Pr. (picking a black card) = $\frac{1}{2}$ (i.e from $\frac{26}{52}$ since there are also 26 black cards)

The probability of picking two cards of different colours means that:

Either the first is a black card AND the second is a red card, OR the first is a red card AND the second is a black card.

This can be calculated by putting x in place of AND and + in place of OR in the above statement as follows:

Pr. (picking a black card) x Pr. (picking a red card) + Pr. (picking a red card) x Pr. (picking a black card)

$$= (\frac{1}{2} \times \frac{1}{2}) + (\frac{1}{2} \times \frac{1}{2})$$

$$= \frac{1}{4} + \frac{1}{4}$$

$$= \frac{2}{4}$$

$$= \frac{1}{2}$$

Therefore, Pr. (picking two cards of different colours) = $\frac{1}{2}$

(f) Pr. (picking two cards of the same colours) = 1 - Pr. (picking two cards of different colours)

$$= 1 - \frac{1}{2}$$

$$= \frac{1}{2}$$

Note that this can also be solved by using the logical process which is:

Either the first is red AND the second is red OR the first is black AND the second is black. This will also give $\frac{1}{2}$

9. Two cards are picked at random one after the other without replacement from a pack of playing cards. What is the probability of picking:

(a) a 5 and a 7

(b) a king and a jack

(c) two aces

(d) two diamond cards

(e) two black cards

(f) a red and a black card

(g) two cards of the same colours

Solution

This problem involves picking a card without replacement. This means that when one card is picked out, the total number of cards remaining in the pack become reduced to 51. That number of that particular type of card also reduces by 1.

(a) There are four cards that are 5. There are also four cards that are 7.

Hence the probability of picking a 5 and a 7 means that:

Either first picking a 5 AND then a 7, OR first picking a 7 AND then a 5.

Now, let us calculate each of the probabilities as follows:

Pr. (first card is a 5) = $\frac{4}{52}$ (There are four cards that are 5)

$\quad = \frac{1}{13}$ (In its lowest term)

We now have 51 cards left in the pack.

Therefore, Pr. (second card is a 7) = $\frac{4}{51}$ (There are four cards that are 7, and a total of 51 cards remaining in the pack)

Or,

Pr. (first card is a 7) = $\frac{4}{52}$ (There are four cards that are 7)

$\quad = \frac{1}{13}$ (In its lowest term)

We now have 51 cards left in the pack.

Therefore, Pr. (second card is a 5) $= \frac{4}{51}$ (There are four cards that are 5, and a total of 51 cards remaining in the pack)

Hence the probability of picking a 5 and a 7 means that:

Either first picking a 5 AND then a 7, OR first picking a 7 AND then a 5. Which is computed as:

Pr. (picking a 5 and a 7) = Pr. (first card is a 5) x Pr. (second card is a 7) + Pr. (first card is a 7) x Pr. (second card is a 5)

$$= (\frac{1}{13} \times \frac{4}{51}) + (\frac{1}{13} \times \frac{4}{51})$$

$$= \frac{4}{663} + \frac{4}{663}$$

$$= \frac{8}{663}$$

(b) There are four cards that are kings. There are also four cards that are jacks.

Now, let us calculate each of the probabilities as follows:

Pr. (first card is a king) $= \frac{4}{52}$ (There are four cards that are kings)

$= \frac{1}{13}$ (In its lowest term)

We now have 51 cards left in the pack.

Therefore, Pr. (second card is a jack) $= \frac{4}{51}$ (There are four cards that are jack, and a total of 51 cards remaining in the pack)

Or,

Pr. (first card is a jack) $= \frac{4}{52}$ (There are four cards that are jack)

$= \frac{1}{13}$ (In its lowest term)

We now have 51 cards left in the pack.

Therefore, Pr. (second card is a king) $= \frac{4}{51}$ (There are four cards that are king, and a total of 51 cards remaining in the pack)

Hence the probability of picking a king and a jack means that:

Either first picking a king AND then a jack, OR first picking a jack AND then a king. This is computed as:

Pr. (picking a king and a queen) = Pr. (first card is a king) x Pr. (second card is a jack) + Pr. (first card is a jack) x Pr. (second card is a king)

$$= (\frac{1}{13} \times \frac{4}{51}) + (\frac{1}{13} \times \frac{4}{51})$$

$$= \frac{4}{663} + \frac{4}{663}$$

$$= \frac{8}{663}$$

(c) There are 4 cards that are aces.

Hence the probability of picking two aces means that:

The first is an ace, and the second is an ace.

Now, let us calculate each of the probabilities as follows:

Pr. (first card is an ace) = $\frac{4}{52}$ (There are 4 cards that are aces)

$\quad = \frac{1}{13}$ (In its lowest term)

We now have 3 aces left in the pack, and a total of 51 cards left in the pack.

Therefore, Pr. (second card is an ace) = $\frac{3}{51}$

Hence the probability of picking two aces is given by:

Pr. (picking two aces) = Pr. (first card is an ace) x Pr. (second card is an ace)

$\quad = \frac{1}{13} \times \frac{3}{51}$

$\quad = \frac{3}{663}$

(d) There are 13 cards that are diamonds.

Hence the probability of picking two diamonds means that:

The first is a diamond, and the second is a diamond.

Now, let us calculate each of the probabilities as follows:

Pr. (first card is a diamond) = $\frac{13}{52}$ (There are 13 cards that are diamonds)

$\quad = \frac{1}{4}$ (In its lowest term)

We now have 12 diamonds left in the pack, and a total of 51 cards left in the pack.

Therefore, Pr. (second card is a diamond) = $\frac{1}{13}$

$\quad = \frac{4}{17}$ (In its lowest term)

Hence the probability of picking two diamonds is given by:

Pr. (picking two diamonds) = Pr. (first card is a diamond) x Pr. (second card is a diamond)

$\quad = \frac{1}{4} \times \frac{4}{17}$

$\quad = \frac{4}{68}$

$\quad = \frac{1}{17}$ (In its lowest term)

(e) There are 26 black cards.

Hence the probability of picking two black cards means that:

The first is a black card, and the second is a black card.

Now, let us calculate each of the probabilities as follows:

Pr. (first card is a black card) = $\dfrac{26}{52}$

$\qquad = \dfrac{1}{2}$ (In its lowest term)

We now have 25 black cards left in the pack, and a total of 51 cards left in the pack.

Therefore, Pr. (second card is a black card) = $\dfrac{25}{51}$

Hence the probability of picking two black cards is given by:

Pr. (picking two black cards) = Pr. (first card is a black card) x Pr. (second card is a black card)

$\qquad = \dfrac{1}{2} \text{ x } \dfrac{25}{51}$

$\qquad = \dfrac{25}{102}$

(f) The logical explanation for this situation is that:

Either the first card is red AND the second is black OR the first card is black and the second is red.

There are 26 red cards and also 26 black cards.

Now, let us calculate each of the probabilities as follows:

Pr. (first card is a red card) = $\dfrac{26}{52}$

$\qquad = \dfrac{1}{2}$ (In its lowest term)

We now have 51 cards left in the pack.

Therefore, Pr. (second card is a black card) = $\dfrac{26}{51}$ (There are 26 black cards, and a total of 51 cards remaining in the pack)

Or,

Pr. (first card is a black card) = $\dfrac{26}{52}$

$\qquad = \dfrac{1}{2}$ (In its lowest term)

We now have 51 cards left in the pack.

Therefore, Pr. (second card is a red card) = $\dfrac{26}{51}$ (There are 26 red cards, and a total of 51 cards remaining in the pack)

Hence the probability of picking a red card and a black card means that:

Either first picking a red card AND then a black card, OR first picking a black card AND then a red card. This is computed as:

Pr. (picking a red and black cards) = Pr. (first card is a red card) x Pr. (second card is a black card) + Pr. (first card is a black card) x Pr. (second card is a red card)

$$= (\frac{1}{2} \times \frac{26}{51}) + (\frac{1}{2} \times \frac{26}{51})$$

$$= \frac{26}{102} + \frac{26}{102}$$

$$= \frac{52}{102}$$

$$= \frac{26}{51}$$

(g) The logical explanation for this situation is that:

Either the first card is red AND the second is red OR the first card is black and the second is black.

There are 26 red cards and also 26 black cards.

Now, let us calculate each of the probabilities as follows:

Pr. (first card is a red card) = $\frac{26}{52}$

$$= \frac{1}{2} \text{ (In its lowest term)}$$

We now have 25 red cards left and a total of 51 cards left in the pack.

Therefore, Pr. (second card is a red card) = $\frac{25}{51}$

Or,

Pr. (first card is a black card) = $\frac{26}{102}$

$$= \frac{1}{2} \text{ (In its lowest term)}$$

We now have 25 black cards left and a total of 51 cards left in the pack.

Therefore, Pr. (second card is a black card) = $\frac{25}{51}$

Hence the probability of picking two cards of the same colour means that:

Either picking a red card AND then another red card, OR picking a black card AND then another black card. This is computed as:

Pr. (picking two cards of the same colour) = Pr. (first card is a red card) x Pr. (second card is a red card) + Pr. (first card is a black card) x Pr. (second card is a black card)

$$= (\frac{1}{2} \times \frac{25}{51}) + (\frac{1}{2} \times \frac{25}{51})$$

$$= \frac{25}{102} + \frac{25}{102}$$

$$= \frac{50}{102}$$

$$= \frac{25}{51}$$

Alternatively, this question can also be solved as follows:

Recall that question (f) above gives the probability of picking a red and a black card. This also means the probability of picking two cards of different colours.

Hence the probability of picking two cards of different colours as given in (f) above = $\dfrac{26}{51}$

Therefore, Pr. (picking two cards of the same colour) = 1 - Pr. (picking two cards of different colours)   (Note that they are opposite statements)

$$= 1 - \dfrac{26}{51}$$
$$= \dfrac{51 - 26}{51}$$
$$= \dfrac{25}{51} \quad \text{(As obtained before)}$$

10. If three cards are picked from a pack of playing cards with replacement, what is the probability if getting:
(a) at least two clubs
(b) at most two clubs

Solution
I am going to be using a special type of tree diagram without actually drawing the diagram. Now, the total outcome in a selection of three items involving two events (i.e. a club or not a club) is given by:

2^n,

where n is the number of selection made.
In the question, n = 3, since three cards were picked.
Hence total outcome = 2^3 = 2 x 2 x 2
        = 8.

Now, in order to write out the outcomes, let us use the letter C to represent a club and letter N to represent not a club. Note that in tree diagrams like this, only two letters should be used in writing the outcomes since the question involves the picking of only one type of item (club). Hence the outcome is written as follows:

   (CCC), (CCN), (CNC), (CNN), (NCC), (NCN), (NNC), (NNN)

Note that there are 8 ways of arranging the two letters in the brackets. There is no fast rule in carrying out the arrangement. You just have to make sure that no two brackets have the same arrangement of the letters. Also make sure the number of brackets is complete.

(a) In order to determine the probability of getting at least two clubs, we need to compute the probabilities of the brackets that contain at least 2 clubs. They are, (CCC), (CCN), (CNC), and (NCC). Note that at least two, means two and above, (i.e. two and three clubs in this case). Hence the probability of getting at least two clubs = (CCC) or (CCN) or (CNC) or (NCC)
Now, let us compute each of the probabilities.

There are 13 clubs in a pack of cards, and there are 39 cards that are not club. Note that this is a case of with replacement, which means that the total number of cards in the pack is always complete. Hence:

(CCC) = Pr. (first card is a club) x Pr. (second card is a club) x Pr. (third card is a club)

$$= \frac{13}{52} \times \frac{13}{52} \times \frac{13}{52}$$

$$= \frac{1}{4} \times \frac{1}{4} \times \frac{1}{4}$$

$$= \frac{1}{64}$$

(CCN) = Pr. (first card is a club) x Pr. (second card is a club) x Pr. (third card is not a club)

$$= \frac{13}{52} \times \frac{13}{52} \times \frac{39}{52}$$ (Note that there are 39 cards that are not club)

$$= \frac{1}{4} \times \frac{1}{4} \times \frac{3}{4}$$

$$= \frac{3}{64}$$

(CNC) = Pr. (first card is a club) x Pr. (second card is not a club) x Pr. (third card is a club)

$$= \frac{13}{52} \times \frac{39}{52} \times \frac{13}{52}$$

$$= \frac{1}{4} \times \frac{3}{4} \times \frac{1}{4}$$

$$= \frac{3}{64}$$

(NCC) = Pr. (first card is not a club) x Pr. (second card is a club) x Pr. (third card is a club)

$$= \frac{39}{52} \times \frac{13}{52} \times \frac{13}{52}$$

$$= \frac{3}{4} \times \frac{1}{4} \times \frac{1}{4}$$

$$= \frac{3}{64}$$

Therefore, Pr. (getting at least two clubs) = (CCC) or (CCN) or (CNC) or (NCC)

$$= (CCC) + (CCN) + (CNC) + (NCC)$$

$$= \frac{1}{64} + \frac{3}{64} + \frac{3}{64} + \frac{3}{64}$$

$$= \frac{10}{64}$$

$$= \frac{5}{32}$$

(b) In order to determine the probability of getting at most two clubs, we need to compute the probabilities of the brackets that contain at most 2 clubs. From the outcome brackets given above, the ones that contain at most two clubs are, (CCN), (CNC), (CNN), (NCC), (NCN), (NNC), (NNN). Note that at most two, means two and below, (i.e. two, one and zero clubs in this case). Hence the probability of getting at most two clubs = (CCN) or (CNC) or (CNN) or (NCC) or (NCN) or (NNC) or (NNN)

Now, let us compute each of the probabilities. Hence:

(CCN) = $\dfrac{3}{64}$ (As calculated in (a) above)

(CNC) = $\dfrac{3}{64}$ (As calculated in (a) above)

(CNN) = Pr. (first card is a club) x Pr. (second card is not a club) x Pr. (third card is not a club)

$= \dfrac{13}{52}$ x $\dfrac{39}{52}$ x $\dfrac{39}{52}$

$= \dfrac{1}{4}$ x $\dfrac{3}{4}$ x $\dfrac{3}{4}$)

$= \dfrac{9}{64}$

(NCC) = $\dfrac{3}{64}$ (As calculated in (a) above)

(NCN) = Pr. (first card is not a club) x Pr. (second card is a club) x Pr. (third card is not a club)

$= \dfrac{39}{52}$ x $\dfrac{13}{52}$ x $\dfrac{39}{52}$

$= \dfrac{3}{4}$ x $\dfrac{1}{4}$ x $\dfrac{3}{4}$

$= \dfrac{9}{64}$

(NNC) = Pr. (first card is not a club) x Pr. (second card is not a club) x Pr. (third card is a club)

$= \dfrac{39}{52}$ x $\dfrac{39}{52}$ x $\dfrac{13}{52}$

$= \dfrac{3}{4}$ x $\dfrac{3}{4}$ x $\dfrac{1}{4}$

$= \dfrac{9}{64}$

(NNN) = Pr. (first card is not a club) x Pr. (second card is not a club) x Pr. (third card is not a club)

$= \dfrac{39}{52}$ x $\dfrac{39}{52}$ x $\dfrac{39}{52}$

$= \dfrac{3}{4}$ x $\dfrac{3}{4}$ x $\dfrac{3}{4}$

$= \dfrac{27}{64}$

Therefore, Pr. (getting at most two clubs) = (CCN) or (CNC) or (CNN) or (NCC) or (NCN) or (NNC) or (NNN)

$= $ (CCN) + (CNC) + (CNN) + (NCC) + (NCN) + (NNC) + (NNN)

$= \dfrac{3}{64} + \dfrac{3}{64} + \dfrac{9}{64} + \dfrac{3}{64} + \dfrac{9}{64} + \dfrac{9}{64} + \dfrac{27}{64}$

$= \dfrac{63}{64}$

Alternatively, a shorter method of solving this problem is as follows:

Pr. (getting at most two clubs) = 1 - Pr. (getting three clubs)   (This is because the only item not in (b) is CCC. All 8 outcomes in brackets make a total probability of 1, but 7 of the outcomes are in (b), hence 1 minus the outcome not in (b) gives (b). This can also be expressed in fraction as, (b) = $\frac{7}{8}$ = $\frac{8}{8}$ - $\frac{1}{8}$). Therefore:

Pr. (getting at most two clubs) = 1 - Pr. (getting three clubs)

= 1 – CCC

= 1 - $\frac{1}{64}$ (Note that CCC computed in (a) = $\frac{1}{64}$)

= $\frac{63}{64}$ (As obtained before)

11. If three cards are chosen from a pack of playing cards without replacement, what is the probability of getting:

(a) at least two diamonds

(b) at most one diamond?

<u>Solution</u>

In order to write out the outcomes, let us use the letter D to represent a diamond and letter N to represent not a diamond.

Hence the outcomes are written as follows:

(DDD), (DDN), (DND), (DNN), (NDD), (NDN), (NND), (NNN)

(a) In order to determine the probability of getting at least two diamonds, we need to compute the probabilities of the brackets that contain at least 2 diamonds. They are, (DDD), (DDN), (DND), and (NDD).

Hence the probability of getting at least two diamonds = (DDD) or (DDN) or (DND) or (NDD)

Now, let us compute each of the probabilities.

There are 13 diamonds in a pack of cards, and there are 39 cards that are not diamonds. Note that this is a case of without replacement, which means that after each selection, both the total number of cards left and the number of the particular card picked, are reduced by 1. Hence:

(DDD) = Pr. (first card is a diamond) x Pr. (second card is a diamond) x Pr. (third card is a diamond)

= $\frac{13}{52}$ x $\frac{12}{51}$ x $\frac{11}{50}$ (Note that the number of diamond and the total number of card left, keep reducing by 1 after each selection)

= $\frac{1}{4}$ x $\frac{12}{51}$ x $\frac{11}{50}$

= $\frac{132}{10200}$

= $\frac{11}{850}$ (In its lowest term, after equal division by 12)

(DDN) = Pr. (first card is a diamond) x Pr. (second card is a diamond) x Pr. (third card is not a diamond)

$$= \frac{13}{52} \times \frac{12}{51} \times \frac{39}{50} \quad \text{(Note that there are 39 cards that are not diamond)}$$

$$= \frac{1}{4} \times \frac{4}{17} \times \frac{39}{50}$$

$$= \frac{156}{3400}$$

$$= \frac{39}{850} \quad \text{(After equal division by 4)}$$

(DND) = Pr. (first card is a diamond) x Pr. (second card is not a diamond) x Pr. (third card is a diamond)

$$= \frac{13}{52} \times \frac{39}{51} \times \frac{12}{50}$$

$$= \frac{1}{4} \times \frac{13}{17} \times \frac{6}{25}$$

$$= \frac{78}{1700}$$

$$= \frac{39}{850}$$

(NDD) = Pr. (first card is not a diamond) x Pr. (second card is a diamond) x Pr. (third card is a diamond)

$$= \frac{39}{52} \times \frac{13}{51} \times \frac{12}{50}$$

$$= \frac{3}{4} \times \frac{13}{51} \times \frac{6}{25}$$

$$= \frac{234}{5100}$$

$$= \frac{39}{850}$$

Therefore, Pr. (getting at least two diamonds) = (DDD) or (DDN) or (DND) or (NDD)

$$= (DDD) + (DDN) + (DND) + (NDD)$$

$$= \frac{11}{850} + \frac{39}{850} + \frac{39}{850} + \frac{39}{850}$$

$$= \frac{128}{850}$$

$$= \frac{64}{425}$$

(b) In order to determine the probability of getting at most one diamond, we need to compute the probabilities of the brackets that contain at most one diamond. From the outcome brackets given above, the ones that contain at most one diamond are, (DNN), (NDN), (NND), (NNN). Note that at most one, means one and below, (i.e. one and zero diamond in this case). Hence the probability of getting at most one diamond = (DNN) + (NDN) + (NND) + (NNN) Now, let us compute each of the probabilities. Hence:

(DNN) = Pr. (first card is a diamond) x Pr. (second card is not a diamond) x Pr. (third card is not a diamond)

$$= \frac{13}{52} \times \frac{39}{51} \times \frac{38}{50}$$

$$= \frac{1}{4} \times \frac{13}{17} \times \frac{19}{25}$$

$$= \frac{247}{1700}$$

(NDN) = Pr. (first card is not a diamond) x Pr. (second card is a diamond) x Pr. (third card is not a diamond)

$$= \frac{39}{52} \times \frac{13}{51} \times \frac{38}{50}$$

$$= \frac{3}{4} \times \frac{13}{51} \times \frac{19}{25}$$

$$= \frac{741}{5100}$$

$$= \frac{247}{1700} \quad \text{(In its lowest term after equal division by 3)}$$

(NND) = Pr. (first card is not a diamond) x Pr. (second card is not a diamond) x Pr. (third card is a diamond)

$$= \frac{39}{52} \times \frac{38}{51} \times \frac{13}{50}$$

$$= \frac{3}{4} \times \frac{38}{51} \times \frac{13}{50}$$

$$= \frac{1482}{10200}$$

$$= \frac{247}{1700} \quad \text{(After equal division by 6)}$$

(NNN) = Pr. (first card is not a diamond) x Pr. (second card is not a diamond) x Pr. (third card is not a diamond)

$$= \frac{39}{52} \times \frac{38}{51} \times \frac{37}{50}$$

$$= \frac{3}{4} \times \frac{38}{51} \times \frac{37}{50}$$

$$= \frac{4218}{10200}$$

$$= \frac{703}{1700} \quad \text{(After equal division by 6)}$$

Therefore, Pr. (getting at most one diamond) = (DNN) or (NDN) or (NND) or (NNN)

$$= (DNN) + (NDN) + (NND) + (NNN)$$

$$= \frac{247}{1700} + \frac{247}{1700} + \frac{247}{1700} + \frac{703}{1700}$$

$$= \frac{1444}{1700}$$

$$= \frac{361}{425}$$

Exercise 42

1. A card is picked at random from a pack of playing cards. Find the probability of picking a jack.

2. A card is picked at random from a pack of playing cards. Find the probability of picking a black 4.

3. A card is picked at random from a pack of playing cards. Find the probability of picking a red king.

4. A card is picked at random from a pack of playing cards. Find the probability of picking a either a black or red card.

5. A card is picked at random from a pack of playing cards. Find the probability of picking a black Queen.

6. A card is picked at random from a pack of playing cards. Find the probability of picking a card that is not an Ace.

7. A card is picked at random from a pack of playing cards. Find the probability of picking

(a) a queen or a king

(b) a 3 or a 9

(c) either a jack or the queen of diamonds

(d) a spade or a black 7

(e) a club or a red king

(f) a 2 or a red card

8. A card is picked at random from a pack of playing cards and then replaced. A second card is picked. What is the probability of picking:

(a) an 8 and a 5

(b) a black card and a 4

(c) two cards between 2 and 9 that have odd numbers

(d) two black cards

(e) two cards with the same number on them

(f) two cards with different number on them

9. Two cards are picked at random one after the other without replacement from a pack of playing cards. What is the probability of picking:

(a) a 4 and an ace

(b) a 2 and a 7

(c) two 8s

(d) two clubs

(e) two red cards

(f) a club and a diamond

(g) two cards that are queens

10. If three cards are picked from a pack of playing cards with replacement, what is the probability if getting:

(a) at least two 9s

(b) at most two 9s

11. If three cards are chosen from a pack of playing cards without replacement, what is the probability of getting:

(a) at least two kings

(b) at most one king?

CHAPTER 43
PROBABILITY ON TOSSING OF COINS

When a coin is tossed, the outcome can either be a head or a tail. However when two or more coins are tossed, the total outcome is obtained from 2^n, where n is the number of times the coin is tossed, or the number of coins tossed together.

Note that 'head' is the part of the coin that shows the person drawn on the coin, while the opposite side of the coin is called the 'tail'

Examples

1. A fair coin is tossed. What is the probability of getting:

(a) a head

(b) a tail

Solution

(a) There are only two possible outcomes. Head or tail.

Therefore, Pr. (getting a head) $= \dfrac{\text{Number of heads}}{\text{Total outcomes}}$

$\quad = \dfrac{1}{2}$

(b) Pr. (getting a tail) $= \dfrac{\text{Number of tails}}{\text{Total outcomes}}$

$\quad = \dfrac{1}{2}$

2. A coin is tossed two times. What is the probability of getting:

(a) a head and a tail

(b) at least a tail

(c) two heads

(d) two tails

(e) a head on the first toss, and a tail on the second toss.

Solution

The outcomes are written by using H for head and T for tail. The total number of outcomes will be $2^2 = 4$ (i.e. from 2^n, and n = 2 in this case)

The outcomes are: (HH), (HT), (TH), (TT).

(a) The outcomes with head and tail are (HT) and (TH). This gives 2 outcomes.

Therefore, Pr. (getting a head and tail) $= \dfrac{\text{Number of outcome with head and tail}}{\text{Total outcomes}}$

$$= \frac{2}{4}$$

$$= \frac{1}{2}$$

(b) The outcomes with at least a tail are (HT), (TH) and (TT). This gives 3 outcomes.

Therefore, Pr. (getting at least a tail) = $\dfrac{\text{Number of outcomes with at least a tail}}{\text{Total number of outcomes}}$

$$= \frac{3}{4}$$

(c) The outcome with two heads is (HH). This gives 1 outcome.

Therefore, Pr. (getting two heads) = $\dfrac{\text{Number of outcomes with heads}}{\text{Total number of outcomes}}$

$$= \frac{1}{4}$$

(d) The outcome with two tails is (TT). This gives 1 outcome.

Therefore, Pr. (getting two tails) = $\dfrac{\text{Number of outcomes with two tails}}{\text{Total number of outcomes}}$

$$= \frac{1}{4}$$

(e) The outcome with a head on the first toss, and a tail on the second toss is (HT). This gives 1 outcome

Therefore, Pr. (getting a head on the first toss, and a tail on the second toss) = $\dfrac{1}{4}$

3. A coin is tossed three times. What is the probability of getting:
(a) two heads and one tail
(b) at least one head
(c) three tails
(d) at least two heads
(e) a tail, a head and a tail

Solution
(a) The total number of outcomes will be $2^3 = 8$ (i.e. from 2^n, and n = 3 in this case)
The outcomes are: (HHH), (HTH), (HTT), (HHT), (THH), (THT), (TTH), (TTT). This gives a total of 8 outcomes

(a) The outcomes with two heads and one tail are (HTH), (HHT) and (THH). This gives 3 outcomes.
Therefore, Pr. (getting two heads and one tail) =

410

Number of outcomes with two heads and one tail
───
      Total number of outcomes

$$= \frac{3}{8}$$

(b) The outcomes with at least one head are (HHH), (HTH), (HTT), (HHT), (THH), (THT) and (TTH). This gives 7 outcomes.

Therefore, Pr. (getting at least one head) = $\dfrac{\text{Number of outcomes with at least one head}}{\text{Total number of outcomes}}$

$$= \frac{7}{8}$$

(c) The outcome with three tails is (TTT). This gives 1 outcome.

Therefore, Pr. (getting three tails) = $\dfrac{1}{8}$

(d) The outcomes with at least two heads are (HHH), (HTH), (HHT) and (THH). This gives 4 outcomes.

Therefore, Pr. (getting at least two heads) = $\dfrac{\text{Number of outcomes with at least two heads}}{\text{Total number of outcomes}}$

$$= \frac{4}{8}$$
$$= \frac{1}{2}$$

(e) The outcome with a tail, a head and a tail is (THT). This is 1 outcome

Hence, Pr. (getting a tail, a head and a tail) = $\dfrac{1}{8}$

4. Four coins are tossed together. Find the probability of getting:
(a) two heads and two tails
(b) four tails
(c) at least three heads
(d) at least two heads
(e) one head

Solution
The total number of outcomes will be $2^4 = 16$ (i.e. from 2^n, and n = 4 in this case)
The outcomes are: (HHHH), (HHHT), (HHTT), (HTTT), (THHH), (TTHH), (TTTH), (THTH), (HTHT), (HHTH), (THHT), HTTH), (TTHT), (THTT), (HTHH), (TTTT). This gives a total of 16 outcomes.

(a) The outcomes with two heads and two tails are (HHTT), (TTHH), (THTH), (HTHT), (THHT), (HTTH). This gives 6 outcomes.

Therefore, Pr. (getting two heads and two tails) =

$$\frac{\text{Number of outcomes with two heads and two tails}}{\text{Total number of outcomes}}$$

$$= \frac{6}{16}$$

$$= \frac{3}{8}$$

(b) The outcome with four tails is (TTTT). This gives 1 outcomes.

Therefore, Pr. (getting four tails) = $\frac{1}{16}$

(c) The outcomes with at least three heads are (HHHH), (HHHT), (THHH), (HHTH), (HTHH). This gives 5 outcomes.

Therefore, Pr. (getting at least three heads) = $\frac{\text{Number of outcomes with at least three heads}}{\text{Total number of outcomes}}$

$$= \frac{5}{16}$$

(d) The outcomes with at least two heads are (HHHH), (HHHT), (HHTT), (THHH), (TTHH), (THTH), (HTHT) (HHTH), (THHT), (HTTH), (HTHH). This gives 11 outcomes.

Therefore, Pr. (getting at least two heads) = $\frac{\text{Number of outcomes with at least two heads}}{\text{Total number of outcomes}}$

$$= \frac{11}{16}$$

(e) The outcomes with one head are, (HTTT), (TTTH), (TTHT), (THTT), . This gives 4 outcomes.

Therefore, Pr. (getting one head) = $\frac{\text{Number of outcomes with one head}}{\text{Total number of outcomes}}$

$$= \frac{4}{16}$$

$$= \frac{1}{4}$$

(5) A coin is tossed five times. Find the probability of getting at least one tail.

Solution

The total number of outcomes will be $2^5 = 32$

The only outcome without a tail is (HHHHH). This is an outcome of 1

Pr. (getting no tail, i.e. all head) = $\frac{1}{32}$

Therefore, Pr. (getting at least one tail) = 1 - Pr. (getting no tail)

$$= 1 - \frac{1}{32}$$

$$= \frac{31}{32}$$

Exercise 43

1. A fair coin is tossed. What is the probability of getting:

    (a) a tail

    (b) a head

    (c) a tail or a head

    2. A coin is tossed two times. What is the probability of getting:

    (a) a tail and then a head

    (b) at least a head

    (c) two tails

    (d) at least a tail

    (e) a head on the first toss, and a tail on the second toss.

    3. Three coins are tossed. What is the probability of getting:

    (a) three heads

    (b) at least one tail

    (c) a head, a tail and then a head

    (d) at least one head

    (e) at least two heads

    (f) at most two tails

    4. Four coins are tossed together. Find the probability of getting:

    (a) at least one head

    (b) four heads

    (c) at least two heads

    (d) at most three tails

    (e) two heads

    (5) A coin is tossed five times. Find the probability of getting at least one head.

CHAPTER 44
PROBABILITY ON THROWING OF DICE

Examples

1. A fair die is thrown. Find the probability of getting:

(a) a 2

(b) a 5

(c) a 7

(d) a 4 or a 5

(e) a number less than 4

(f) an odd number

Solution

Note that a die has six faces numbered 1 to 6. That means that each number appears once.

(a) Pr. (getting a 2) = $\dfrac{\text{Number of faces having 2}}{\text{total number of faces}}$

$= \dfrac{1}{6}$

(b) Pr. (getting a 5) = $\dfrac{\text{Number of faces having 5}}{\text{total number of faces}}$

$= \dfrac{1}{6}$

(c) Pr. (getting a 7) = $\dfrac{\text{Number of faces having 7}}{\text{total number of faces}}$

$= \dfrac{0}{6}$

$= 0$ (This is a case of an impossible event)

(d) Pr. (getting a 4 or a 5) = $\dfrac{\text{Number of faces having 4 and having 5}}{\text{total number of faces}}$

$= \dfrac{2}{6}$

$= \dfrac{1}{3}$

(e) Pr. (getting a number less than 4) = $\dfrac{\text{Number of faces having numbers less than 4}}{\text{total number of faces}}$

$= \dfrac{3}{6}$ (Note that the faces with numbers less than 4 are 3, 2, and 1. This makes a total of 3 faces)

$= \dfrac{1}{2}$

(f) Pr. (getting an odd number) = $\dfrac{\text{Number of faces having odd numbers}}{\text{total number of faces}}$

$= \dfrac{3}{6}$ (Faces with odd numbers are 1, 3 and 5, i.e. three faces)

$= \dfrac{1}{2}$

2. A fair die is rolled once. What is the probability of getting:

(a) a number divisible by 3

(b) a multiple of 2

(c) at least 5

(d) at most 2

(e) a prime number or an even number

(f) either a number greater that 2 or a multiple of 4

Solution

(a) Pr. (getting a number divisible by 3) = $\dfrac{\text{Number of faces having numbers divisible by 3}}{\text{total number of faces}}$

$= \dfrac{2}{6}$ (Faces with numbers divisible by 3 are 3 and 6, i.e. 2 faces)

$= \dfrac{1}{3}$

(b) Pr. (getting a multiple of 2) = $\dfrac{\text{Number of faces having numbers that are multiple of 2}}{\text{total number of faces}}$

$= \dfrac{3}{6}$ (Faces with numbers that are multiple of 2 are 2, 4 and 6, i.e. 3 faces)

$= \dfrac{1}{2}$

(c) Pr. (getting at least 5) = $\dfrac{\text{Number of faces having numbers that are at least 5}}{\text{Total number of faces}}$

$= \dfrac{2}{6}$ (Faces with numbers that are at least 5 are 5 and 6, i.e. 2 faces)

$= \dfrac{1}{3}$

(d) Pr. (getting at most 2) = $\dfrac{\text{Number of faces having numbers that are at least 2}}{\text{Total number of faces}}$

$= \dfrac{2}{6}$ (Faces with numbers that are at most 2 are 1 and 2, i.e. 2 faces)

$= \dfrac{1}{3}$

(e) Pr. (getting a prime number or an even number) =

$$\frac{\text{Number of faces having prime numbers and even numbers}}{\text{Total number of faces}}$$

$= \frac{5}{6}$ (Faces with prime numbers are 2, 3, and 5. Faces with even numbers are 2, 4, 6. This will give a total of 5 faces because 2 which is both a prime and even number should be counted once)

Therefore, Pr. (getting a prime number or an even number) = $\frac{5}{6}$

(f) Pr. (getting either a number greater that 2 or a multiple of 4) =

$$\frac{\text{Number of faces having numbers greater than 2 and numbers that are multiple of 4}}{\text{Total number of faces}} = \frac{4}{6}$$ (Faces with numbers greater than 2 are 3, 4, 5 and 6. Faces with multiple of 4 is 4. This will give a total of 4 faces because 4 which appears in both events should be counted once)

Therefore, Pr. (getting either a number greater that 2 or a multiple of 4) = $\frac{4}{6}$

$= \frac{2}{3}$

3. A die is thrown and a coin is tossed. What is the probability of getting:

(a) a 3 and a head

(b) a tail and a prime number

Solution

(a) From the die, Pr. (getting a 3) = $\frac{1}{6}$

From the coin, Pr. (getting a head) = $\frac{1}{2}$

Since AND means multiplication in probability:

Therefore, Pr. (getting a 3 and a head) = Pr. (getting a 3) x Pr. (getting a head)

$= \frac{1}{6} \times \frac{1}{2}$

$= \frac{1}{12}$

(b) (a) From the coin, Pr. (getting a tail) = $\frac{1}{2}$

From the die, Pr. (getting a prime number) = $\frac{3}{6}$ (The prime numbers are 3, i.e. 2, 3 and 5)

$= \frac{1}{2}$

Therefore, Pr. (getting a tail and a prime number) = Pr. (getting a tail) x Pr. (getting a prime number)

$$= \frac{1}{2} \times \frac{1}{2}$$

$$= \frac{1}{4}$$

4. Two fair dice are thrown at the same time. Find the probability of getting:

(a) at least one six

(b) a sum of at least 10

(c) a sum of at most 5

(d) a sum less than 3

(e) a total of seven

(f) a sum that is either a prime number or a multiple of 3

(g) a sum that is either divisible by 3 or a multiple of 2

Solution

The outcome table is as shown below. The numbers in the bracket give the outcome from the first and second die respectively. Adding the numbers in the bracket will give the respective sum that will be obtained.

Number on second die

+	1	2	3	4	5	6
1	(1,1)	(1,2)	(1,3)	(1,4)	(1,5)	(1,6)
2	(2,1)	(2,2)	(2,3)	(2,4)	(2,5)	(2,6)
3	(3,1)	(3,2)	(3,3)	(3,4)	(3,5)	(3,6)
4	(4,1)	(4,2)	(4,3)	(4,4)	(4,5)	(4,6)
5	(5,1)	(5,2)	(5,3)	(5,4)	(5,5)	(5,6)
6	(6,1)	(6,2)	(6,3)	(6,4)	(6,5)	(6,6)

Number on first die

The outcome table above can be presented in a more direct form by adding the values in the brackets above to obtain the sum. This is as shown below. In the table below, the numbers in the brackets represent the numbers on each die. The numbers that are not in bracket are the outcomes from the sum of numbers on first and second dice.

Number on second die

+	(1)	(2)	(3)	(4)	(5)	(6)
(1)	2	3	4	5	6	7
(2)	3	4	5	6	7	8
(3)	4	5	6	7	8	9
(4)	5	6	7	8	9	10
(5)	6	7	8	9	10	11
(6)	7	8	9	10	11	12

Number on first die

Note that any of the tables above can be used to answer the questions asked above.

(a) The outcomes that can be obtained from getting at least a six are (6,1), (6,2), (6,3), (6,4), (6,5), (6,6), (1,6), (2,6), (3,6), (4,6), (5,6). They are from the first table. They are the outcomes from the 6 on the first die, and 6 on the second die respectively. The number of brackets from this outcome is 11 (when the brackets are counted). Note that the total outcomes from any of the two outcome tables above is 36. This is easily obtained from the second table by counting the numbers that are not in bracket.

Therefore, Pr. (getting at least a six) = $\dfrac{\text{Number of outcomes obtained when at least a six shows}}{\text{Total number of outcomes on the table}}$

$= \dfrac{11}{36}$

(b) A sum of at least 10 as shown on the second table above are, 10, 10, 10, 11, 11, and 12. This gives a total of 6 outcomes.

Therefore, Pr. (getting a sum of at least 10) = $\dfrac{6}{36}$ (Note that 36 is the total outcome)

$= \dfrac{1}{6}$

(c) A sum of at most 5 as shown on the second table above are, 5, 5, 5, 5, 4, 4, 4, 3, 3, and 2. This gives a total of 10 outcomes.

Therefore, Pr. (getting a sum of at most 5) = $\dfrac{10}{36}$

$= \dfrac{5}{18}$

(d) A sum less than 3 as shown on the second table above 2 only. This gives a total of 1 outcome.

Therefore, Pr. (getting a sum less than 3) = $\dfrac{1}{36}$

(e) A total of 7 as shown on the second table above appears 6 times. This gives a total of 6 outcomes.

Therefore, Pr. (getting a total of 7) = $\dfrac{6}{36}$

$\qquad = \dfrac{1}{6}$

(f) Sums which are prime numbers are 2, 3, 5, 7 and 11. Sums which are multiple of 3 are 3, 6, 9 and 12. Hence we are to count the outcomes from 2, 3, 5, 6, 7, 9, 11, and 12 (3 should be counted once). Hence, from the table, 2 appears 1 time, 3 appears 2 times, 5 appears 4 times, 6 appears 5 times, 7 appears 6 times, 9 appears 4 times, 11 appears 2 times, 12 appears 1 time. This gives a total outcome of 1 time + 2 times + 4 times + 5 times + 6 times + 4 times + 2 times + 1 time = 25. This is easier done on the table by counting all 2, 3, 5, 6, 7, 9, 11 and 12. It will also give a total of 25 outcomes.

Therefore, Pr. (getting a sum that is either a prime number or a multiple of 3) = $\dfrac{25}{36}$

(g) Sums which are divisible by 3 are 3, 6, 9 and 12. Sums which are multiples of 2 are 2, 4, 6, 8, 10 and 12. Hence we are to count the outcomes from 2, 3, 4, 6, 8, 9, 10 and 12 (6 and 12 which appear in both events should be counted once each). Hence, we go to the second table above and count all 2, 3, 4, 6, 8, 9, 10 and 12. It will give a total of 24 outcomes.

Therefore, Pr. (getting a sum that is either divisible by 3 or a multiple of 2) = $\dfrac{24}{36}$

$\qquad = \dfrac{2}{3}$

5. An unbiased die with faces numbered 1 to 6 is rolled twice. Find the probability that the product of the numbers obtained is:
(a) odd
(b) even
(c) 12
(d) prime
(e) either odd or a multiple of 5

Solution

The outcome table is as shown below. The numbers in brackets are the numbers on the die.

Number on second die

	x	(1)	(2)	(3)	(4)	(5)	(6)
	(1)	1	2	3	4	5	6
Number on	(2)	2	4	6	8	10	12
first die	(3)	3	6	9	12	15	18
	(4)	4	8	12	16	20	24
	(5)	5	10	15	20	25	30
	(6)	6	12	18	24	30	36

(a) All the odd numbers from the outcome table above are 1, 3, 3, 5, 5, 9, 15, 15, 25. This gives a total of 9 outcomes.

Therefore, Pr. (product of numbers is odd) = $\dfrac{9}{36}$ (Note that the total outcomes is 36)

$$= \dfrac{1}{4}$$

(b) Pr. (product of numbers is even) = 1 - Pr. (product of numbers is odd)

$$= 1 - \dfrac{1}{4}$$

$$= \dfrac{3}{4}$$

This can also be obtained by counting all the outcomes that are even numbers in the table above. Total even numbers is 27.

Hence, Pr. (product of numbers is even) = $\dfrac{27}{36}$

$$= \dfrac{3}{4} \quad \text{(As obtained before)}$$

(c) Pr. (product of numbers is 12) = $\dfrac{4}{36}$ (12 appears 4 times in the table)

$$= \dfrac{1}{9}$$

(d) All the prime numbers are, 2, 2, 3, 3, 5, 5. This gives a total outcomes of 6.

Therefore, Pr. (product of numbers is prime) = $\dfrac{6}{36}$

$= \dfrac{1}{6}$

(e) All products that are odd numbers are, 1, 3, 3, 5, 5, 9, 15, 15, 25. All products that are multiples of 5 are, 5, 5, 10, 10, 15, 15, 20, 20, 25, 30, 30.
They will both give a total outcome of 15. Note that, 5, 5, 15, 15, 25 are counted once under odd number. They should not be counted under multiples of 5, as this will result to double counting. Hence with this total outcome of 15,

Pr. (product of numbers is either odd or a multiple of 5) = $\dfrac{15}{36}$

$= \dfrac{5}{12}$

6. Three dice are thrown together. What is the probability of getting a total score of 10?

<u>Solution</u>.
If a die is thrown once, the total outcome is given by $6^1 = 6$. If two dice are thrown, the total outcome is $6^2 = 36$. Similarly, if three dice are thrown, the total outcome will be $6^3 = 216$.
Now, for us to draw a table with 216 outcomes will be very tedious. So, a direct way of solving this problem will be to select the outcomes from each die that will result in a total score of 10. These outcomes are:
(6, 3, 1), (6, 2, 2), (5, 4, 1), (5, 3, 2), (4, 4, 2), (4, 3, 3)
Each of the brackets above can give us 6 outcomes. For example, the first bracket above can give us the following 6 outcomes:
(6, 3, 1): which means - First die shows 6, second die shows 3, third die shows 1
(6, 1, 3): which means - First die shows 6, second die shows 1, third die shows 3
(1, 6, 3): which means - First die shows 1, second die shows 6, third die shows 3
(1, 3, 6): which means - First die shows 1, second die shows 3, third die shows 6
(3, 1, 6): which means - First die shows 3, second die shows 1, third die shows 6
(3, 6, 1): which means - First die shows 3, second die shows 6, third die shows 1
Similarly, each of the other brackets can give us 6 outcomes.
Let us write out our outcome brackets again. They are, (6, 3, 1), (6, 2, 2), (5, 4, 1), (5, 3, 2), (4, 4, 2), (4, 3, 3)
When each of these brackets give us 6 outcomes, then we will obtain a total of 36 (i.e. 6 x 6) outcomes. Recall that our overall outcome table will give us a total of 216 (i.e. 6^3) outcomes.

Therefore, Pr. (getting a total score of 10) $= \dfrac{36}{216}$

$= \dfrac{1}{6}$

Exercise 44

1. A fair die is thrown once. Find the probability of getting:

(a) a 5

(b) a 1

(c) a 9

(d) a 2 or 3 or 6

(e) a number less than 6

(f) a prime or an even number

2. A fair die is rolled once. What is the probability of getting:

(a) a number divisible by 2

(b) a multiple of 3

(c) at least 2

(d) at most 3

(e) a perfect square or an odd number

(f) either a number greater that 5 or a multiple of 3

3. A die is thrown and a coin is tossed. What is the probability of getting:

(a) a 5 and a head

(b) a tail and a perfect cube

4. Two fair dice are thrown at the same time. Find the probability of getting:

(a) at least one four

(b) a sum of at least 6

(c) a sum of at most 10

(d) a sum less than 8

(e) a total of 12

(f) a sum that is either a perfect square or a multiple of 5

(g) a sum that is either divisible by 6 or a multiple of 4

5. An unbiased die with faces numbered 1 to 6 is rolled twice. Find the probability that the product of the numbers obtained is:

(a) prime

(b) divisible by 6

(c) 9

(d) a factor of 10

(e) either perfect cube or a multiple of 8

6. Three dice are thrown together. What is the probability of getting a total score of 11?

CHAPTER 45
MISCELLANEOUS PROBLEMS ON PROBABILITY

Examples

1. A box contains two green balls, three yellow balls and four white balls. A ball is picked at random from the box. What is the probability that it is:

(a) green

(b) yellow

(c) white

(d) blue

(e) not white

(f) either yellow or green

<u>Solution</u>

Total number of balls in the box = 2 + 3 + 4 = 9

(a) Pr. (that it is green) = $\dfrac{\text{Number of green balls}}{\text{Total number of balls in the box}}$

$= \dfrac{2}{9}$

(b) Pr. (that it is yellow) = $\dfrac{\text{Number of yellow balls}}{\text{Total number of balls in the box}}$

$= \dfrac{3}{9}$

$= \dfrac{1}{3}$

(c) Pr. (that it is white) = $\dfrac{\text{Number of white balls}}{\text{Total number of balls in the box}}$

$= \dfrac{4}{9}$

(d) There is no blue ball in the box.

Therefore, Pr. (that it is blue) = 0

(e) Pr. (that it is not white) = 1 - Pr. (that it is white)

$= 1 - \dfrac{4}{9}$

$= \dfrac{5}{9}$

(f) Pr. (that it is either yellow or green) = $\dfrac{\text{Number of yellow and green balls}}{\text{Total number of balls in the box}}$

$$= \dfrac{3+2}{9}$$

$$= \dfrac{5}{9}$$

Or,

Pr. (that it is either yellow or green) = Pr. (that it is yellow) + Pr. (that it is green)   (Since OR means addition)

$$= \dfrac{1}{3} + \dfrac{2}{9}$$

$$= \dfrac{3+2}{9}$$

$$= \dfrac{5}{9} \quad \text{(As obtained before)}$$

2. A letter is chosen at random from the word COMPUTER. What is the probability that it is:

(a) either in the word MORE or in the word CUT

(b) either in the word COPE or in the word CUTE

(c) neither in the word ROT nor in the word CUP

Solution

(a) The total number of letters in COMPUTER is 8 letters.

In the word MORE, the number of letters is 4, while in the word CUT, the number of letters is 3. They both give a total of 7 letters.

Therefore, Pr. (that it is either in the word MORE or in the word CUT) = $\dfrac{7}{8}$

(b) In the word COPE, the number of letters is 4, while in the word CUTE, the number of letters is 4. Without counting any letter twice (i.e. C and E), the two words give a total of 6 letters (i.e. C, O, P, E, U, T).

Therefore, Pr. (that it is either in the word COPE or in the word CUTE) = $\dfrac{6}{8}$ (The total number of letters in COMPUTER is 8 letters).

$$= \dfrac{3}{4}$$

(c) Out of the 8 letters in COMPUTER, the letters that are neither in the word ROT nor in the word CUP are letters M and E. They are 2 letters.

Therefore, Pr. (that it is neither in the word ROT nor in the word CUP) = $\dfrac{2}{8}$

$$= \dfrac{1}{4}$$

(3) In a college 80% of the boys and 45% of the girls can drive a car. If a boy and a girl are chosen at random, what is the probability that:

(a) both of then can drive a car |

(b) the boy cannot drive a car and the girl can drive a car

(c) neither of them can drive a car?

(d) one of them can drive a car

Solution

The probabilities are given in percentage. Hence the total for each probability is 100%

Therefore, Pr. (a boy can drive a car) $= \dfrac{80}{100}$

$= \dfrac{4}{5}$

Pr. (a boy cannot drive a car) $= \dfrac{20}{100}$ (i.e. 100 - 80 = 20)

$= \dfrac{1}{5}$ (Can also be obtained from $1 - \dfrac{4}{5}$)

Similarly, Pr. (a girl can drive a car) $= \dfrac{45}{100}$

$= \dfrac{9}{20}$ (After equal division by 5)

Pr. (a girl cannot drive a car) $= 1 - \dfrac{9}{20}$)

$= \dfrac{11}{20}$

(a) Therefore, Pr. (both of them can drive a car) = Pr. (a boy can drive a car) AND Pr. (a girl can drive a car)

= Pr. (a boy can drive a car) x Pr. (a girl can drive a car)

$= \dfrac{4}{5} \text{ x } \dfrac{9}{20}$

$= \dfrac{36}{100}$

$= \dfrac{9}{25}$

(b) Pr. (the boy cannot drive a car and the girl can drive a car) = Pr. (a boy cannot drive a car) AND Pr. (a girl can drive a car)

= Pr. (a boy cannot drive a car) x Pr. (a girl can drive a car)

$= \dfrac{1}{5} \text{ x } \dfrac{9}{20}$

$= \dfrac{9}{100}$

(c) Pr. (neither of them can drive a car) = Pr. (a boy cannot drive a car) AND Pr. (a girl cannot drive a car)

= Pr. (a boy cannot drive a car) x Pr. (a girl cannot drive a car)

$$= \frac{1}{5} \times \frac{11}{20}$$

$$= \frac{11}{100}$$

(d) Since we do not know which of then can drive a car, then this case is logically explained as follows:

Pr. (one of them can drive a car) = either the boy can drive a car AND the girl cannot drive a car OR the girl can drive a car AND the boy cannot drive a car.

This in now calculated as follows:

Pr. (one of them can drive a car) = Pr. (the boy can drive a car) x Pr. (the girl cannot drive a car) + Pr. (the girl can drive a car) x Pr. (the boy cannot drive a car)

$$= (\frac{4}{5} \times \frac{11}{20}) + (\frac{9}{20} \times \frac{1}{5})$$

$$= \frac{11}{25} + \frac{9}{100}$$

$$= \frac{44 + 9}{100}$$

$$= \frac{53}{100}$$

4. The probability of a seed germinating is $\frac{2}{5}$. If three of the seeds are planted, what is the probability that:

(a) none will germinate

(b) at least one will germinate

(c) at least one will not germinate

(d) only one will germinate

Solution

This is a case of selection of three items from two possible events. We are going to write our outcomes in bracket like a tree diagram method. In order to write out the outcomes, let us use the letter G to represent germinate and letter N to represent not germinate.

Hence the outcomes are written as follows:

(GGG), (GGN), (GNG), (GNN), (NGG), (NGN), (NNG), (NNN)

(a) The probability that none will germinate is given by (NNN).

From the question, the probability that a seed germinate, G = $\frac{2}{5}$. Therefore the probability that it will

not germinate, N = 1 - G = $1 - \frac{2}{5} = \frac{3}{5}$

Hence, $G = \dfrac{2}{5}$, $N = \dfrac{3}{5}$

Therefore, Pr. (that none will germinate) = (NNN)

$$= \frac{3}{5} \times \frac{3}{5} \times \frac{3}{5}$$

$$= \frac{27}{125}$$

(b) The outcomes of the probability that at least one will germinate are, (GGG), (GGN), (GNG), (GNN), (NGG), (NGN), (NNG). Hence we can compute each of the outcomes and add them together. But this will be tedious. An easier way of solving this problem is as explained below. The difference between the outcome in question (a) and (b) is (NNN). This shows that subtracting (NNN) from the total probability will give us the outcomes in question (b). Recall that the total of any probability is 1. Therefore, 1 - (NNN) = outcomes in (b)

Hence, Pr. (that at least one will germinate) = 1 - (NNN)

$$= 1 - \frac{27}{125} \quad \text{[Note that (NNN)} = \frac{27}{125} \text{ as calculated in question (a)]}$$

$$= \frac{108}{125}$$

(c) The outcomes of the probability that at least one will not germinate are, (GGN), (GNG), (GNN), (NGG), (NGN), (NNG), (NNN). Similar to (b) above, the difference between this outcomes of this question and the overall outcomes is (GGG).

Therefore, Pr. (that at least one will not germinate) = 1 - (GGG)

Let us calculate (GGG) as follows:

Pr. [that all three will germinate, i.e. (GGG)] $= \dfrac{2}{5} \times \dfrac{2}{5} \times \dfrac{2}{5}$

$$= \frac{8}{125}$$

Therefore, Pr. (that at least one will not germinate) = 1 - (GGG)

$$= 1 - \frac{8}{125}$$

$$= \frac{117}{125}$$

(d) The outcomes of the probability that only one will germinate are, (GNN), (NGN), (NNG). Hence we will calculate each of these outcomes and add them together.

(GNN) = Pr. (that the first will germinate) x Pr. (that the second will not germinate) x Pr. (that the third will not germinate)

$$= \frac{2}{5} \times \frac{3}{5} \times \frac{3}{5}$$

$$= \frac{18}{125}$$

(NGN) $= \dfrac{3}{5} \times \dfrac{2}{5} \times \dfrac{3}{5}$

$$= \frac{18}{125}$$

$$(NNG) = \frac{3}{5} \times \frac{3}{5} \times \frac{2}{5}$$

$$= \frac{18}{125}$$

Therefore, Pr. (that only one will germinate) $= \frac{18}{125} + \frac{18}{125} + \frac{18}{125}$

$$= \frac{54}{125}$$

5. When children are born, they are equally likely to be boys or girls. What is the probability that in a family of four children:

(a) three are boys and one is a girl

(b) at least two are girls

(c) two are boys and two are girls

(d) the first and second born are girls

Solution

Since children are equally likely to be boys or girls, it means that the probability of having a boy is $\frac{1}{2}$,

and the probability of having a girl is also $\frac{1}{2}$. This is similar to the case of tossing a coin (i.e. $\frac{1}{2}$ for head and $\frac{1}{2}$ for tail).

Therefore, the case of a family of four children is like when four coins are tossed. Refer to the example on tossing four coins in our previous chapter.

Let us use B for boy and G for girl to write out the total outcomes of 16 (i.e. 2^4 = 16) as shown below.

The outcomes are: (BBBB), (BBBG), (BBGG), (BGGG), (GBBB), (GGBB), (GGGB), (GBGB), (BGBG), (BBGB), (GBBG), (BGGB), (GGBG), (GBGG), (BGBB), (GGGG). This gives a total of 16 outcomes.

(a) The outcomes that the children are three boys and one girl are, (BBBG), (GBBB), (BBGB), (BGBB). This gives 4 outcomes.

Therefore, Pr. (three are boys and one is a girl) $= \frac{4}{16}$

$$= \frac{1}{4}$$

(b) The outcomes that the children are at least two girls are, (BBGG), (BGGG), (GGBB), (GGGB), (GBGB), (BGBG), (GBBG), (BGGB), (GGBG), (GBGG), (GGGG). This gives 11 outcomes.

Therefore, Pr. (at least two are girls) = $\dfrac{11}{16}$

(c) The outcomes that the children are two boys and two girls are, (BBGG), (GGBB), (GBGB), (BGBG), (GBBG), (BGGB). This gives 6 outcomes.

Therefore, Pr. (two are boys and two are girls) = $\dfrac{6}{16}$

$= \dfrac{3}{8}$

(d) The outcomes that the first and second born are girls are, (GGBB), (GGGB), (GGBG), (GGGG). This gives 4 outcomes.

Therefore, Pr. (the first and second born are girls) = $\dfrac{4}{16}$

$= \dfrac{1}{4}$

6. A bag contains three blue balls, four red balls and five white balls. Three balls are removed from the bag without replacement. What is the probability of getting:
(a) a white, blue and red balls in that order
(a) one of each colour
(c) at least two white balls

Solution
The total number of balls in the bag = 3 + 4 + 5 = 12

(a) A white, blue and red balls in that order means that the first is white, the second is blue and the third is red. This can be represented as (WBR).
Note that this is a case of without replacement. Hence after each ball is removed, the total number of ball remaining and the number of the particular ball removed are both reduced by one.

Therefore, Pr. (getting a white, blue and red balls, i.e. WBR) = $\dfrac{5}{12} \times \dfrac{3}{11} \times \dfrac{4}{10}$. (Notice how the total balls is reduced by 1 after each ball is removed from the bag.

$= \dfrac{60}{1320}$

$= \dfrac{1}{22}$ (After equal division by 60)

(b) Let B represent blue, R represent red and W represent white. Then the outcomes for getting one of each colour are given by: (BRW), (BWR), (RBW), (RWB), (WBR), (WRB).

Let us now calculate each of them.

(BRW) = Pr. (First is blue) x Pr. (Second is red) x Pr. (Third is white)

$$= \frac{3}{12} \times \frac{4}{11} \times \frac{5}{10}$$

$$= \frac{1}{4} \times \frac{4}{11} \times \frac{1}{2}$$

$$= \frac{4}{88}$$

$$= \frac{1}{22}$$

Similarly, each of the other five outcomes, i.e. (BWR), (RBW), (RWB), (WBR), (WRB), will each give us a value of $\frac{1}{22}$ when calculated. This is because each is obtained by multiplying 3 x 4 x 5, to give the numerator, and 12 x 11 x 10, to give the denominator, which simplifies to $\frac{1}{22}$.

Therefore, Pr. (getting one of each colour) $= \frac{1}{22} + \frac{1}{22} + \frac{1}{22} + \frac{1}{22} + \frac{1}{22} + \frac{1}{22}$

$$= \frac{6}{22}$$

$$= \frac{3}{11}$$

(c) Let us write out a different outcome for this problem. Since we are concerned about one colour, we are going to use W to represent white colour, and N to represent not a white colour. This will give us 8 outcomes in brackets as usual. The outcomes are:

(WWW), (WWN), (WNW), (WNN), (NWW), (NWN), (NNW), (NNN).

The outcomes representing at least two white balls are: (WWW), (WWN), (WNW), (NWW).

Number of white balls is 5. Therefore number of balls that are not white = 12 - 5 = 7, or blue + red = 3 + 4 = 7. (Blue and red ball are the balls that are not white balls).

Let us now calculate each of the outcomes above as follows:

(WWW) = Pr. (first is white) x Pr. (second is white) x Pr. (third is white)

$$= \frac{5}{12} \times \frac{4}{11} \times \frac{3}{10}$$ (Take note of the reduction in the white balls and total number of

balls as each ball is removed from the bag)

$$= \frac{60}{1320}$$

$$= \frac{1}{22}$$

(WWN) $= \frac{5}{12} \times \frac{4}{11} \times \frac{7}{10}$ (Note that there are 7 balls that are not white)

$$= \frac{140}{1320}$$

$$= \frac{7}{66}$$

(WNW) $= \frac{5}{12} \times \frac{7}{11} \times \frac{4}{10}$

431

$$= \frac{140}{1320}$$

$$= \frac{7}{66}$$

$$(NWW) = \frac{7}{12} \times \frac{5}{11} \times \frac{4}{10}$$

$$= \frac{140}{1320}$$

$$= \frac{7}{66}$$

Therefore, Pr. (getting at least two white balls) = (WWW) or (WWN) or (WNW) or (NWW)

$$= (WWW) + (WWN) + (WNW) + (NWW)$$

$$= \frac{1}{22} + \frac{7}{66} + \frac{7}{66} + \frac{7}{66}$$

$$= \frac{3 + 7 + 7 + 7}{66}$$

$$= \frac{24}{66}$$

$$= \frac{4}{11}$$

7. A committee consist of 6 men and 4 women. A subcommittee made up of three members is randomly chosen from the committee members. What is the probability that:

(a) they are all men

(b) two of them are women?

Solution

Let us write out the outcome for this problem. Let M represent man, and W represent woman. This will give us 8 outcomes in brackets as usual. The outcomes are:

(WWW), (WWM), (WMW), (WMM), (MWW), (MWM), (MMW), (MMM).

(a) The total members in the committee are: 6 + 4 = 10.

The outcomes representing all men is (MMM)

Therefore, Pr. (they are all men, i.e. MMM) = Pr. (first is a man) x Pr. (second is a man) x Pr. (third is a man)

$$= \frac{6}{10} \times \frac{5}{9} \times \frac{4}{8}$$ (Notice the reduction in the number of men and people left, as each member is chosen from the committee).

$$= \frac{130}{720}$$

$$= \frac{13}{72}$$

432

(b) The outcomes showing that two of them are women are: (WWM), (WMW), (MWW)

Let us calculate each of them as follows:

(WWM) = Pr. (the first is a woman) x Pr. (the second is a woman) x Pr. (the third is a man)

$$= \frac{4}{10} \times \frac{3}{9} \times \frac{6}{8}$$

$$= \frac{72}{720}$$

$$= \frac{1}{10}$$

(WMW) = $\frac{4}{10} \times \frac{6}{9} \times \frac{3}{8}$

$$= \frac{72}{720}$$

$$= \frac{1}{10}$$

(MWW) = $\frac{6}{10} \times \frac{4}{9} \times \frac{3}{8}$

$$= \frac{72}{720}$$

$$= \frac{1}{10}$$

Therefore, Pr. (two of them are women) = (WWM) or (WMW) or (MWW)

$$= (WWM) + (WMW) + (MWW)$$

$$= \frac{1}{10} + \frac{1}{10} + \frac{1}{10}$$

$$= \frac{3}{10}$$

8. A box contains seven blue pens and three red pens. Three pens are picked one after the other without replacement. Find the probability of picking:

(a) two blue pens

(b) at least two red pens

(c) at most two blue pens

Solution

Let B represent blue pen, and R represent red pen. The outcomes are:

(BBB), (BBR), (BRB), (BRR), (RBB), (RBR), (RRB), (RRR).

The total number of pens = 7 + 3 = 10

(a) The outcomes showing two blue pens are: (BBR), (BRB), (RBB)

Let us calculate each of them as follows:

(BBR) = Pr. (the first is a blue pen) x Pr. (the second is a blue pen) x Pr. (the third is a red pen)

$$= \frac{7}{10} \times \frac{6}{9} \times \frac{3}{8}$$

$$= \frac{126}{720}$$

$$= \frac{7}{40} \quad \text{(In its lowest term after equal division by 18)}$$

$$(BRB) = \frac{7}{10} \times \frac{3}{9} \times \frac{6}{8}$$

$$= \frac{126}{720}$$

$$= \frac{7}{40}$$

Also, $(RBB) = \frac{7}{40}$ (Similar to the once above)

Therefore, Pr. (picking two blue pens) $= \frac{7}{40} \times \frac{7}{40} \times \frac{7}{40}$

$$= \frac{21}{40}$$

(b) The outcomes representing at least two red pens are: (RRR), (RRB), (RBR), (BRR)

Let us now calculate each of the outcomes as follows:

(RRR) = Pr. (first is a red pen) x Pr. (second is a red pen) x Pr. (third is a red pen)

$$= \frac{3}{10} \times \frac{2}{9} \times \frac{1}{8} \quad \text{(Take note of the reduction in the red pens and total number of pens as}$$

each pen is picked from the box)

$$= \frac{6}{720}$$

$$= \frac{1}{120}$$

$$(RRB) = \frac{3}{10} \times \frac{2}{9} \times \frac{7}{8}$$

$$= \frac{42}{720}$$

$$= \frac{7}{120}$$

Hence, $(RBR) = \frac{7}{120}$ (This is similar to the one above)

And, $(BRR) = \frac{7}{120}$ (Same reason as above)

Therefore, Pr. (picking at least two red pens) $= \frac{1}{120} + \frac{7}{120} + \frac{7}{120} + \frac{7}{120}$

$$= \frac{1 + 7 + 7 + 7}{120}$$

$$= \frac{22}{120}$$

$$= \frac{11}{60}$$

(c) The outcomes that represent picking at most two blue pens are: (BBR), (BRB), (BRR), (RBB), (RBR), (RRB), (RRR). Note that at most two blue pens means 2, 1 or 0 blue pens.

Notice that there is only (BBB) missing from this outcome. This shows that it can be obtained by: total probability - (BBB). Which is: 1 - (BBB).

Let us calculate (BBB) as follows:

(BBB) = Pr. (first is a blue pen) x Pr. (second is a blue pen) x Pr. (third is a blue pen)

$$= \frac{7}{10} \times \frac{6}{9} \times \frac{5}{8}$$

$$= \frac{210}{720}$$

$$= \frac{7}{24} \quad \text{(After equal division by 30)}$$

Therefore, Pr. (picking at most two blue pens) = 1 - (BBB)

$$= 1 - \frac{7}{24}$$

$$= \frac{17}{24}$$

Exercise 45

1. A box contains 5 green balls, 8 yellow balls and 7 white balls. A ball is picked at random from the box. What is the probability that it is:

(a) green

(b) yellow

(c) white

(d) blue

(e) not white

(f) either yellow or green

2. A letter is chosen at random from the word NORMADIC. What is the probability that it is:

(a) either in the word MAD or in the word CORN

(b) either in the word NORM or in the word DAM

(c) neither in the word RID nor in the word CAN

(3) In a college 20% of the boys and 8% of the girls who had graduated from the college, graduated with distinction since the inception of the college. If a boy and a girl are chosen at random, what is the probability that:

(a) both of them will graduate with distinction

(b) the boy will not and the girl will graduate with distinction

(c) neither of them will graduate with distinction?

(d) one of them will graduate with distinction

4. The probability of a seed germinating is $\frac{1}{4}$. If three of the seeds are planted, what is the probability that:

(a) none will germinate

(b) at least one will germinate

(c) at least one will not germinate

(d) only one will germinate

5. When parents who are carriers of sickle cell disorder get married, they are equally likely to give birth to normal child and sick child. What is the probability that in a family of three children:

(a) two are normal and one is sick

(b) at least two are sick

(c) one is normal and two are sick

(d) the first is sick

(e) at most one is normal

6. A box contains six blue balls, three red balls and five white balls. Three balls are removed from the bag without replacement. What is the probability of getting:

(a) a white, blue and red balls in that order

(a) one of each colour

(c) at least two white balls

7. A committee consist of 4 men and 2 women. A subcommittee made up of two members is randomly chosen from the committee members. What is the probability that:

(a) they are all men

(b) one of them is a woman?

8. A bag contains 5 blue balls and seven red balls. Three balls are picked one after the other without replacement. Find the probability of picking:

(a) two blue balls

(b) at least two red balls

(c) at most two blue balls

ANSWERS TO EXERCISES

Exercise 1

1. (a) -15   (b) -11   (c) -5   (d) -12   (e) 19   (f) 6   (g) -82

2. (a) -10   (b) -22   (c) 36   (d) 30   (e) -4   (f) -6   (g) 8   (h) 4

3. (a) 12   (b) 24   (c) $\frac{71}{3}$ or $23\frac{2}{3}$ (d) $\frac{19}{8}$ or $2\frac{3}{8}$ (e) -1910

Exercise 2

1. 1000y metres   2. $(38 + m)$years, $(38 + m + y)$years   3. $(b - c)$years

4. (a) 100y cents   (b) $\frac{n}{60}$ minutes, $\frac{n}{60 \times 60}$ 5. $(22x - x^2)$cm^2 6. $(x - 1)(x + 10)$cm^2

7. $10x - 5y$ 8. $8x - 3y - 5$ 9. $2a - 3b$ 10. $11x^2 + 15x$ 11. $2a^2 - 7ab - 15b^2$

12. $15x^2 - 13xy + 2y^2$ 13. $16a^2 - 40a + 25$ 14. 24   15. 3   16. (a) lmn (b) 6cde

(c) $30a^3x^2y^2$ (d) 20mn   (e) $(x + y)(3x - y)$ 17. (a) $\frac{3x}{5}$ (b) $\frac{4x + 3}{9x^2}$ (c) $\frac{3ax - 14b}{30bx}$

(d) $\frac{10bx + 25b + 8ab - 24a}{40ab}$ (e) $\frac{25a^3 - 15a^2 - 15a^2b - 3a + 15b}{15a^2}$ 18. (a) $\frac{-15x + 8}{30}$ (b) $\frac{3}{2(3a + 7)}$

(c) $\frac{-7a(a+2) + 6}{(3a - 1)(a - 2)}$ 19. (a) 5   (b) 2   (c) $6a^2b^2$ (d) ab^2c^2 (e) $12x^2yz^2$ 20. $2(2m - 4n)$

21. $4(3y - 2)$ 22. $-10y(2 + 5z)$ 23. $m^2(5m^2 - 10 - m)$ 24. $(2 + 5m)(x + 3y)$ 25. $(a - 1)(3b + 7)$

26. $(5b + n)(x - y)$ 27. $(5 + y)(3 - x)$ 28. $(10 + m)(10 - m)$ 29. $(5x + 3y(5x - 3y)$

30. $7(1 + b)(1 - b)$ (31) $(6p + 2)(6p - 2)$ 32. 2600   33. 8400   34. $(x + 3)(x + 3)$

35. $(b - 5)(b + 4)$ 36. $(n - 8)(n - 6)$ 37. $(5a + 4)(a - 3)$ 38. (a) -6   (b) -5   39. (a) 23   (b) -78

(c) $\frac{6}{7}$ 40. -12   41. (a) 0   (b) -24

Exercise 3

1. (a) $-3t^4e^{12}$ (b) $64a^3b^9$ (c) $-(a^4)$ (d) g^{20} (e) $-m$ 2. (a) $\frac{1}{3a}$ (b) $\frac{1}{a}$ (c) $7x^{\frac{3}{2}}$ (d) $9x$

3. (a) $x = \frac{1}{25}$ (b) $a = \frac{1}{3}$ (c) $x = 3\frac{1}{2}$ (d) $x = 4$

Exercise 4

1. (a) $x = \frac{4}{3}$ (b) $a = \frac{5}{16}$ (c) $b = -\frac{4}{7}$ 2. (a) $x = \frac{5}{11}$ (b) $x = 1\frac{11}{20}$ (c) $\frac{113}{62}$ 3. $p = \frac{md^2}{3(md - t)}$

4. $C = \sqrt{\frac{P^2}{I^2V^2} - \frac{E^2}{I^2}}$ or $C = \frac{1}{I}\sqrt{\frac{P^2}{V^2} - E^2}$ 5. $x = \frac{R^2T - S}{b - aR^2}$ 6. $m = 0$ 7. $Y = \sqrt[3]{\frac{TC^2}{L} - B}$

or $Y = \sqrt[3]{\frac{TC^2 + BL}{L}}$ 8. $x = \sqrt{\frac{M}{2Q^2}}$ or $x = \frac{1}{Q}\sqrt{\frac{M}{2}}$

Exercise 5

1. 8   2. 4   3. Man = 100years, daughter = 20years   4. 10years   5. 150km/hr   6. 150km/hr

7. Eighteen $10 notes and six $1 notes   8. 15days   9. 1050litres   10. 2250km   11. 53.3km

12. 12   13. 13, 15 and 17   14. 5 and 24   15. 12   16. 15 and 25   17. 32years   18. $-\frac{5}{4}$

Exercise 6

1. $x = -1, y = 1$ 2. $x = 6$ 3. m = 2, n = 2   4. a = 10, b = 5   5. P = 4, q = -3   6. $x = -2, y = 2$

7. $x = 3, y = 6$ 8. $x = \frac{31}{23}, y = \frac{3}{23}$ 9. m = 1, n = -1   10. p = $-\frac{11}{23}$, q = $-\frac{3}{23}$ 11. c = $3\frac{1}{2}$, d = -$1\frac{1}{2}$

12. x = $-\frac{60}{53}$, y = $\frac{250}{53}$

Exercise 7

1. Book = $10, pen = $5   2. Mathematics = 30, Biology = 70   3. 18.75 and 56.25   4. m =
5hours,

n = 3hours   5. $x = -2, y = 3$. Sides are 14cm, 18cm and 18cm   6. (a) $x = 6, y = 7.5$ (b) 30cm
each

7. 47   8. 92   9. 35   10. $\frac{3}{8}$

Exercise 8

1. (a) P: (Closed, True)   (b) Q: (Open, Impossible to say)   (c) R: (Neither, not applicable)
2. (a) Stanley is not neat   (b) Jane is a doctor   (c) m ≠ -5   3. (a) Y ⇒ X   (b) ~Z ⇒ Y
4. (a) If r is negative then r^3 is negative   (b) If y^3 is equal to -64 then y is equal to -4
5. Abel is intelligent ⇒ Abel is clever   6. (a) If you do not travel by road, it will not take you 11
hours.
(b) Those students who do not offer mathematics will not score above 80
7.

p	q	~p	~q	~p ∧ ~q	~(~p ∧ ~q)	p ∨ q
T	T	F	F	F	T	T
T	F	F	T	F	T	T
F	T	T	F	F	T	T
F	F	T	T	T	F	F

8. Valid   9. Valid   10. Not valid   11. Not valid

Exercise 9

1. (a) a(7a + 9)   (b) (m + 4)(m - 4)   (c) (3p + 11)(3p - 11)   (d) 8(2b + 5)(2b - 5)   2. (a) (3x +
5)(x - 3)

(b) $(x + 9)(x - 8)$ (c) $(5x + 2)(x + 4)$ (d) $(10x + 6)(x - 1)$ 3. (a) Yes   (b) Yes   (c) No   (d) No

(e) No   4. (a) a = 0, or -4   (b) b = -6 or 6   (c) y = $\frac{7}{2}$ or -$\frac{7}{2}$ (d) $x = 4$ or 1   5. (a) $x = \frac{4}{3}$ or -1

438

(b) $y = 2$ or -3 (c) $y = 2$ or 18 (d) $m = \frac{1}{2}$ or $-\frac{4}{5}$ (e) $x = \frac{6}{7}$ or -1 6. (a) $x^2 - 3x - 40 = 0$

(b) $4x^2 + 27x - 7 = 0$ (c) $5x^2 + 32x + 12 = 0$ 7. (a) $\frac{15}{2}$ (b) -4 8. (a) $-\frac{4}{3}$ (b) $-\frac{11}{3}$

9. $M = 5$. The other root is $-\frac{9}{2}$ 10. $K = -8$. The other root is $-\frac{11}{3}$ 11. $K = -6$, $P = -27$

12. $A = 9$, $B = 30$ 13. $K = 12$ or -12. The two possible equations are $m^2 - 12m + 27 = 0$ and $m^2 +$

$12m + 27 = 0$ 14. 64 15. $\frac{169}{4}$ 16. $a^2 - 6a + 9$ 17. $36m^2 + 60m + 26$ 18. $x - 4$ 19. $x + 0.6$

20. (a) $y = -11$ or 2 (b) $x = 3$ or 8 (c) $x = 4$ or $-\frac{3}{2}$ (d) $x = 1$ or $\frac{7}{3}$ (e) $y = -1$ or $-\frac{6}{5}$ (f) $m = 2$

or $-\frac{7}{2}$

(g) $x = 5.56$ or 1.44 21. (a) $x = 3$ or -4 (b) $x = 1$ or $\frac{10}{7}$ (c) $x = -\frac{1}{2}$ or 4 (d) $x = 2.76$ or -1.09

(e) $x = 0.52$ or -1.92 (f) $n = \frac{3}{2}$ twice

Exercise 10

1. -4 and -18 or 4 and 18   2. -3 and 7 or -7 and 3   3. 11years and 18years   4. 2.89cm   5. 2m
6. 11years   7. 4 years   8. -16 and -18 or 16 and 18   9. 10   10. 12years and 36years

Exercise 11

1. (a) $x = \frac{5y}{4}$ (b) $x = 12.5$ (c) $y = 12.8$ 2. (a) $h = \frac{5}{3}\sqrt{P}$ (b) $p = 144$ (c) $h = \frac{25}{6}$ 3. (a) $p = \frac{36}{q}$

(b) $q = \frac{9}{5}$ (c) $p = 7\frac{1}{5}$ 4. (a) $m = \frac{15}{\sqrt[3]{n}}$ (b) $m = 7\frac{1}{2}$ (c) $n = 24{,}146$ 5. (a) $a = \frac{8b^2}{3c}$ (b) $a = \frac{20}{3}$

(c) $b = 1.225$ 6. (a) $w = \sqrt{\frac{9h}{5L}}$ (b) $w = 2.32$ (c) $L = 0.576$ 7. (a) $x = 2 + 3y$ (b) $x = 26$

8. (a) $E = 5 + 10F$ (b) $E = 30$ (c) $F = 3\frac{1}{2}$

Exercise 12

1. (a) $(-12, 73)$ or $(2, 3)$ (b) $(-\frac{23}{9}, -\frac{10}{9})$ or $(-1, 2)$ (c) $(-\frac{13}{2}, -\frac{19}{2})$ or $(-2, 1)$ (d) $(-\frac{3}{5}, -\frac{5}{6})$ or $(\frac{1}{2}, 1)$

(e) $(\frac{23}{5}, -\frac{3}{5})$ or $(3, -1)$ (f) $(-5, 3)$ (g) $(2, 1)$ or $(\frac{14}{17}, -\frac{33}{17})$ 2. Boy = 30years old, Father = 60 years
old

3. $(\frac{47}{4}, -\frac{13}{2})$ or $(5, 7)$ 4. 6years and 11years

Exercise 13

1. $x > 2$

2. $x < -1$

3. $x < 2\frac{32}{43}$

4. $x \le -4$

5. -1, -2 and -3   6. 1, 2 and 3   7. 65, 66 and 67   8. 3, 4 and 5

9. $-1 \le y \le 2$

10. $-2 < y < 3$

11. $y > -1$ or $y \le -7$

12. $4 < m < 5$ 13. (a) $x \ge 5$, $y \ge 2$, $x + y \le 10$ and $2x + y \le 15$, where x = drivers and y = supervisors

(b) 5 supervisors   (c) 5 drivers and 5 supervisors   14. (a) 50 items consisting of 40 bicycles and 10 tricycles   (b) 11 tricycles and 33 bicycles   15. (a) 7 shirts and 5 trousers   (b) $10

16. (a) $320f + 80c \ge 1680$, $8f + 6c \ge 62$ and $f + c \le 20$, where f = fux tin and c = clusion tin

(b) 4 fux tins and 5 clusion tins   17. (a) $4500   (b) 5 manual and 3 electric mixing machines

18. (a) A = 62units, B = 55 units   (b) $3440   19. 57.45   20. 94.93

Exercise 14

1. $-5 < x < 6$ 2. $x \ge \frac{2}{3}$ or $x \le -4$ 3. $\frac{4}{5} \le x \le 1$ 4. $-\frac{3}{5} < x < \frac{3}{2}$ 5. $-1.95 \le x \le 2.95$ 6. $-2 < x < \frac{9}{2}$

7. $x \ge \frac{6}{7}$ or $x \le -1$ 8. $-2 \le x \le -\frac{7}{5}$ 9. $-3 < x < 5$ 10. $-1.5 \le x \le 3.7$

Exercise 15

1. 10 units   2. $x = 12$ 3. (a) $\begin{pmatrix} -13 \\ 27 \end{pmatrix}$ (b) $\begin{pmatrix} -1 \\ 1 \end{pmatrix}$ (c) $\begin{pmatrix} -2\frac{1}{2} \\ 8 \end{pmatrix}$ (d) 59.1   4. (a) $p = -\frac{41}{23}$, $q = -\frac{7}{23}$

(b) 37.6   5. (a) $\begin{pmatrix} -11 \\ 38 \end{pmatrix}$ (b) $(\sqrt{50}, 81.9°)$ 6. $\theta = 33.7°$, (7.2, 33.7°)   7. (a) $\begin{pmatrix} -2 \\ 4 \end{pmatrix}$, $\sqrt{20}$ (b) $\begin{pmatrix} -6 \\ -2 \end{pmatrix}$, $\sqrt{40}$

(c) $\begin{pmatrix} 6 \\ -3 \end{pmatrix}$, 6.7   (d) $\begin{pmatrix} -4 \\ -6 \end{pmatrix}$, 7.2   8. (a) Q = (-1, 6), R = (8, 9)   (b) $\begin{pmatrix} 1\frac{1}{2} \\ 2\frac{1}{2} \end{pmatrix}$ 9. 29.4units   10. 14.3

11. (a) $\begin{pmatrix} 12 \\ -8 \end{pmatrix}$ (b) $\begin{pmatrix} -7 \\ 9 \end{pmatrix}$ (c) $\begin{pmatrix} -5 \\ 8\frac{3}{4} \end{pmatrix}$ (d) 42.2   12. (a) $p = \frac{22}{7}$, $q = \frac{26}{21}$ (b) 19.6

13. (a) $PR = \begin{pmatrix} -4 \\ 5 \end{pmatrix}$, $|PR| = \sqrt{41}$ (b) $RQ = \begin{pmatrix} 4 \\ 2 \end{pmatrix}$, $|RQ| = \sqrt{20}$ (c) $-5QS = \begin{pmatrix} 15 \\ -10 \end{pmatrix}$, $|-5QS| = \sqrt{325}$

(d) $2PQ = \begin{pmatrix} 16 \\ -6 \end{pmatrix}$, $|2PQ| = \sqrt{292}$ or 17.09

Exercise 16

1. 9　　2. $-\dfrac{8}{5}$　　3. 9　　4. 14　　5. −14　　6. $\dfrac{2}{5}$　　7. $\dfrac{1}{3}$　　8. −1　　9. −15

10. 10　　11. $\dfrac{1}{4}$　　12. 27　　13. 12　　14. $\dfrac{1}{10}$　　15. $\dfrac{3}{7}$　　16. Continuous

17. Continuous　　18. Discontinuous　　19. Not continuous (Discontinuous

20. Not continuous　　21. Continuous　　22. Not continuous　　23. Not continuous

24(a) 8　　(b) 5　　26. $\dfrac{7}{2}$　　27. $\dfrac{1}{3}$　　28. 1　　29. $-\dfrac{1}{2}$　　30. $2\dfrac{1}{4}$

Exercise 17

1. 2　　2. $2x$　　3. $-3x^{-4}$ or $-\dfrac{3}{x^4}$　　4(a) $10x + 5h$　　(b) $10x$

5(a) $27x^2 + 27xh + 9h^2$　　(b) $27x^2$　　6(a) $\dfrac{2x^3 + 3x^2\Delta x + x(\Delta x)^2 + 2}{x(x + \Delta x)}$　　(b) $2x + \dfrac{2}{x^2}$

7. $6x - 10$　　8. $1 - \dfrac{3}{x^2}$　　9. $5 - 6x$　　10. $2 + \dfrac{1}{5} = 2\dfrac{1}{5}$

Exercise 18

1(a) $40x^4$　　(b) $2x^4$　　(c) $\dfrac{1}{3x^{\frac{2}{3}}}$ or $\dfrac{1}{3\sqrt[3]{x^2}}$　　(d) $\dfrac{1}{x^{\frac{6}{7}}}$ or $\dfrac{1}{\sqrt[7]{x^6}}$　　(e) $-\dfrac{5}{8x^{\frac{13}{8}}}$ or $-\dfrac{5}{8\sqrt[8]{x^{13}}}$

(f) $-\dfrac{5}{x^{\frac{7}{2}}}$ or $-\dfrac{5}{\sqrt{x^7}}$　　2. (a) $10x^4 - 12x^3 - 12x^2 + 10x - 6$　　(b) $7x^6 + 8x^3 + \dfrac{3}{x^2}$

(c) $18x^5 - 4x^3 - 5 - \dfrac{2}{x^2} + \dfrac{3}{x^4}$　　(d) $\dfrac{5}{4\sqrt[4]{x^3}} - \dfrac{5}{3\sqrt[3]{2x^4}}$　　3. $8(2x-5)^3$　　4. $\dfrac{-6}{(x^3-7)^3}$

5. $\dfrac{6x^2 + 7}{2(2x^3 + 7x)^{\frac{1}{2}}}$　　6. $5(21x^2 - 2x)(7x^3 - x^2 + 3)^4$　　7. $\left(9 + \dfrac{2}{x^2}\right)\left(3x - \dfrac{2}{3x}\right)^2$

8. $1 + \dfrac{9(6x - 1)}{(3x^2 - x - 10)^2}$　　9. $\dfrac{-4x^3}{\sqrt{1 - 2x^4}}$　　10. $\dfrac{-5x^2}{\sqrt[3]{(5x^3 - 1)^4}}$

Exercise 19

1. $12x - 1$　　2. $12x^3 + 50x$　　3. $\dfrac{15x + 30}{2\sqrt{3 + x}}$　　4. $(7x^2 + 30 - 7)(x^2 - 7)^2$

5. $48x^3 + 24x^2 - 6x - 20$　　6. $\dfrac{\sqrt{2}\,[(3x^2 - 1)^3 + 36x^2(9x^4 - 6x^2 + 1)]}{2\sqrt{x}}$

7. $\dfrac{15 - 7x}{4(x+3)^{\frac{1}{4}}}$ or $\dfrac{15 - 7x}{4\sqrt[4]{x + 3}}$　　8. $12x^3 + 9x^2 - 46x - 5$　　9. $15x^4 - 12x^3 - 3x^2 + 6x - 2$

10. $\dfrac{12x + 1}{2}$　　11. $\dfrac{9x^{\frac{7}{2}}}{2}$ or $\dfrac{9\sqrt{x^7}}{2}$　　12. $\dfrac{4x^3(7x - 33)}{3\sqrt[3]{2x - 11}}$

13. $-30x^5 - 175x^4 + 20x^3 + 102x^2 - 14x + 1$　　　14. $75x^4 + 28x^3 + 3x^2 + 28x - 7$

15. $-\dfrac{2}{x^2} + \dfrac{12}{x^5} - \dfrac{25}{x^6}$

Exercise 20

1. $\dfrac{x^2 - 2x + 4}{(x-1)^2}$　　2. $\dfrac{8x^2 + 24x - 3}{(2x+3)^2}$　　3. $\dfrac{42x}{(3x^2+1)^2}$　　4. $\dfrac{1}{(1-x)^{\frac{3}{2}}\sqrt{x+1}}$

5. $\dfrac{4(x^3-2)(x^3+1)}{x^3}$　　6. $\dfrac{-6(x^2+1)}{x^4\sqrt{3x^2+2}}$　　7. $\dfrac{-8x^2 + 12x + 24}{3(2x+1)^3(x^2-x-4)^{\frac{2}{3}}}$　　8. $\dfrac{10x^3 - 2x^2 - x + 4}{2(2-x)^{\frac{3}{2}}}$

9. $\dfrac{-6x}{(1-3x^2)^2}$　　10. $\dfrac{4}{(x-2)^2}$　or　$\dfrac{4}{(2-x)^2}$

Exercise 21

1. $\dfrac{3t}{2}$　　2. $\dfrac{2(4-t^3)^2}{3t}$　　3. $\dfrac{2r}{3(l+2r)}$　　4. $\dfrac{2t-1}{10t-1}$　　5. $\dfrac{-2m(v-u)}{(u+v)t^2}$　　6. $\dfrac{5(t^4+1)^2}{3(t^2-1)^2}$

7. $\dfrac{1-12t^2}{2t}$　　8. $\dfrac{15t^2}{2}$　　9. $\dfrac{8}{r}$　　10. $\dfrac{2s^3}{3}$

Exercise 22

1. $\dfrac{-15x^2}{3-2y}$　or　$\dfrac{15x^2}{2y-3}$　　2. $\dfrac{-4x}{9y^2}$　　3. $\dfrac{3y^2 + 12x^2 - y}{x - 6xy}$　　4. $\dfrac{3x^2 - y^2 - 4x}{2xy}$　　5. $\dfrac{x^2}{y}$

6. $\dfrac{4xy}{2x^2 + 5}$　　7. $\dfrac{-2xy}{x^2 + 6y^2 - 1}$　　8. $\dfrac{4y}{10y - 4x - 1}$　　9. $\dfrac{2xy^2 - 1}{2y(x^2-1)}$　　10. $\dfrac{-5x^4 y^2}{3}$

Exercise 23

1. $2\sec^2 2x$　　2. $-\dfrac{1}{5}\sin\dfrac{1}{5}x$　　3. $-50\sin 5x$　　4. $3\sin^2 x\cos x$　　5. $2\sec^2 x\tan x$

6. $60x^4\sin^3 3x^5\cos 3x^5$　　7. $-12x\cot 6x^2\csc 6x^2$　　8. $10x^4\sec 2x^5\tan 2x^5$　　9. $6x\sec^2 3x^2$

10. $-x(3x\sin 3x - 2\cos 3x)$　　11. $\dfrac{6x\cos 2x - 9\sin 2x}{x^4}$　　12. $48x^3\sec x^4\tan x^4$

13. $\dfrac{-5\tan x\cos 5x - \sec^2 x(2 - \sin 5x)}{\tan^2 x}$　　14. $\dfrac{\sin^2 x(3x\cos x - \sin x)}{2x^2}$

15. $\dfrac{-(2x+6)\csc^2 2x - \cot 2x)}{(x+3)^2}$　　16. $\dfrac{-\sin\sqrt[3]{3}}{5x^2}$　　17. $\dfrac{2[x - (x^2-3)\tan 2x]}{\sec 2x}$

18. $6x\cos 3x^2 - 18x^3\sin 3x^2$　　19. $-\cos x(2\cot x\sin x + \cos x\csc^2 x)$

20. $\dfrac{2(\sin^2 x + \sin^2 x)}{(\sin x + \cos x)^2}$　or　$\dfrac{2}{(\sin x + \cos x)^2}$　　21. $\dfrac{4\sec 4x\sin(4x-2) + \tan 4x\cos(4x-2)}{\cos^2(4x-2)}$

22. $-3\sin 3x - x^2\sec x\tan x - 2x\sec x$　　23. $27x^2\cos x^3\sin^8 x^3$　　24. $\dfrac{3\cos 6\sqrt{x}}{\sqrt{x}}$

25. $3\cos x^3 - \dfrac{2\sin x^3}{x^3}$　　26. $-30x^4\cos^2 2x^5\sin 2x^2$　　27. $-10\sin 2x\sin 10x - 2\cos 2x\cos 10x$

28. $6x^2\cos x^4 - \dfrac{3\sin x^4}{2x^2}$　　29. $\dfrac{(10x - 5)\sec^2 5x - 2\tan 5x}{(2x-1)^2}$　　30. $\dfrac{\sec^2 x\tan x}{x} - \dfrac{\tan^2 x}{2x^2}$

442

Exercise 24

1. $\dfrac{1}{3\sqrt[3]{7y^2}}$

2. $\dfrac{3y^2}{2\sqrt{y^3-1}}$

3. $\dfrac{1}{5\sqrt[5]{2(y+3)^4}}$

4. $\dfrac{1}{\sqrt[3]{2(3y+27)^2}}$

5. $\dfrac{-1}{(y-2)^2}$

6. $\dfrac{-3}{2y^2\sqrt{\frac{3}{y}+5}}$

7. $2y^3$

8. $5y^4$

9. $\dfrac{-5}{(y-4)^2}$

10. $\dfrac{-1}{3y^2\sqrt[3]{(\frac{1}{y}+8)^2}}$

Exercise 25

1. $\dfrac{2}{\sqrt{1-4x^2}}$

2. $\dfrac{-1}{\sqrt{1-x^2}}$

3. $\dfrac{-6x}{9x^4+1}$

4. $\dfrac{-4x^3}{x^8+1}$

5. $3\sec^{-1}x+\dfrac{3}{\sqrt{x^2-1}}$

6. $2x+\dfrac{3}{x^2+1}$

7. $\dfrac{1}{(x+5)^2+1}$ or $\dfrac{1}{x^2+10x+26}$

8. $\dfrac{1}{x^2\sqrt{1-\frac{1}{x^2}}}$

9. $\dfrac{-1}{5\sqrt{1-y^2}}$

10. $\dfrac{1}{2y\sqrt{y-1}}$

11. $5\tan^{-1}3x+\dfrac{15x}{9x^2+1}$

12. $2x+\dfrac{20x^4}{\sqrt{1-x^{10}}}$

13. $\dfrac{-3}{x\sqrt{x^6-1}}$

14. $\dfrac{1}{2(y^2+1)}$

15. $\dfrac{1}{15y^{\frac{2}{3}}\sqrt{1-y^{\frac{2}{3}}}}$

Exercise 26

1. $3\cosh 3x - 2\sinh x$

2. $-5x\operatorname{sech}x(x\tanh x - 2)$

3. $\dfrac{2(x\cosh 2x - \sinh 2x)}{3x^3}$

4. $6\cosh 3x\sinh 3x$

5. $60x^4\cosh^2 4x^5\sinh 4x^5$

6. $-10x^4\operatorname{cosech}2x^5$

7. $-12x\operatorname{sech}^3 2x^2\tanh 2x^2$

8. $x(5x\sinh 5x + 2\cosh 5x)$

9. $\dfrac{3\operatorname{cosech}^2 3x\tanh 5x + 5\coth 3x\operatorname{sech}^2 5x}{\coth^2 3x}$

10. $\dfrac{2\cosh^2 x(3x\sinh x - \cosh x)}{3x^2}$

Exercise 27

1. $\dfrac{1}{x}\log_a e$

2. $\dfrac{3x^2}{x^3+5}\log_a e$

3. $\dfrac{36x^2}{2x^3-5}\log_a e$

4. $\dfrac{2}{3x}\log_a e$

5. $\dfrac{4x}{(x^2+1)(1-x^2)}\log_a e$

or $\dfrac{-4x}{(x^2+1)(x^2-1)}\log_a e$ or $\dfrac{-4x}{(x^4-1)}\log_a e$

6. $\dfrac{15x^2}{5x^3-1}\log_5 e$

7. $\dfrac{-2}{x}\log_2 e$

8. $\dfrac{2x}{x^2+3}$

9. $\dfrac{3\ln^2 x}{x}$

10. $\dfrac{5x^4}{2x^5-1}$

11. $2x^3(4\ln x + 1)$

12. $\dfrac{-28}{3-7x}$ or $\dfrac{28}{7x-3}$

13. $6x\ln(4x^3+1)+\dfrac{36x^4}{4x^3+1}$

14. $\dfrac{2-4\ln x}{x^3}$

15. $\dfrac{-30x}{1-5x^2}$ or $\dfrac{30x}{5x^2-1}$

Exercise 28

1. $10a^{2x}\log_e a$

2. $a^{5x^2-x}\ln a(10x-1)$

3. $a^{3x}x^3(3x\ln a + 4)$

4. $-2e^{-2x}-3e^{-x}$ or $-e^{-2x}(2+3e^x)$

5. $\dfrac{\sqrt{3}\,e^{\sqrt{3x}}}{\sqrt{x}}$

6. $6x^2e^{x^2}(2x^2+3)$

7. $\dfrac{5\ln a(a^{10x}-1)}{a^{5x}}$

8. $\dfrac{6e^{4x}(2x\ln 2x^2+1)}{x}$

9. $\dfrac{3x\sqrt{e^{5x}}(5x+4)}{2a}$

10. $x^3(\log_{10}e + 4\log_{10}7x)$

11. $3\ln6(6^{3x})$ 12. $2x - 3(e^{x^2 - 3x})$ 13. $\dfrac{-(5x + 3)e^{\frac{3}{x}}}{5x^7}$ 14. $6e^{-3x}$

15. $3(2x + 1)(x^2 + x)^2 e^{(x^2 + x)^3}$ 16. $2^{x^2} 2x\ln2$ 17. $2a^{2x}\ln a$ 18. $\dfrac{e^x(x\ln 10x^3 + 3)}{x}$

19. $\dfrac{\sqrt{5}(4x + 1)e^{2x}}{2\sqrt{x}}$ 20. $2x^4(4\log_{10}e + 5\log_{10}5x^4)$

Exercise 29

1. $x^{2x}(2 + 2\ln x)$ 2. $\dfrac{3x^{\ln 2x}(\ln 2x + \ln x)}{x}$ 3. $\ln(1 - 4x^2) - \dfrac{8x^2}{1 - 4x^2}$

4. $(6x^x - 5)^{3x}\left[3\ln(6x^2 - 5) + \dfrac{36x^2}{6x^2 - 5}\right]$ 5. $\dfrac{(2x^3 + 1)(x - 2)^3}{3x^2(x^3 - 1)^2}\left(\dfrac{6x^2}{2x^3 + 1} + \dfrac{3}{x - 2} - \dfrac{2}{x} - \dfrac{6x^2}{x^3 - 1}\right)$

6. $\dfrac{(2x - 1)(x^2 - 2)}{(1 - x)(x - 3)^2}\left(\dfrac{2}{2x - 1} + \dfrac{2x}{x^2 - 2} + \dfrac{1}{1 - x} - \dfrac{2}{x - 3}\right)$

7. $15x^3x^{3x^4}(4\ln x + 1)$ or $15x^{3x^4 + 3}(4\ln x + 1)$ 8. $3e^{3x}(e^{e^{3x}})$ or $3e^{e^{3x} + 3x}$

9. $2^{e^x}e^x\ln2$ 10. $\dfrac{x^{\ln(x^2 - 4x)}[(x - 4)\ln(x^2 - 4x) + (2x - 4)\ln x]}{x(x - 4)}$ 11. $\ln(2x^2 + 10) + \dfrac{4x^2}{2x^2 + 10}$

12. $(3x^2 - 8)^{2x}\left[2\ln(3x^3 - 8) + \dfrac{18x^3}{3x^3 - 8}\right]$ 13. $\dfrac{(x + 3)^2(x - 3)^2}{x^3 - 1}\left(\dfrac{2}{x + 3} + \dfrac{2}{x - 3} - \dfrac{3x^2}{x^3 - 1}\right)$

14. $15x^{(\ln x)^2}(\ln x)^2$ 15. $30x^{3x^2}x(2\ln x + 1)$ or $30x^{3x^2 + 1}(2\ln x + 1)$

Exercise 30

1. $\dfrac{3}{5x^2}$ 2. $\dfrac{5xe^{5x}}{2}$ 3. $\dfrac{-2x\sin x^2}{\cos x}$ 4. $\dfrac{x(6x - 5)}{x - 1}$ 5. $\dfrac{14x^2}{3}$ 6. $\dfrac{2a^x\ln a}{e^x}$

7. $\dfrac{-2\tan x \sec^2 x}{5\sin 5x}$ 8. $\dfrac{1}{(x - 1)(e^{x - 1})}$ 9. $\dfrac{1}{xe^{5x}}$ 10. $\dfrac{e^x + e^{-x}}{(e^x - e^{-x})(1 + \ln x)}$

Exercise 31

1. $\dfrac{dy}{dx} = 2x(2x^2 - 3x - 7),\quad \dfrac{d^2y}{dx^2} = 2(6x^2 - 6x - 7),\quad \dfrac{d^3y}{dx^3} = 2(12x - 6)$ 2. $-\dfrac{3}{x^2}$

3. $20x^2e^{5x^4}(20x^4 + 3)$ 4. $-50x^8\sin^2x^5 + 40x^3\cos x^5\sin x^5 + 50x^8\cos x^5$ 5. $\dfrac{128}{(4x - 10)^2}$

6. $\dfrac{-(x^2 - 2)\sin x - 2x\cos x}{x^3}$ 7. $\dfrac{4\ln x - 6}{x^3}$ 8. $-2\cosec^2x(\cosec^2x + 2\cot^2x)$

9. $2\cos^2x - 3\sin^2x - 5\cos2x$ 10. 0

11. $-8x^2\sin x^2 + 4\cos x^2 + 2\sin^2x - 2\cos^2x - 4x\cos x^2 + 2\cos x\sin x$

12. $\dfrac{2[(x^2\sec^2x + 1)\tan x - x\sec^2x]}{3x^3}$ 13. $e^{-x}(4\sin2x - 3\cos2x)$ 14. 0

15. $\dfrac{-(9x^2 - 2)\sin 3x - 6x\cos 3x}{3x^3}$

444

Exercise 32

1. $2x(3x^4 - 5)^3 + 36x(3x^4 - 5)^3$ or $2x(3x^4 - 5)^2(21x^4 - 5)$ 2. 209

3(a) $\dfrac{\sqrt{3x^2 - 1}\,(3x^2 + 2)}{x^3}$ (b) $5\sqrt{2}$ 4(a) $\dfrac{12x^2 - 3y^3}{9xy^2 - 2y - 5}$ (b) $-\dfrac{9}{16}$ 5. $-\dfrac{2}{y}$

6. $3x(x^2 + 5)^2(3x^3 + 5x - 2)$ 7. $e^{\sin x + \tan x}(\sec^2 x + \cos x)$ 8. $\dfrac{2x(\sin x - x\cos x)}{\sin^3 x}$

9. $\dfrac{2\cos x \sin x - 3\sin 3x}{\sin^2 x + \cos 3x}$ 10. $-5a^{\cos 5x}\sin 5x \ln a$ 11. $5x^2 e^{-3x} - 10xe^{-3x} + 2e^{-3x}$

12. $-(2x^3 + 2)\cos\left(\dfrac{1 - 2x^3}{x^2}\right)$ 13. $\dfrac{4x^4 - 6x^3 + 12x^2 - 18x + 6}{x^3}$ 14. 0

15. $5\left(\dfrac{2}{x^2} - \dfrac{2}{x^3}\right)\left(\dfrac{1}{x^2} - \dfrac{2}{x}\right)^4$ or $\dfrac{10(x-1)(1-2x)^4}{x^{11}}$ 16(a) $\dfrac{3x^2 e^{x^3 - y^3} - y}{3y^2 e^{x^3 - y^3} + x}$ (b) 1

17. 0   18. $\dfrac{2\sin\frac{1}{x^2}}{x^2} - \dfrac{4\cos\frac{1}{x^2}}{x^4} + 2\cos\dfrac{1}{x^2}$ 19(a) $\dfrac{2e^x(e^x + 1)}{(e^x - 1)^3}$

(b) $\dfrac{4(e^{2x} - e^{-x}) + (1 + e^{-x})(e^x - 1)^3}{(e^x - 1)^3(1 - e^{-x})}$ 20. $2\tan x$ 21. 0

22. $3\left(\dfrac{\sqrt{5}}{4\sqrt{x}} + 3\right)\left(3x + \dfrac{\sqrt{5x}}{2}\right)^2$ 23. $\dfrac{-10x - 18}{x^4}$ or $\dfrac{-2(5x + 9)}{x^4}$

24. $9x^2(x^3 + 1)^2 + \dfrac{5 + 4x - 10x^2}{(x^2 + 1)^2}$ 25. 0

Exercise 33

1.

Mark	65	66	67	68	69	70	71	72	73	74
Frequency	4	2	6	5	6	3	3	2	3	5

2.

Score	5	6	7	8	9
Frequency	5	7	2	2	4

3.

No of Seed	20	21	22	23	24	25	26	27	28	29	30
Frequency/No of Pod	6	4	3	5	2	2	1	4	7	3	3

4.

Age	10	11	12	13	14	15
No of School	5	4	5	5	6	5

Exercise 34

1. Mean = 4   Median = 3   Mode = 2
2. (a) 24   (b) 25   (c) 25
3. Mean = 111.2   Median = 111.5   Mode = 105
4. Mean = 4.3   Median = 4.5   Mode = 5
5. (a) 5.1   (b) 5   (c) 4
6. (a) 13.2   (b) 13.5   (c) 14
7. (a) 11   (b) 58kg

Exercise 35

1.

Score	40 − 44	45 − 49	50 - 54	55 - 59	60 − 64
Frequency	8	14	13	9	6

2.

Weight	21 - 25	26 - 30	31 - 35	36 - 40	41 - 45	46 - 50	51 − 55
Frequency	2	9	8	2	8	2	9

3.

Class interval	Frequency	Class Boundary	Class width	Class mid-value
5 − 9	2	4.5 − 9.5	20	7
10 − 14	5	9.5 − 14.5	20	12
15 − 19	5	14.5 − 19.5	20	17
20 − 24	7	19.5 − 24.5	20	22
25 − 29	1	24.5 − 29.5	20	27

4.

Class interval	Frequency	Class Boundary	Class width	Class mid-value
1 – 20	1	0.5 – 20.5	20	10.5
21 – 40	4	20.5 – 40.5	20	30.5
41 – 60	7	40.5 – 60.5	20	50.5
61 – 80	3	60.5 – 80.5	20	70.5
81 – 100	5	80.5 – 100.5	20	90.5

5.

Class interval	Frequency	Class Boundary	Class width	Class mid-value
0 – 90	2	-5 – 95	100	45
100 – 190	4	95 – 195	100	145
200 – 290	1	195 – 295	100	245
300 – 390	7	295 – 395	100	345
400 – 490	1	395 - 495	100	445

Exercise 36

1. (a) 72.5   (b) 70.9   (c) 67.9

2. (a) 90.4   (b) 91.5   (c) 92.7

3. (a) 44.1   (b) 41.7   (c) 36.9

Exercise 37

1. 1.67   2. 1.82   3. 1.6   4. 16.27   5. 12.5   6. 9.1   7. 2.7

8. (a) The frequency table is as shown below.

No on Die	1	2	3	4	5	6
No of Times	7	7	9	11	7	9

(b) 1.41

Exercise 38

1. Variance = 2   Standard Deviation = 1.41

2. Variance = 6   Standard Deviation = 2.45

3. (a) 199.04   (b) 14.11

4. (a) 40.69   (b) 6.38

5. (a) 1.049   (b) 1.024

Exercise 39

1. The frequency table is as shown below.

Mark	11 - 20	21 – 30	31 - 40	41 – 50	51 - 60	61 - 70	71 - 80	81 - 90	91-100
No of Student	4	5	4	5	8	3	5	3	3

(a) 53   (b) 33   (c) 72.5   (d) 39.5   (e) 19.75   (f) 38   (g) 64.5

2. (a) 32.36   (b) 25.18   (c) 40.25   (d) 15.07   (e) 7.54   (f) 41.6

(g) 37.33   (h) 22.06

3. (a) 1.7   (b) 1.23   (c) 2.31   (d) 1.60   (e) 0.98

4. (a) 17.5   (b) 11.29   (c) 22.56   (d) 11.27   (e) 5.63   (f) 20.06   (g) 29.5

5. (a) 1.54   (b) 0.83   (c) 2.08   (d) 0.83   (e) 1.01

Exercise 40   There is no exercise in chapter 40.

Exercise 41

1. (a) $\frac{1}{10}$ (b) $\frac{9}{20}$ (c) $\frac{1}{4}$ (d) $\frac{3}{4}$ (e) $\frac{7}{20}$

2. $\frac{1}{4}$ 3. $\frac{4}{13}$ 4. $\frac{2}{5}$ 5. $\frac{4}{5}$

6. (a) $\frac{3}{5}$ (b) $\frac{2}{5}$

7. $\frac{11}{25}$ 8. $\frac{7}{10}$

9. (a) $\frac{2}{5}$ (b) $\frac{1}{2}$ (c) $\frac{9}{10}$ (d) $\frac{3}{10}$ (e) $\frac{9}{20}$

10. (a) $\frac{1}{26}$ (b) $\frac{3}{26}$ (c) $\frac{2}{13}$ (d) $\frac{4}{13}$ (e) $\frac{4}{13}$

11. $\frac{2}{11}$

Exercise 42

1. $\frac{1}{13}$ 2. $\frac{1}{26}$ 3. $\frac{1}{26}$ 4. 1   5. $\frac{1}{26}$ 6. $\frac{12}{13}$

7. (a) $\frac{2}{13}$ (b) $\frac{2}{13}$ (c) $\frac{5}{52}$ (d) $\frac{7}{26}$ (e) $\frac{15}{52}$ (f) $\frac{7}{13}$

8. (a) $\frac{1}{169}$ (b) $\frac{1}{13}$ (c) $\frac{6}{169}$ (d) $\frac{1}{4}$ (e) $\frac{1}{13}$ (f) $\frac{12}{13}$

9. (a) $\frac{8}{663}$ (b) $\frac{8}{663}$ (c) $\frac{1}{221}$ (d) $\frac{1}{17}$ (e) $\frac{25}{102}$ (f) $\frac{13}{102}$ (g) $\frac{1}{221}$

10. (a) $\frac{37}{2197}$ (b) $\frac{2196}{2197}$

11. (a) $\frac{73}{5525}$ (b) $\frac{5452}{5525}$

Exercise 43

1. (a) $\frac{1}{2}$ (b) $\frac{1}{2}$ (c) 1

2. (a) $\frac{1}{4}$ (b) $\frac{3}{4}$ (c) $\frac{1}{4}$ (d) $\frac{3}{4}$ (e) $\frac{1}{4}$

3. (a) $\frac{1}{8}$ (b) $\frac{7}{8}$ (c) $\frac{1}{8}$ (d) $\frac{7}{8}$ (e) $\frac{1}{2}$ (f) $\frac{7}{8}$

4. (a) $\frac{15}{16}$ (b) $\frac{1}{16}$ (c) $\frac{11}{16}$ (d) $\frac{15}{16}$ (e) $\frac{3}{8}$

5. $\frac{31}{32}$

Exercise 44

1. (a) $\frac{1}{6}$ (b) $\frac{1}{6}$ (c) $\frac{1}{6}$ (d) $\frac{1}{2}$ (e) $\frac{5}{6}$ (f) $\frac{5}{6}$

2. (a) $\frac{1}{2}$ (b) $\frac{1}{2}$ (c) $\frac{5}{6}$ (d) $\frac{1}{2}$ (e) $\frac{2}{3}$ (f) $\frac{1}{3}$

3. (a) $\frac{1}{12}$ (b) $\frac{1}{12}$

4. (a) $\frac{11}{36}$ (b) $\frac{13}{18}$ (c) $\frac{11}{12}$ (d) $\frac{7}{12}$ (e) $\frac{1}{36}$ (f) $\frac{7}{18}$ (g) $\frac{7}{18}$

5. (a) $\frac{1}{6}$ (b) $\frac{5}{12}$ (c) $\frac{1}{36}$ (d) $\frac{1}{6}$ (e) $\frac{1}{6}$

6. $\frac{1}{6}$

Exercise 45

1. (a) $\frac{1}{4}$ (b) $\frac{2}{5}$ (c) $\frac{7}{20}$ (d) 0 (e) $\frac{13}{20}$ (f) $\frac{13}{20}$

2. (a) $\frac{7}{8}$ (b) $\frac{3}{4}$ (c) $\frac{1}{4}$

3. (a) $\frac{2}{125}$ (b) $\frac{8}{125}$ (c) $\frac{92}{125}$ (d) $\frac{31}{125}$

4. (a) $\frac{27}{64}$ (b) $\frac{37}{64}$ (c) $\frac{63}{64}$ (d) $\frac{27}{64}$

5. (a) $\frac{3}{8}$ (b) $\frac{1}{2}$ (c) $\frac{3}{8}$ (d) $\frac{1}{2}$ (e) $\frac{1}{2}$

6. (a) $\frac{15}{364}$ (b) $\frac{145}{182}$ (c) $\frac{25}{91}$

7. (a) $\frac{2}{5}$ (b) $\frac{8}{15}$

8. (a) $\frac{7}{22}$ (b) $\frac{7}{11}$ (c) $\frac{21}{22}$

Please if you found this book well simplified enough for easier understanding, kindly give it a five star rating on amazon so as to encourage people to buy this book, thereby putting more money in my pocket and helping students improve on their skills on algebra and differential calculus. Thank you.

If you want to see other books written by the author, just simply search for the author's name, Kingsley Augustine on amazon.com

If you have any enquiry, suggestion or information concerning this book, please contact the author through the email below.

KINGSLEY AUGUSTINE

kingzohb2@yahoo.com

Printed in Great Britain
by Amazon

45933369R00249